# GENETIC CLASSIFICATION OF GLACIGENIC DEPOSITS

FINAL REPORT OF THE COMMISSION ON GENESIS AND LITHOLOGY OF GLACIAL
QUATERNARY DEPOSITS OF THE INTERNATIONAL UNION FOR QUATERNARY
RESEARCH (INQUA)

# Genetic Classification of Glacigenic Deposits

*Edited by*
R.P.GOLDTHWAIT
*Ohio State University*

C.L.MATSCH
*University of Minnesota-Duluth*

A.A.BALKEMA / ROTTERDAM / BROOKFIELD / 1989

CIP-DATA KONINKLIJKE BIBLIOTHEEK, DEN HAAG

Genetic

Genetic classification of glacigenic deposits: final report of the Commission on Genesis and Lithology of Glacial Quaternary Deposits of the International Union for Quaternary Research (INQUA). – Rotterdam [etc.]: Balkema. – Ill.
ISBN 90 6191 694 1 bound
SISO 568.5 UDC 551.332
Subject heading: glacigenic deposits; genetic classification.

Published by

A.A.Balkema, P.O.Box 1675, 3000 BR Rotterdam, Netherlands

A.A.Balkema Publishers, Old Post Road, Brookfield, VT 05036, USA

ISBN 90 6191 694 1

© 1988 A.A.Balkema, Rotterdam

Printed in the Netherlands

*Genetic Classification of Glacigenic Deposits, Goldthwait & Matsch (eds)*
*© 1988 Balkema, Rotterdam. ISBN 90 6191 694 1*

# Contents

Genetic Classification of Glacigenic Deposits, Goldthwait & Matsch (eds)
© 1988 Balkema, Rotterdam. ISBN 90 6191 694 1

# Preface

## 1 CONCERNING THIS BOOK

This is a report by a Commission -- not
the usual 1- or 2-day symposium. After a
12-year-long gathering of material by key
commission members, this volume tackles
specific glacial problems concerning the
genesis of glacial deposits and a
resulting classification. The personnel of
each group changed some at each meeting,
and the field of inspection of each
meeting changed every time. As a consequence, all thirty meetings aired lots
of opinion, evidence, and controversy.
These were outdoor conferences, indoor
papers on specific areas, and Commission
C-2 symposia at larger meetings. With at
least 10 to 20 countries represented at
each of the several meetings, this final
report is truly international in scope.

The fact that English was the official
language of the Commission, and that work
was presented orally most often in
English, did exclude a few participants.
Of course many INQUA participants chose
other commissions as their particular
interest. There are a few countries where
Quaternary studies are common, but few or
no representatives came forward with this
commission's interest -- e.g., China or
Chile, New Zealand, Australia. Foreign
language equivalent terms are discussed
somewhat in Chapter 20, and more
thoroughly in Chapter 2.

The nine working groups that complete
their work here are reported most often
thoroughly by each final chairman of that
group (shortened titles):

1 or 2A, Genetic classification of till.
Chapter 2 by Dreimanis.

2 or 2B, Field and laboratory methods.
Chapter 15 by Raukas, Haldorsen, and
Mickelson.

3 or 2C, Glaciofluvial deposits. Chapter
17 by Jurgaitis and Juozapavičius.

4 or 2D, Glaciolacustrine deposits.

Chapter 18 by Ashley.

5 or 2E, Glaciomarine deposits. Chapter
19 by Borns and Matsch.

6 or 2F, Glacigenic landforms. Chapter
20 by Goldthwait.

7 or 2G, Engineering-geological investigation of tills. Chapter 14 by White and
Ellis.

8 or 2H, Standards for computer processing of glacigenic data. Chapter 15 by
Raukas, Haldorsen, and Mickelson.

9 or 2I, Glacigenic deposits as indicators of glacial movements. Chapter 12 by
Hirvas, Kujansuu, Nenonen, and Saarnisto.

You will note that most of these are in
the second "half" of the book. The first
"half" is devoted to till because so many
symposia since 1969 involved its origin.
Work group 1 or 2A was led by the
Commission Chairman, Aleksis Dreimanis.
Its report is given the place of promi-
nence in Part I right after a general
introductory paper by the next Chairman
(from 1987 on), Jan Lundqvist. Although
origins and terms are exhaustively treated
by Dreimanis (Chapter 2), he wished some
of the main arguments to be given in
detail by their proponents. For this
reason, Chapters 3, 4, 5, 8, 9, and 10 are
very short controversial viewpoints.
Chapters 6 and 7 are longer field studies
that present specific examples. Chapter
11 indicates the strong trend toward sedimentation in glacial studies.

Part II concerns the application of till
studies to two prominent and lucrative
areas: ore sources for mining, and foundations or materials for building. Other
materials, such as gravel and sand resources for roads and concrete aggregate, are
only implied. Note that every basic main
chapter has abundant references, especially Chapters 2 and 14.

Finally, Part III comprises the elements
of the sorted meltwater-handled glacial
deposits in Chapters 17, 18, and 19.

Chapter 15 concerns methods for standardized analysis and statistical handling of these deposits, and Chapter 16 demonstrates one classic area of application. Chapter 20 presents a genetic classification of all glacially produced morphologic features. After repeated modifications during 12 years, one short critique of this is given in Chapter 21. The book winds up with one examination of glaciotectonic landforms -- a study of ice-made structures that affect surface form. It is one rapidly developing new approach to ice action and depositional genetics.

Such a book would be almost unreadable were it not for editors and peer reviewers. In nearly all cases pertinent changes were accepted by the author. We tried to pick peers in the same specialty. At least one commission editor and one outside reader had a hand in every chapter. For these many hours of effort we wish to thank:

| | |
|---|---|
| J.T. Andrews | W.H. Johnson |
| R.L. Bates | P.F. Karrow |
| J.J. Clague | T.J. Kemmis |
| L. Clayton | C.L. Matsch |
| R.N. DiLabio | G.D. McKenzie |
| L.D. Drake | D.M. Mickelson |
| L. Dredge | M. Parent |
| A. Dreimanis | F. Pessl |
| E.B. Evenson | N. Potter, Jr. |
| M.M. Fenton | R.W. Ojakangas |
| R.P. Goldthwait | W.W. Shilts |
| S. Haldorsen | J.T. Teller |

The formidable and challenging task of final manuscript preparation was accomplished by two members of the staff of the Dean's Office, College of Science and Engineering, University of Minnesota, Duluth: Judy Holz supervised the entire final editing and lay-out; Avis Hedin converted all of the individual manuscripts into a two-column format using a CPT Phoenix Jr. Their skillful ministrations permeate this book. To them we express our admiration and our appreciation.

2 THE COMMISSION

The Commission on Genesis and Lithology of Quaternary Deposits (C-2) was established at INQUA VI (International Union for Quaternary Research) in Spain. At that time, most commission members were interested in glacial deposits, their landforms, and in the development of methods that would effectively decipher the origin of glacial deposits. During a reorganization of the Commission in 1973

(INQUA IX in New Zealand), two main objectives were declared for two intercongress periods: genetic classification of all glacial deposits and their landforms, and the review of their field and laboratory investigation methods. In order to achieve these objectives, seven autonomous work groups were established during the 1973-77 period, and increased to ten during 1977-82.

During the discussions at the 16 field conferences and symposia held during the 1973-82 period (Dreimanis, this volume), it became obvious that genetic classification was not the main object of most Commission members, but instead, a better understanding of the processes of formation of glacial deposits. The field and laboratory criteria for the identification of varieties of till appeared to be most important. Therefore most of the papers presented and most of the field discussions dealt with individual case studies of glacial deposits, as may be seen in the 26 publications issued mainly by the field conference organizers during 1973-82.

Classifications of tills, glaciofluvial deposits, and glacigenic landforms also were proposed, but others were still pending. Therefore the XI INQUA Congress (USSR) extended the mandate of the Commission for one more intercongress period (1982-87). It also restricted the Commission's name by adding "Glacial" before "Quaternary." During this period 13 more meetings were held and 16 more volumes published, with 4 still in preparation. Various individual case studies dominated at the meetings. The previously proposed classifications of tills, glaciofluvial deposits, and all glacigenic landforms were reevaluated and modified. The first proposals for the genetic classification of glaciolacustrine and glaciomarine deposits did develop, but were too late to be integrated with the other classifications.

By 1987 when all the genetic classifications of glacial deposits and landforms had to be completed for the XIIth INQUA Congress (Canada) and for publication in the present volume, we had long summary reports on the genetic classifications of tills and of glacigenic landforms. These were based on repeated discussions and reappraisals by about 100 Commission members from five continents. There was a genetic classification of glaciofluvial deposits with considerable petrographic dependence, based mainly on regional investigations east of the Baltic Sea. Other work groups had shorter, less-discussed presentations with maximum interest shown in indicator tracing in the search for ore deposits in Finland. These were all presented at the Special Session 31 of the XIIth INQUA Congress (Canada). There was obviously an imbalance in the scope and details of the classifications; the emphasis was on tills and landforms. Although all varieties of glacigenic deposits had been investigated by one or another of almost 200 Commission members from 30 countries, most of the actively participating members were discussing or demonstrating their research results on glacial sedimentation and glacial tectonics.

Any classification, whether descriptive or interpretive, draws some arbitrary boundaries or emphasizes typical endmembers. Nature usually produces a continuum from one endmember to another, from one facies to another. The selection of the most typical endmembers, and the proper combination of them, is accomplished by considering the natural facies associations and multiple criteria, which in turn are based on the depositional processes and environments. Pre-depositional transport, derivation, and provenance of materials are also critical for glaciers. With the application of new methodologies, and with more detailed investigations, existing interpretations will change in time. It is easier to make changes and improvements, however, if you have a proposal to start from. Therefore this volume presents several alternatives whenever our Commission members were willing to present their alternative views, and alternate opinions are mentioned in the main classification reports.

Richard P. Goldthwait
Ohio State University
March, 1988


Charles L. Matsch
University of Minnesota-Duluth
March, 1988


Aleksis Dreimanis
University of Western Ontario
February, 1988

1 Tills and related diamictons

Genetic Classification of Glacigenic Deposits, Goldthwait & Matsch (eds)
© 1988 Balkema, Rotterdam. ISBN 90 6191 694 1

# Glacigenic processes, deposits, and landforms

Jan Lundqvist
*Department of Quaternary Research, University of Stockholm, Sweden*

ABSTRACT: A review of glacigenic processes, deposits and landforms, grouped according to their environment of formation. Deposits are controlled by derivation of the debris, its transport by and release from the ice, further transport by other agents, and sedimentation processes. Sedimentation is controlled by mode of deposition and position in relation to the ice, state of the ice, and hydrography. The characteristics of the deposits and their landforms are discussed and considered essential for the interpretation of former environments.

## 1 INTRODUCTION

Glacier ice, especially when it is moving, may exert erosive activity as well as accumulative. The following discussion will be restricted to processes of deposition. We may distinguish between processes directly related to the ice and processes induced by the ice. For all these processes, the general term glacigenic is used, whereas those directly related to the ice are called glacial. This is in accordance with the classification proposed by the Commission's Work Group 6.

Processes and deposition of related deposits similar to the ones described as glacigenic will often take place without the presence of glacier ice. Glacigenic processes require the presence of ice and are controlled by the activity of the ice and the variations in its melting. The sediments deposited were transported by the glacier before deposition took place by means of other agents.

Processes and deposits that require cold climate but not necessarily the presence of ice are not included as glacigenic. They may be called periglacial. We thus get a broad classification:

1. glacial features (directly related to glacier ice)
2. glacigenic features (requiring glacier ice but not necessarily directly related to it)
3. periglacial features (caused by cold climate and thus often occurring outside a glacier but not requiring its presence)
4. nonglacial (nonglacigenic) features.

A number of the processes and deposits related to groups 1 and 2 will be discussed in detail in the following volume. This introductory article aims at a general discussion and classification of the first two groups of features. These are controlled by a number of parameters, as shown in Figure 1 (cf. Lundqvist 1984). In addition to these parameters, others like lithology, granulometry, bedding, etc. characterize the deposits, but they will not be considered here.

## 2 DERIVATION OF DEBRIS

The debris in the glacier, which will subsequently form different types of deposits, may be supplied to the ice in different ways, essentially either subglacially or supraglacially. Subglacial derivation is the result of erosion by the glacier. During its movement it will break off fractions from the substratum by direct pressure or by freeze-thaw processes. Debris at the base of the ice will have a grinding effect on the substratum. These processes will not be discussed in detail here, but through them debris of different granulometry is accumulated in the basal part of the glacier or moved entirely below the ice.

Supraglacial debris is transported to the glacier surface by different extraglacial processes, of which slumping and flowing from adjacent hillsides in broken

Fig. 1. During the annual winter advance of an ice front, small moraines may be bulldozed by the glacier. When the ice margin (left) passes stagnant water, fine-grained sediments may be incorporated in the moraine. Fjallsjökull, Iceland (photo J. Lundqvist 1977).

terrain are the most important. Occasionally, sediments may be transported to the ice by streams from adjacent high ground or by wind. In volcanic areas tephra of varying grain sizes may be deposited upon the ice in large quantities.

The mode of derivation has a great influence upon the petrography of the deposits to be formed later. The subglacially derived debris may at least partly be rounded or subangular, whereas the debris falling to the glacier as a result of frost action in its vicinity is mostly sharp angular.

## 3 TRANSPORT BY ICE

The debris accumulated in or upon the ice is carried along during its movement. From the position in relation to the ice body we may distinguish between different types of transport.

Debris that stays on the glacier surface during transport is considered supraglacial or subaerial. When further accumulation of snow causes the glacier to grow upward, some of this debris will be incorporated in the ice body and further transported englacially. Englacial debris may also derive from the base of the ice because of upward transport, which takes place where there is compressive flow and sheets of ice are moved over each other. In this way a large amount of debris may be carried to the ice surface to form supraglacial debris. This mainly happens toward the ice margin and forms the end of the glacial transport.

Debris resulting from basal erosion will stay mainly at the base, to be transported along it as a debris-rich sole of the ice. Debris falling to the bottom through crevasses or moulins may eventually be incorporated in the basal debris. During transport, basal debris contributes to further erosion through grinding of the substratum. In this way fine-grained material will be added to the basal debris, making it more fine-textured in most cases than the supraglacial.

The directed pressure of the moving ice will cause glaciotectonic disturbances in its substratum. There is a gradation from slight disturbances like folds and fractures to transport of fractured blocks or floes with or without addition of debris transported a longer distance. In the first case we should consider the result to be tectonized bedrock, whereas in the second case we could consider it a new deposit -- for instance, a till or floes embedded in till and so-called deformation till (Elson 1961).

Glaciotectonic disturbances of a similar type are also formed outside the ice margin as a result of the bulldozing effect of an advancing ice front (Fig. 1). Naturally, only a narrow extramarginal zone is affected, but zones with a morphology formed in this way may occur (Christiansen and Whitaker 1976).

## 4 RELEASE FROM ICE

Debris transported by the ice is released during its wastage or earlier. The particular release process determines the deposits formed.

The simplest case is release by melting. The debris is released in its original position in the ice, which it may retain or lose during the process, which we call melt-out. It takes place from the moving ice as well as after its stagnation, from the base as well as the surface (Fig. 2). Lawson (1979) considers till formed by this process to be the main till.

Usually melt-out implies the transformation of ice into water, which may affect the deposit under formation to a varying extent. In very dry areas the ice may evaporate directly, in which case release from the surface occurs by sublimation. The deposit then will be less affected by water action.

At the base of the ice, increased pressure during movement may cause a release of debris when the pressure-melting point is reached. The debris will be deposited either as individual grains or as sheets or slabs. Because of changes in pressure and temperature, debris may be reincorporated in the ice and deposited repeatedly. This process will affect the deposit strongly, often making it compact and producing a strong fabric. The effect is increased by the fact that the basal debris is often fine-grained.

Where sheets of ice are moved over each other along shear-planes or by compressive flow, some release of debris may take place at the front of these planes. The importance of the shearing process has probably been somewhat exaggerated in the past. A limited amount of material moved

along the shear-planes will accumulate where they end at the ice margin, but most of the debris is situated just above a plane and will be released by melt-out.

Some debris will be released by other agents of erosion, mainly water. Except for large boulders the debris released in this way will be carried away by the eroding agent to be deposited later as sorted glacigenic sediments.

Some debris is carried on top of the glacier or moved by it without actually being frozen to it. In these cases it is irrelevant to discuss release.

## 5 FURTHER TRANSPORT

Debris released from the ice may be deposited directly (glacial deposits) or carried away by some agent before deposition.

The simplest case is further transport by the moving glacier. It may seem illogical to consider this a special case, but it means only that debris could melt out from the ice surface and then be carried on as supraglacial debris not frozen to the ice. At the ice front, debris may be released and pushed further by advancing ice before it comes to rest as a deposit.

A more important further transport is caused by slumping and flow from the glacier. We may use the term slope processes for all this activity. Because a glacial deposit is accumulated directly by the ice, and other deposits are called glacigenic, but not glacial, this transport will constitute the difference between the two groups. Strictly speaking, only the subglacially accumulated material released by pressure-melting is deposited directly by the glacier. In all other instances some settling or movement takes place at the deposition. However, for practical purposes it may be convenient to disregard an insignificant secondary movement of this type (especially if it consists only of a settling) and consider these deposits glacial unless a "more important" transport has taken place (Dreimanis and Lundqvist 1984). The definition is certainly vague, and gradual transitions are numerous -- or even dominant.

Further transport by water is common and most important. This water is mainly meltwater that forms streams upon, within, or underneath the ice (Fig. 3). Occasionally other streams may also erode the glacier margin, and glaciers terminating in water may be abraded by wave action. The debris is carried away by the water and deposited as waterlaid (cf. Francis 1975) sediments.

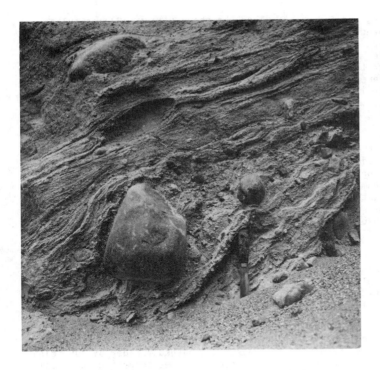

Fig. 2. At basal melt-out from a glacier, the bedding of
debris in the ice is to some extent preserved. According
to Shaw this applies to the Sveg till, here seen near
Sveg, Härjedalen, Sweden (photo J. Lundqvist 1963).

These are called glaciofluvial (cf.
Lundqvist 1985) if they are deposited by
streaming water, otherwise glaciolacustrine or glaciomarine.

Especially in dry areas, wind action may
erode the released debris, which will be
transported and deposited as eolian
sediments. In general this process is of
minor importance. The bulk of the eolian
sediments deposited in periglacial regions
derive their origin from previously deposited sediments. The portion deriving
directly from the glacier can hardly be
identified as a glacioeolian sediment, and
will not be further discussed here.

6 SEDIMENTATION

Debris deriving from a glacier and transported by various agents will be deposited
in different environments. The sediment
character is controlled by a number of
parameters, especially the place of deposition in relation to the ice and to water
bodies of various types. The state of the

ice has a great influence, as well as the
mode of deposition.

6.1 Mode of deposition

The debris transported can accumulate in
different ways to form sediments. The
simplest way is settling controlled only
by gravity. If this settling occurs in a
water body or in the air, the process will
follow common sedimentation rules, which
will not be further discussed here. If the
settling occurs directly from ice it
implies a removal of interstitial ice supporting the grains and clasts of the
debris. The particles retain their position relative to each other but become
more closely packed. Theoretically the
only move is a lowering due to gravity. In
reality the withdrawal of interstitial
water implies transport of the finest particles in the pores between the coarser
ones (cf. Muller 1983). The deposit in
this way may become either more compact,
by the addition of fines, or less compact

Fig. 3. Meltwater from a glacier will deposit sediments
in its channel in the ice, in this case a tunnel, and
proglacially as an outwash plain. Gulkana Glacier,
Alaska (photo J. Lundqvist 1965).

by their removal.

Most commonly the entire mass will move
some distance directly upon its release
from the ice before it is deposited (see
Boulton 1968). Accumulation takes place as
a flow or dumping. Flow (solifluction) is
a very common process in the periglacial
environment. Theoretically, a deposit
should be considered glacial only if there
is no sedimentation before the flow
process. If a sediment is first deposited
and later on redeposited by flow it should
no longer be considered glacial, but
rather a slope deposit, periglacial deposit, solifluction deposit, or the like,
formed by reworking of a glacial deposit.
However, in reality it is virtually
impossible to distinguish between slope
deposits etc. formed shortly after the
primary deposition and strictly glacial
deposits. Therefore we probably have to
allow some flow in close connection with
the primary formation even of glacial
deposits. This problem especially concerns
the definition of some diamictons, and
whether they should be considered tills.
Dumping may take place directly from a

steep ice margin, but more important
perhaps are subglacial dumping and flow
(Evenson, Dreimanis, and Newsome 1977)
from floating ice margins, ice shelves, or
even isolated icebergs. Where many icebergs are formed, extensive dumping of
debris may occur far outside an ice
margin.

Lodgement is a term for the deposition
of debris released at the glacier base by
pressure-melting. In this case the processes of transport, release, and deposition almost coincide, affecting the
deposit strongly, and producing a strong
fabric, compactness and structure.
According to an extreme opinion a deposit
(lodgement till) formed in this way is the
only true till.

Finally, by the pressure of the ice,
mainly in crevasses but also at its base
or margin, some debris will be deposited
by squeezing. This effect is of minor
importance, forming rather small deposits,
with one exception. Some truly subglacial
features related to glaciotectonics may be
considered a type of squeeze phenomena.
The same applies to some marginal

features. Otherwise this mode of deposition is mainly of theoretical interest.

## 6.2 Position in relation to glacier

Large-scale accumulation of debris transported to the surface in zones of compressive flow or falling to the ice from its surroundings (Boulton 1972) may take place on the glacier surface behind the ice margin. The debris forms diamictons, but in this position meltwater may also accumulate sorted sediments. It may seem illogical to discuss formation of deposits while there is still ice underneath them. The final settling of the deposits will not occur until the ice is gone. However, the deposits are essentially formed in the supraglacial position, where individual grains and clasts obtain their position relative to each other, and sedimentary structures as well as textures are also formed. Melting of the underlying ice may result only in a lowering of the entire sediment unit, without further sedimentation processes. If a true redeposition with rearrangement of particles takes place, the deposit will be classified in a different way.

Debris released at the ice margin in different ways will partly accumulate to form ice-marginal deposits, as well as debris released behind the margin and transported to it by different agents. Obviously virtually any kind of deposit may be formed in this position, but we define true ice-marginal deposits as those requiring an ice margin for their formation.

Subglacial deposition takes place where debris is released at or falls to the base of the ice, with or without further transport along the base. Sedimentation occurs both directly at the base and, especially where the ice is floating, at some distance below it (cf. Evenson, Dreimanis, and Newsome 1977).

Large amounts of glacigenic sediments, especially those which are transported by water from the ice, are deposited far outside the ice margin (pro- or extramarginally). Floating icebergs may also deposit glacial sediments far from a glacier front.

## 6.3 State of the ice

The state of the glacier greatly influences its deposits. Some deposits need active ice. It may be less important whether the ice is climatically active or if it is just moving mechanically through deformation under its own weight. However, climatically active ice will certainly be more efficient in terms of glacial sedimentation. Active ice is a prerequisite of the process of lodging, and in general it favours sedimentation through entrainment and formation of new debris. Very rapidly moving, or surging, glaciers should possibly be distinguished as a special category, although it is questionable whether their deposits are different from those of other glaciers. Conditions in terms of advance and retreat may also affect the deposition (Boulton and Eyles 1979).

Stagnant (passive, dead) ice favours the formation of a number of deposits. The significance of so-called dead-ice deposits, however, has often been misinterpreted in previous literature. It is not always recognized that their formation mostly requires a rather active glacier to produce a sufficient amount of debris, in particular on the ice surface (Eyles 1983). The downwasting of dead-ice then takes place only in a limited marginal zone. On the other hand, deposits indicating active ice and formed subglacially may require dead-ice downwasting in order to be preserved and uncovered by "dead-ice deposits."

The temperature at the base of the ice is also important. Generally, a warm base is required to release debris and produce meltwater. Therefore most glacigenic deposits indicate a warm-based (temperate) ice. If the ice is cold-based, there seems to be a consensus of opinion that few or no subglacial sediments can be formed. Deposition implies either some release at the ice margin, or supraglacial formation at the ice surface. However, if the ice is cold-based the amount of debris transported to its surface is probably small.

## 6.4 Hydrography

The hydrographic conditions at the time of deposition greatly influence not only the formation of waterlaid sediments but also that of other deposits, such as till. According to Francis (1975) and Dreimanis (1979), we should distinguish between waterlaid sediments (deposited by water) and waterlain sediments (deposited in water). Deposition may take place either above (supra-aquatically) or below (subaquatically) the surface of the sea, lake, or temporary glacial lake. Some deposits are formed at the water surface in the sense that they require a water surface for their formation.

Above water bodies, streaming as well as

standing, deposition takes place sub-aerially without significant contribution of water. It is controlled only by ice, wind, or slumping, etc. Waterlaid sediments in this environment are deposited by flowing water (glaciofluvial).

Where the flowing water reaches standing water and its current ceases, water-level deposits (deltas) are formed. If the current is strong and the amount of debris is large, a Gilbert-type delta forms. If the transition is more gradual, the delta type is different and its sediments pass more gradually into deep-bottom sediments.

Deposition in standing water takes place either by sedimentation of the material transported by the streams, mainly in suspension, or by material dumped into the water by slumping, flow or directly from glacier ice. Usually some sorting takes place during this process; at least the deposition of suspended material is greatly affected by the salinity of the water. We therefore distinguish between marine and lacustrine subaquatic deposits.

## 7 CHARACTERISTICS OF GLACIGENIC DEPOSITS

Mode of deposition as well as earlier glacial history control the deposits in terms of texture, bedding, fabric, etc. Deposits laid down in water are controlled by the common laws of sedimentation and do not differ principally from other waterlain sediments. A thorough sedimentological study will reveal their sedimentological history, although the character of their glacigenic deposits is not always clear. In general, the glacigenic sediments indicate a higher flow regime than that corresponding to their present environment. Especially the fine-textured sediments often indicate a cyclic process of sedimentation (Fig. 4; De Geer 1940; Stevens 1985). However, a high flow regime is not exclusively related to the glacial environment, and rhythmites can be formed without the presence of a glacier with annual variations in melt-water outflow.

Therefore, other criteria must be used to identify waterlain sediments as glacigenic. The presence of dropstones, particularly of striated rocks, is one indication. Others are larger amounts of diamicton beds dropped from icebergs. The topographic position may often give good evidence for the former presence of a glacier. Generally, the most reliable evidence is the environment and the relationship between different deposits. We may conclude that in the case of waterlain -- as well as eolian -- sediments, no

single evidence can be considered conclusive. Environmental studies and the combined evidence of different facts are most important. Thorough sedimentological investigations are among the most useful tools in this respect.

Glacial deposits (direct deposits from glacier ice) are often more easily defined as glacial in general, but their detailed history is more difficult to interpret. Glacial deposits may be misinterpreted as, for instance, mudflows and other deposits from mass movements, lahar flows, etc. or vice versa. Studies of the paleoenvironment are also essential for a correct interpretation. Existence of striated boulders and some sedimentary structures are indications, but no single piece of evidence is conclusive.

One complication in the identification of glacial deposits is that they do not follow sedimentary laws. From the sedimentological point of view any kind of sediment can be deposited by a glacier. Lithology, texture, grain and clast shape and similar parameters are all to a great extent controlled by the origin of the debris. For instance, it is often claimed that glacial deposits are poorly sorted. This is mostly true, but it is not uncommon that a glacier erodes and entrains a well-sorted deposit and redeposits it as a glacial sediment with inherited good sorting (e.g., Rabassa, Rubulis, and Suárez 1979). The same applies to clast roundness and a number of other parameters.

The sediment, especially diamicton, directly deposited by glacier ice is generally referred to as till. The definition of this term, however, is controversial. According to the definition adopted by the Till Work Group of the INQUA Commission on Genesis and Lithology of Quaternary Deposits "Till is a sediment that has been transported and is subsequently deposited by or from glacier ice, with little or no sorting by water" (Dreimanis 1982). This is a very general definition, which seems to be the only one that can be reasonably widely accepted. There is one vague point in it: how much is "little sorting"? Because interstitial ice has to be removed at deposition, water action will inevitably have some effect during the deposition of most glacial deposits. Consequently, according to the most orthodox till researchers, only the deposit formed by the process of lodging, lodgement till, should be considered till. Other types of till -- melt-out till, flow till etc. (see Dreimanis 1969, 1983) -- are more or less affected by other processes and therefore are not true tills

Fig. 4. Glacial clay with a bedding, varves, representing the annual cyclicity of the drainage from the glacier. Sundsvall, Sweden (photo J. Lundqvist 1971).

and should be referred to as diamictons. From a practical point of view this is not a convenient definition of till. Dreimanis (1982, 1983) therefore introduced the terms ortho-tills (cf. Harland, Herod, and Krinsley 1966) for the "true," unaffected types, and allo-tills for types slightly affected by other processes. Boulton and Deynoux (1981) similarly used the terms primary and secondary tills.

If this definition of till is accepted, we find a great variety of tills. Dreimanis (1969, 1982) defined the main groups -- lodgement, melt-out and flow-till as well as a few less common types. If we consider all the parameters controlling till formation we find that there are theoretically at least thirty varieties (J. Lundqvist 1984). However, it is virtually impossible to identify all these, so for practical purposes we could limit the number to the main types defined by Dreimanis. Very often we will just be able to distinguish between basal (or subglacial) till and supraglacial till.

The restricted use of the term till, mentioned above, may be justified in very

detailed sedimentological investigations. In a section, where such work can be done, the exact sedimentological characterization of the strata is important (see Eyles, Eyles, and Miall 1983). For all other work, especially regional investigations in vast areas with few or no sections, this definition is too restricted. The problem is particularly obvious in other languages, where the word for till is "moraine" -- in slightly different form in different languages. This word is generally accepted and can hardly be replaced by, for instance, a term like "diamicton."

The term diamicton is just as debatable as till. Diamicton is a sedimentological term referring to a specific texture (poor sorting), compatible with gravel, sandy silt, silty clay, etc. Till is usually, but not always, a diamicton. Consequently, the term diamicton should not be used unless we really know the texture. We have to accept that we are dealing with two separate terminologies, one sedimentological-textural, and one genetic. Till is a genetic term and diamicton a sedimen-

10

tological one, and they should not be mixed up. If we accept this view, we can define till in different ways. For practical use, a fairly wide definition seems to be the only one possible to apply.

The problem of the definition of till exemplifies a general problem: ubiquitous gradual transitions between the different types of deposits. Till grades into slope deposits and glaciofluvial deposits, the latter grade into glaciomarine and glaciolacustrine deposits, etc. Glacigenic deposits grade into nonglacial through interference with sediments deriving from nonglacial environments.

It is easy to find numerous such examples. We may argue forever about problems of where to put the boundaries between different deposits, but we will probably never reach a consensus of opinion. Usage is different in different countries, and solutions will be different for different purposes. For instance, regional aspects may favour one solution, stratigraphical aspects another. Therefore, instead of arguing in vain about the definition of terms, we should try to understand and explain the conditions and environment of formation of different deposits. This does not mean that it is useless to define existing terms and to find out the characteristics of the deposits.

In the interpretation of deposits, sedimentological analysis is essential. It helps to explain the formation and origin of different strata. However, such an analysis requires good sections and a knowledge that they show representative sequences of strata. In vast areas this is not the case. To some extent sections may be replaced by drilled cores, but for practical and economic reasons only limited possibilities of this type are available. Then we have to trust the surface and landforms. Therefore a good understanding of glacial landforms is essential in glacial geology. The following discussion deals with this issue.

8 GLACIGENIC LANDFORMS

The landforms of glacigenic deposits formed outside the glacier margin -- and independent of isolated ice bodies -- do not differ principally from those of nonglacigenic deposits. Fluvial and other waterlaid sediments (as well as eolian sediments) show the same forms with banks, dunes, stream channels, etc. whether they are glacigenic or not. The only difference is that glacigenic deposits are often more coarse-grained and indicate a higher flow regime. The forms give information about the process of sedimentation but (unless we also consider the general environment) they do not indicate whether a sediment is glacigenic or not. These landforms will not be discussed here.

Glacial deposits formed directly from or by glacier ice and glacigenic deposits formed in contact with the ice both show a morphology highly influenced by the ice, its movements, and melting. Results differ greatly depending on whether deposition took place upon or underneath the ice, in its crevasses, or at its margin. There is a great difference between landforms controlled by actively moving ice or stagnant ice. A number of morphological features are not fully understood, even if their glacial origin is beyond any doubt.

Landforms from the base of active ice are often very even plains of "ground moraine." Genetically, the deposit is a basal till, lodgement or melt-out. Where other moraine forms are developed, they are mostly controlled by the ice movement, that is, they are extended in the direction of flow. The land surface may be striated, grooved, or fluted (Boulton 1976). The relative differences in height of this landform amount to only a few centimeters or decimeters. The striation is clearly visible in barren periglacial areas, but is often more or less invisible on ground covered with vegetation. Even in that case, air photos may reveal the striation (Svensson 1970). Large-scale fluting forms a transition to drumlins.

Flutes are normally formed of till that has been squeezed into open spaces in leeside positions (Boulton 1976, Åmark 1980), but we must observe that similar landforms may be erosional. In the latter case, they do not give information about the nature of the underlying deposits. No matter what their origin, flutes give important information about the direction of ice movement.

Flutings may grade into the larger drumlins or drumlinoid forms of different types in other ways. Rock drumlins, crag-and-tails, precrags and related landforms (cf. Gillberg 1976; Dionne 1984) belong to the same group. Their formation is not understood in all respects, although it seems clear that there are both depositional and erosional forms within this group. Drumlins may consist of till -- with or without a core of bedrock -- but in many cases the bulk of their interior is waterlaid sediments, for instance glaciofluvial (Whittecar and Mickelson 1979). Even organic deposits have been found in them (Hillefors 1969). This observation is important from the point of view of glacial transport: even if there is till in a drumlin, it does not necessarily correspond to the direction of transport indicated by the drumlin form.

Drumlins and flutings are often interpreted as indicating active ice. This is true to some extent, but does not necessarily apply to the stage of deglaciation. Where the ice above the drumlins was clean and lacking supraglacial debris it may well have stagnated and wasted down without changing the landforms of the preceding active ice.

In concave positions in the terrain, drumlins tend to become incomplete, their distal ends being more or less concave (Lundqvist 1969). The incomplete drumlins often join side by side to form ridges transverse to the ice movement. They form a transition to Rogen moraine. The latter is a type of transverse ridge, consisting of either complete, small drumlins or crescentic parts. There is a consensus of opinion about the morphological relationship between drumlins and Rogen moraine, but the genetic relationship is not completely understood.

Crevasse fillings in a limited sense are restricted to stagnant ice. They may certainly be formed also in active ice, but they will then almost inevitably be transformed to drumlinoid forms. In a wider sense, for instance, flutes are also a type of crevasse filling. In stagnant ice we may find crevasses that are longitudinal as well as transverse, and often winding. They may extend all through the ice body, occur only at the base, or reach some distance down from the surface.

In such positions different types of sediments may accumulate. Till melts out from the adjacent walls or falls from the surface. Basal till may be squeezed in. Naturally enough, meltwater follows the crevasse systems and sorting will take place. Consequently, the crevasse fillings show all gradations from till ridges to ridges of glaciofluvial sediments. The latter are a type of esker. It seems appropriate to use the term esker also for these forms, and not try to separate eskers from crevasse fillings (Flint 1971:211), because most eskers are rather complex features with sediments accumulated in different stages and varying in size within wide limits.

The so-called De Geer moraines, which have formerly often been interpreted as end moraines (De Geer 1940), are probably a type of transverse crevasse fillings (Elson 1957; Hoppe 1957; Strömberg 1965).

I have suggested (Lundqvist, this volume) that they were formed in basal crevasses some distance behind the margin of a rapidly advancing glacier in zones of extending flow. We may infer that the moraines represent a zone mainly of transverse crevasses similar to the one observed in surging glaciers (Robin and Barnes 1969). Evidence in support of this interpretation is the pattern, which is sometimes very similar to the pattern of crevasse fillings described by Haselton (1979).

Deposits formed below a moulin are a special type of crevasse landform. The vertically falling water will efficiently remove all fine-grained debris and leave only the largest clasts to form a hummock, which may be considered a special case of subglacial kame.

Giant floes are another special type of subglacial landform of active ice (Petersen 1924; Clayton and Moran 1974; Moran et al. 1980). Under certain conditions, among others, a high hydrostatic pressure below the ice, such floes or slabs can be moved a considerable distance to form hills with varying internal structure. A corresponding erosion scar is sometimes seen behind the hill.

At the margin of the active ice the characteristic landform is the moraine. A moraine may be formed of any material occurring in front of the glacier and bulldozed by it to form a ridge. Commonly, however, a moraine is formed of debris transported to the glacier front and accumulated there. It may consist of subglacial debris transported to the front under compressive flow to be accumulated as imbricated beds. Debris released in the ice front or upon the ice surface may flow down the front to be deposited as flow till or to fall down a steep front as a dump. It appears that the bulk of most moraines, however, consists mainly of waterlaid sediments -- glacio-fluvial gravel and sand. This may form the entire ridge or be interbedded with diamictons. Even when the surface layer is typical till, the interior of the moraine is often a glaciofluvial sediment, more or less disturbed by glacier movements. One gets the impression that glaciofluvial accumulation is essential in the formation of a moraine. Where the glacier terminated in open water the sediments are usually more fine-grained than where it had a supraaquatic margin.

It must be noted that the term moraine is used in different senses. The description above refers to a single ridge. However, the term is often used in a chronological sense (e.g., Thwaites 1943).

Fig. 5. Esker formed as crevasse or tunnel filling in Lake Rörströmssjön, Ångermanland, Sweden (photo J. Lundqvist 1970).

Then it may refer to a broad zone including different landforms and representing a specific stage of deglaciation. The landforms are mostly those of stagnant ice discussed below.

The term moraine usually refers to a frontal position. At a glacier tongue this moraine will curve around the glacier front and extend along its sides as a lateral moraine. In a valley the moraine may form a terrace along the valley side, referred to as a kame terrace or lateral terrace. Like the frontal moraine, it consists of till, glaciofluvial sediments, and slope deposits in varying proportions. The morphology of the terrace is influenced by the adjacent ice margin, either even or pitted by kettle-holes or furrowed by stream channels.

A very special type of marginal formation is the large subaquatic esker first interpreted in this way by De Geer (1897). According to his theory these eskers were formed as incomplete deltas at the mouths of subglacial meltwater channels in open water. During the retreat of the ice margin such deltas were lined up one after the other to form a long and high ridge, a true esker. It is now clear that even though this theory probably is still valid, it does not apply to all eskers (e.g., Tanner 1934), most of which formed in crevasses (Fig. 5). Even the "De Geer eskers" are probably complex features, in which only a part is formed in this classical way (Eriksson 1960).

Stagnant ice wastes down by surface ablation and basal melting. In basal melting subglacial melt-out of debris takes place. A melt-out till is deposited, the landforms of which reflect the forms of the substratum and the distribution of debris in the ice. The landforms consequently lack characteristic features; they are mostly a smooth cover of ground-moraine, to a great extent following the landforms of the substratum. Most probably a large part of the extensive till cover of formerly glaciated regions is of this type. In the stagnant ice, meltwater will easily find its way down to the base, and eskers with winding courses and irregular pattern are formed.

The most characteristic deposit of stagnant ice is the supraglacially formed hummocky moraine (dead-ice moraine), a

Fig. 6. Rounded structures characterize the Veiki moraine of Swedish Lapland, formed either in basal crevasses or, more probably, at the melting of buried, stagnant ice. From the Gällivare region, Swedish Lapland (photo J. Lundqvist 1975).

landform characterized by an irregular landscape of large or small hummocks and ridges and collapse features. The material of the hummocks is any kind of sediment that has accumulated upon the ice surface. These are diamicton (till) melted out by ablation, which either stay in place or move by gravity flow before deposition, further debris released along shear-planes or other debris-rich bands, glaciofluvial deposits and even glaciolacustrine, accumulated in supraglacial ponds (cf. Westergård 1906, Clayton and Cherry 1967). Deposition on an uneven glacier surface causes great differences in thickness, and this is further increased at the differential melting of the ice underneath, followed by slumping and flow. If the sediment is mainly diamicton, the hummocky terrain is referred to as moraine. If glaciofluvial sediments dominate, the term kames is often preferred.

The existence of such a hummocky terrain is often interpreted as evidence that the ice was stagnant at the time of deglaciation, an opinion dominant in the old literature. However, we probably should pay more attention to the fact that in order to get a supraglacial debris cover

thick and extensive enough to form such a landscape, active ice is required before stagnation. The hummocky moraine should therefore be interpreted as indication of rather active ice, the movement of which has ceased successively in only a narrow marginal zone of debris-laden ice (Eyles 1983).

A special case of hummocky moraine, which may indicate regional stagnancy of the ice, may be the Veiki moraine (Fig. 6; Hoppe 1952). It is characterized by numerous rounded depressions at different levels and by plateaus with rim ridges. Its sediments are diamicton (probably flow till) and glaciofluvial or glaciolacustrine deposits. Originally the Veiki moraine was interpreted as subglacial crevasse fillings (Hoppe 1952), but recent investigations (R. Lagerbäck pers. comm.; Minell 1979) point instead to a supraglacial formation. The rounded depressions are then interpreted as kettles, formed at the melting of ice diapirs pressed up from below by the weight of thick debris upon stagnant ice during a fairly long time.

The Veiki forms also occur as isolated ring ridges, very similar to the so-called prairie doughnuts. The formation of these features should consequently be interpreted as a local accumulation of supraglacial debris, temporarily protecting the underlying ice from melting. Slumping and flow from the ice core at the melting create the ring forms (cf. Clayton and Cherry 1967; Parizek 1969).

When hummocky landforms of these types are formed in association with ice that is slightly active or reactivated, they are influenced by the ice movements. More regular ridge systems are created in crevasse systems or by thrusting and pushing. The irregular forms gradually pass into those of the end-moraine zones.

As is clear from the previous discussion, the interpretation of glacigenic landforms is essential for our understanding of the environment and climate at the time of wastage of former glaciers. Some quite new aspects and interpretations may considerably change our opinion in this respect.

## REFERENCES

Åmark, M. 1980. Glacial flutes at Isfallsglaciären, Tarfala, Swedish Lapland. Geol. Fören. Stockh. Förhandl. 102:251-259.

Boulton, G.S. 1968. Flow tills and related deposits on some Vestspitsbergen glaciers. J. Glaciol. 7:391-412.

Boulton, G.S. 1972. Modern Arctic glaciers as depositional models for former ice sheets. J. Geol. Soc. London 128:361-393.

Boulton, G.S. 1976. The origin of glacially fluted surfaces - observations and theory. J. Glaciol. 17:287-309.

Boulton, G.S. and N. Eyles. 1979. Sedimentation by valley glaciers; a model and genetic classification. In Ch. Schlüchter (ed.), Moraines and varves, p. 11-23. Rotterdam, Balkema.

Boulton, G.S. and M. Deynoux. 1981. Sedimentation in glacial environments and the identification of tills and tillites in ancient sedimentary sequences. Precambr. Res. 15:397-422.

Christiansen, E.A. and S.H. Whitaker. 1976. Glacial thrusting of drift and bedrock. Roy. Soc. Can. Spec. Publ. 12:121-130.

Clayton, L. and J.A. Cherry. 1967. Pleistocene superglacial and ice-walled lakes of west-central North America. N. Dakota Geol. Surv. Misc. Ser. 30:47-52.

Clayton, L. and S.R. Moran. 1974. A glacial process-form model. In D.R. Coates (ed.), Glacial geomorphology, p. 89-119. New York, Publ. Geomorph. State Univ.

De Geer, G. 1897. Om rullstensåsarnas bildningssätt. Geol. Fören. Stockh. Förhandl. 19:366-388.

De Geer, G. 1940. Geochronologia Suecica Principles. Kungl. Sven. Vet.-Akad. Handl. Ser. 3, vol. 18(6).

Dionne, J.-C. 1984. Le rocher profilé: une forme d'erosion glaciaire négligée. Géogr. phys. Quat. 38:69-74.

Dreimanis, A. 1969. Selection of genetically significant parameters for investigation of tills. Zesc. Nauk. Univ. A. Mickiewicza w Poznaniu. Geografia 8:15-29.

Dreimanis, A. 1979. The problems of waterlain tills. In Ch. Schlüchter (ed.), Moraines and varves, p. 167-177. Rotterdam, Balkema.

Dreimanis, A. 1982. INQUA-Commission on Genesis and Lithology of Quaternary Deposits. Work Group (1) - Genetic classification of tills and criteria for their differentiation: Progress report on activities 1977-1982, and definitions of glacigenic terms. In Ch. Schlüchter (ed.), p. 12-31. INQUA Commission on Genesis and Lithology of Quaternary Deposits. Report on Activities 1977-1982. Zurich, ETH.

Dreimanis, A. 1983. Precontemporaneous partial disaggregation and/or resedimentation during the formation and deposition of subglacial till. Acta Geol. Hispanica 18:153-160.

Dreimanis, A. and J. Lundqvist. 1984. What should be called till? Striae 20:5-10.

Elson, J.A. 1957. Origin of washboard moraines. Geol. Soc. Am. Bull. 68:1721.

Elson, J.A. 1961. The geology of tills. In E. Penner and J. Butler (eds.). Proceedings of 14th Canadian Soil Mechanics Conference. Nat. Res. Counc. Can. Ass. Comm. Soil and Snow Mech. Techn. Mem. 69:5-36.

Eriksson, K.G. 1960. Studier över Stockholmsåsen vid Halmsjön. Geol. Fören. Stockh. Förhandl. 82:43-125.

Evenson, E.B., A. Dreimanis, and J.W. Newsome 1977. Subaquatic flow tills: A new interpretation for the genesis of some laminated till deposits. Boreas 6:115-133.

Eyles, N. 1983. Modern Icelandic glaciers as depositional models for "hummocky moraine" in the Scottish Highlands. In E.B. Evenson, Ch. Schlüchter and J. Rabassa (eds.), p. 47-59. Tills and related deposits. Rotterdam, Balkema.

Eyles, N., C.H. Eyles, and A.D. Miall 1983. Lithofacies types and vertical profile models; an alternative approach to the description and environmental interpretation of glacial diamict and diamictite sequences. Sedimentology 30:393-410.

Flint, R.F. 1971. Glacial and quaternary geology. New York, John Wiley and Sons.

Francis, E. 1975. Glacial sediments: A selective review. In A.E. Wright and F. Moseley (eds.), p. 43-68. Ice ages: Ancient and modern. Geol. J. Spec. Issue 6.

Gillberg, G. 1976. Drumlins in southern Sweden. Bull. Geol. Inst. Univ. Uppsala. N.S. 6:125-189.

Harland, W.B., K.N. Herod, and D.H. Krinsley 1966. The definition and identification of tills and tillites. Earth Sci. Rev. 2:225-256.

Haselton, G.M. 1979. Some glaciogenic landforms in Glacier Bay National Monument, Southeastern Alaska. In Ch. Schlüchter (ed.), p. 197-205. Moraines and varves. Rotterdam, Balkema.

Hillefors, A. 1969. Västsveriges glaciala historia och morfologi. Naturgeografiska studier. Medd. Lunds Univ. Geogr. Inst. Avh. 60.

Hoppe, G. 1952. Hummocky moraine regions with special reference to the interior of Norrbotten. Geogr. Ann. 34:1-72.

Hoppe, G. 1957. Problems of glacial morphology and the Ice Age. Geogr. Ann. 39:1-18.

Lawson, D.E. 1979. Sedimentological analysis of the western terminus region of the Mantanuska Glacier, Alaska. U.S. Army Corps of Engineers, Cold Regions Research and Engineering, Rep. 79-9.

Lundqvist, J. 1969. Problems of the so-called Rogen moraine. Sver. Geol. Unders. C 648.

Lundqvist, J. 1984. INQUA Commission on Genesis and Lithology of Quaternary Deposits. Striae 20:11-14.

Lundqvist, J. 1985. What should be called Glaciofluvium? Striae 22:5-8.

Minell, H. 1979. The genesis of tills in different moraine types and the deglaciation in a part of central Lappland. Sver. Geol. Unders. C 754.

Moran, S.R., L. Clayton, R. Hooke, M.M. Fenton, and L.D. Andriashek 1980. Glacier-bed landforms of the prairie region of North America. J. Glaciol. 25:457-476.

Muller, H.M. 1983. Dewatering during lodgement of till. In E.B. Evenson, Ch. Schlüchter, and J. Rabassa (eds.), p. 13-18. Tills and related deposits. Rotterdam, Balkema.

Parizek, R.R. 1969. Glacial ice contact rings and ridges. Geol. Soc. Am. Spec. Pap. 123:49-102.

Petersen, G. 1924. Die Schollen der norddeutschen Moränen in ihrer Bedeutung für die diluvialen Krustenbewegungen. Fortschr. Geol. Palaeontol. 9:179-274.

Rabassa, J., S. Rubulis, and J. Suárez 1979. Rate of formation and sedimentology of (1976-1978) push-moraines, Frias Glacier, Mount Tronador (41°10'S; 71°53'W), Argentina. In Ch. Schlüchter (ed.), p. 65-79. Moraines and varves. Rotterdam, Balkema.

Robin, G. de Q. and P. Barnes 1969. Propagation of glacier surges. Can. J. Earth Sci. 6:969-977.

Stevens, R. 1985. Glaciomarine varves in late-Pleistocene clays near Göteborg, southwestern Sweden. Boreas 14:127-132.

Strömberg, B. 1965. Mappings and geochronological investigations in some moraine areas of south-central Sweden. Geogr. Ann. 47A:73-82.

Svensson, H. 1970. Spår av "fluted surfaces" från den senaste nedisningen. Geol. Fören. Stockh. Förhandl. 92:79-85.

Tanner, V. 1934. The problems of the eskers. IV. The glaciofluvial formations of the Rassé muetke valleys, Petsamo, Lapland. A geomorphological study of the origin and development of the shape and configuration of supra-aqueous deposited eskers. Fennia 58:1.

Thwaites, F.T. 1943. Pleistocene of part of northeastern Wisconsin. Bull. Geol. Soc. Am. 54:87-144.

Westergård, A.H. 1906. Platålera, en supramarin hvarfvig lera från Skåne. Sver. Geol. Unders. C 201, 9 pp., and Geol. Fören. Stockh. Förhandl. 28:408-414.

Whittecar, G.R. and D.M. Mickelson 1979. Composition, internal structures, and an hypothesis of formation for drumlins, Waukesha County, Wisconsin, U.S.A. J. Glaciol. 22:357-371.

*Genetic Classification of Glacigenic Deposits, Goldthwait & Matsch (eds)*
*© 1988 Balkema, Rotterdam. ISBN 90 6191 694 1*

# Tills: Their genetic terminology and classification

Aleksis Dreimanis
*Department of Geology, University of Western Ontario, London, Ontario, Canada*

ABSTRACT: The origin and meaning of "till" and other synonymous and/or related terms, such as "moraines," are discussed first. Then the development of genetic classifications of till during the last 150 years is reviewed.

The polygenetic character of till and the synchronism of deposition at various places in the glacier system have to be accommodated by a broad genetic definition of till, and by a broad classification, which can be supplemented by a more detailed step-wise classification in a hierarchical manner.

The genetic classification of tills may begin with their categorization into primary and secondary tills based on the dominant processes of formation and deposition in relation to glacier ice, or by their place of deposition: subglacial, supraglacial, and ice-marginal tills. A more detailed classification is made possible by applying multiple criteria, and by considering such formative parameters as manner of deposition, release of debris from glacier ice, mode of debris transport, and derivation of glacial debris prior to its deposition. Syndepositional and postdepositional glacitectonic deformations also have to be considered, but the terminology of their products is still under debate.

The opinions on diagnostic criteria for the recognition of detailed genetic varieties of till still differ so much that the criteria for only three main varieties, lodgement, melt-out, and flowtill (broad meaning) are given, and even they overlap to a great extent.

The translation of the most commonly used glacial terms in twenty languages and several definitions of terms related to till are given in the Appendixes.

## 1 INTRODUCTION

Till is the most widespread and variable of all glacial or glacigenic deposits, and it is also probably one of the most widespread sediments on earth. It is particularly common in the northern hemisphere, covering over 30 percent of all continents (Goldthwait 1971:3) as unconsolidated Quaternary sediment. Another 10 percent or more of the land may be underlain by its lithified variety, tillite, from glaciations hundreds of millions of years ago (ibid.).

Tills and tillites, being glacial deposits, serve as environmental and climatic indicators for the deciphering of the geologic history of the earth. Since tills are traceable laterally for considerable distances, they are also commonly designed as lithostratigraphic units, even though the International Stratigraphic Guide (Hedberg 1976:43) suggests that lithogene-tic terms should be avoided for formally named lithostratigraphic units. However, the same guide urges a preservation of traditional names.

Both glacial transport and deposition are reflected in tills. They tell us about the paths of former glacial movements, and thus assist in the reconstruction of former glacial flow patterns and distribution of glaciers. Tills are also useful for indicator tracing in search of ore deposits in formerly glaciated areas (Hirvas et al., this volume; Salonen, this volume).

Tills are of considerable concern to hydrogeologists (Grisak et al. 1976), engineering geologists (Milligan 1976; White and Ellis, this volume), and pedologists (St. Arnaud 1976). Tills may behave either as aquicludes, or as aquifers, depending upon hydraulic conductivity. Tills are encountered in all kinds of construction projects, either as foundation material to be excavated, or as material to be used as fill. Where exposed on the land surface, till is the parent material of soils and influences agricultural and forest developments.

The properties of till depend upon a variety of factors, such as the type of material incorporated by glaciers to form each till, and also the processes of genesis, including the formation and deposition of till (see also Lundqvist, this volume) that also influence its composition, structure, and fabric. Postdepositional processes may continue to influence the properties of till, but a discussion of these changes is beyond the scope of this discussion.

This paper has resulted from the task given to the Till Work Group (2-A) by the INQUA Commission on Genesis and Lithology of Quaternary Deposits (further referred to as INQUA Commission C-2) in 1973: to develop a genetic classification of till and to establish criteria that characterize the different genetic varieties of till. The handling of this task during the 1974-86 period will be further discussed in Chapter 6.5. Some background information, particularly on the term "till" and its equivalents, and on the development of the classifications of till, is presented here in Chapters 3-6.4. The results of the Till Work Group's work are given in Chapters 7-14.

Though the Till Work Group's task was to discuss and establish a genetic classification of till, the classification itself and the terminology of the genetic varieties of till were not considered the most important objectives. The names of tills are merely means of communication among those who are dealing with glacial deposits. The means of communication have to be accurate, but linguistically simple and understandable not only to specialists, but also to others who must deal with them. Furthermore, the terms must be interchangeable among various languages or translatable from one language to another without great problems. A separate dictionary and glossary were planned to serve this purpose, but for various reasons had to be postponed. Therefore, a very condensed dictionary, containing the main terms in English and their equivalents in 20 languages, is added as Appendix C. A brief glossary of some terms is given as Appendix B.

As will be discussed in Chapters 8 and 9 (particularly in 8.3 and 8.4), the formation and deposition of till is a complex polygenetic process in four dimensions, three of them in space, the fourth in time. In order to decipher these processes and their products, till investigations

have been made and will continue at selected or merely available and accessible locations in the areas of present-day glaciers and past glaciations. From such studies the investigators develop local depositional models, and many of them propose that their models are universally applicable. While discussing a similar modelling approach to fluvial sedimentation, which is not even as complex as its glacial counterpart, Miall (1985:262) makes the following statement: "It has become clear that these models reflect fixed points on a continuum of variability...The continuum is, in fact, a multidimensional one because of the complexity of partly interdependent controls that govern fluvial sedimentation. A continuation of modelling studies along existing lines will simply result in a proliferation of arbitrary fixed points. Clearly, a new approach is needed." Miall (1985) proposed an architectural-element analysis as a new method of facies analysis for fluvial deposits.

In our case, the deciphering of glacial sedimentation may also need a new approach, and even some of the present approaches have not been tested sufficiently. Therefore, many members of the Till Work Group have proposed that discussion of glacial sedimentation processes has to be continued by the INQUA Commission C-2 during the next intercongress period.

## 2 ACKNOWLEDGEMENTS

Though A. Dreimanis is listed as the author of this paper, the paper is a result of the collective work of over 150 members of the Till Work Group (Appendix A) of INQUA Commission C-2, plus many other scientists who have participated from time to time in the discussions of the Till Work Group, either at its meetings, by correspondence, or by individual exchange of opinions among the Till Work Group members.

Originally it was planned to distribute this report among all the currently active Till Work Group members, and then to list those as co-authors who had participated most actively at revising the submission. However, for various reasons, we ran short of time, and the first draft of this report could only have been reviewed by a small group. Therefore, I am listing all the co-workers as co-authors in Appendix A, with the names of the most active participants in bold print (some were most active at the beginning, others towards the completion of the project). The names

of contributors to the foreign terminology
table are listed in the introduction to
Appendix C.

My sincere thanks to all co-workers for
their ideas, criticism, and patience, particularly to T.J. Kemmis and C.L. Matsch
for the very thorough review of the semi-
final draft of the manuscript.

The expenses of the preparation of this
report have been covered by grants from
INQUA and NSERC.

# 3 ORIGIN AND MEANING OF THE TERM "TILL" IN THE PAST

"Till" is a term that long antedates the
Glacial Theory. It is a Scottish word,
used by countryfolk to describe "...a kind
of coarse obdurate land, the soil developed on the stony clay that covers much
of northern Britain" (Flint 1971:148).
Since the early detailed glacial studies
in Britain were done particularly by
Scots, this term became widely used. Thus
Agassiz (1842:228, according to von Böhm
1901:99) described the Scottish "till" as
follows: "The rounded, polished and
scratched blocks of very various dimensions, are everywhere indiscriminately
mixed together in a marly or clayey
paste." Geikie (1863:185) defined till
more accurately as a "...stiff clay full
of stones varying in size up to boulders
produced by abrasion carried on by the ice
sheet as it moved over the land." This
definition contains both lithologic and
genetic aspects. In the lithologic part of
the definition, the emphasis is on diamictic texture ("clay full of stones") and
compactness ("stiff"). The genetic part
emphasizes till's glacial origin
(transport and abrasion). Since the term
"till" was not meant to be used for any
nonglacial sediments, the double term
"glacial till," still often seen in print,
is redundant (Flint 1971:148).

Till is more variable than any other
sediment known by a single name, as
pointed out independently by two experts
on tills — Flint (1971:154) and
Goldthwait (1971:4) — and many others who
have studied tills. Therefore, the meaning
of "till" varies from one investigator to
another, although everyone agrees that the
word "till" does mean glacial handling
(Goldthwait 1971:5).

Till variability and diversity have
frustrated many students of tills and
till-like sediments. Harland, Herod, and
Krinsley (1966:231) have tried to avoid
frustrations by applying the terms "till"
or "tillite" in a very broad meaning to
any diamictic sediment that contains glacially transported material, for instance.
Others, for instance Lawson (1979a:28),
have gone the opposite way, using very
narrow restrictive definitions, by
selecting the "most glacial" end members
(see also Chapter 7.2 of this book). Some,
for instance Drewry (1986:126), even think
that "...the genetic term 'till' ...is no
longer applicable and new nomenclature is
necessary."

Actually it does not matter what
nomenclature is used for a genetic
designation of glacial sediments, as will
be further discussed in Chapter 7. Glacial
sediments are not formed and deposited
merely by glacier ice: they are polygenetic, produced by glacier ice and its meltwater (except for sublimation till), with
the participation of gravity. If a definition of "till" considers its polygenetic
nature and close relationship to glacier
ice, no new term is needed.

As a convenient short term for glacial
sediment, "till" dominates not only in
English but is also invading many other
languages (see Appendix C). It gradually
replaces the previously dominant French
term "moraine" (see Chapter 4),
restricting "moraine" mainly for
landforms.

Though most till investigators have
emphasized that tills are usually diamictic, they may range from clast-supported
very stony tills to clayey tills with few
boulders visible (see also Lundqvist, this
volume). Their granulometric composition
is usually bimodal or multimodal. Tills
may be massive (nonstratified) or stratified; many are fissile. Lithologically
they may consist of any material that has
been picked up and transported by a
glacier, and the presence of at least some
distantly transported ("erratic" or
"foreign") material has often been cited
as one of the characteristics of till.
(For the characteristics of till and
further references, see Flint 1971;
Goldthwait 1971; Dreimanis 1976; Sugden
and John 1979; Dreimanis and Lundqvist
1984; Dreimanis and Schlüchter 1985; Shaw
1985).

Whatever criteria are chosen for
distinguishing till from other similar
looking sediments, there is no sufficient
single criterion. Multiple criteria have
to be applied, and they will be further
discussed in Chapters 8-11.

# 4 OTHER NAMES EQUIVALENT TO "TILL"

In the Western Alps, the French word
"moraine" has been used for glacial depo-

sits as landforms, along the present-day valley glaciers (for references and more details, see von Böhm 1901; von Klebelsberg 1948:156; Charlesworth 1957:404; Lliboutry 1965:690). This term, in various spellings, began to appear in print in the 1770's.

De Saussure (1779) introduced "moraine" in the currently used spelling in the French scientific literature. In the Swiss German literature, the local term "Gandecke" was used until about the middle of the 19th century. In English, the term "moraine" was applied both to glacial landforms and to glacial materials during the last quarter of the 19th century. In German, two spellings appeared at the beginning: "Moraine" and "Moräne," the latter one taking over gradually.

Up to 1838, the term "moraine" was applied mainly to glacier-deposited landforms, and also to the sediment they contained, but not to the glacial debris on glacier ice called "Guffer" in German in the Alps (von Böhm 1901:328). Agassiz (1838), as pointed out by von Böhm (1901:68-89), was first to introduce the application of "moraine" to glacial debris.

During the 19th century, the French and German languages were used more than any others in the scientific literature and in international communications all over continental Europe. Therefore the name "moraine" and its derivatives — for instance, the French "moraine profonde," the German "Grundmoräne" — became firmly established as the dominant designations for glacial debris, glacial sediments, and glacial landforms in most European languages, with specific linguistic modifications in spelling and pronunciation.

In English, the word "moraine" was and still is used mainly for landforms, with one notable exception. It was used also for debris on the surface of glaciers, such as "lateral moraine," "medial (median) moraine" and "ablation moraine." Recently "ablation moraine" in reference to supraglacial materials was replaced by the terms "supraglacial drift" (Flint 1957:121) or "supraglacial debris," a term used much earlier by Wright (1889).

The prevalent sedimentologic meaning of "moraine," as for "till," has been genetic: a glacial sediment. Both terms, however, imply also a textural and lithologic meaning: a variable particle size matrix with large clasts, many of which are of distant ("erratic") derivation.

Another group of old English terms that has been used synonymously with "till" or "moraine," particularly during the 19th century are: "unstratified drift," "nonstratified drift," "commingled drift." Chamberlin (1883:296) applied them in the structural classification of drift. Although "drift" was introduced in the British geologic literature about 1840 for material drifted by icebergs into the sea and deposited on the sea bottom, this word was changed to a collective term for glacially derived sediments after the Glacial Theory took hold. The dominantly nonstratified character of till was emphasized in the above group of terms, but their glacial derivation was also implied.

Descriptive terms with implied glacial genetic meaning also developed during the 19th century, emphasizing the presence of boulders in a fine-grained matrix, for instance the English "boulder clay," the French "argile a blocaux," the German "Geschiebelehm," "Geschiebemergel." The latter also points out that the matrix is calcareous. Similar terms also developed in other languages. All of them still had genetic meaning — glacial sediment — although this was not expressed verbally.

Some composite terms, with "glacial" as one of the words in them, have been used as substitutes for "till." Thus Rukhina (1973:32) calls tills in Russian "sobstvenno lednikovie otlozhenia," which means "glacial deposits proper" (Rukhina 1980:12, reproduced here as Table 7). Clayton and Moran (1974) use "glacial sediment" instead of "till," and similar terms have been used also in other languages (see Appendix C).


5 NONGENETIC DESCRIPTIVE TERMS USED FOR TILL

The purpose of this report is not to deal with a descriptive classification of till. However, any identification of a sediment as a till and its further, more specific genetic classification has to be preceded by an objective description of the sediment investigated.

During recent decades, sedimentologic investigations of glacigenic deposits became more and more detailed. They demonstrated that not all deposits with a till-like appearance were true tills, nor glacigenic deposits of any kind.

Since the already existing descriptive terms, such as "boulder clay," had the connoted glacial origin, new purely descriptive terms were needed and they began to appear in print. Some examples will be mentioned, mainly those which have been applied to till.

Pettijohn (1949:211) proposed to include

tills in the group of "cataclastic rudites" and described them as "glacial conglomerates." Miller (1953:26-27) applied the term "'conglomeratic' sandy mudstone" to sediments that resembled till but which he interpreted as marine. Wayne (1963) followed his example for describing tills: he first applied the term "conglomeratic mudstone" in the descriptions of the Pleistocene rock stratigraphic units in Indiana, but subsequently interpreted them as glacial sediments and called them "tills." Pettijohn (1957:261) included tills in "paraconglomerates."

About the same time, Flint, Sanders, and Rodgers (1960) proposed the terms "diamicton" and "diamictite" as nongenetic terms for terrigenous sediments and sedimentary rocks consisting of a wide range of particle sizes. Harland, Herod, and Krinsley (1966) further suggested "diamict" as a general term for all poorly sorted sediments, both their lithified and unlithified varieties. Schermerhorn (1966) proposed "mixtite" with a definition similar to "diamict." It is interesting to note that a similar-sounding term, "miktit," had been proposed already by Teodorovich (1939) about 25 years earlier. Flint (1971:Table 7-B) broadened the term "diamicton" to include all possible poorly sorted materials, even volcanic. Since that time, "diamicton," "diamictite," and "diamict" have been accepted by most investigators for poorly sorted sediments of various origins, as a nongenetic term to be used before assigning their origin. IGCP Project 38: Pre-Pleistocene Tillites found that diamictite was preferred by a majority of that group (Hambrey and Harland 1981:23). "Mixtite" is still used by some authors, for instance Martin, Porada, and Walliser (1985) and Spencer (1985).

Whatever descriptive term is used as the initial step, a genetic interpretation is applied as the next step by most authors, either by using "till" or "moraine" or their equivalents in other languages (see Appendix C), and then proceeding with further genetic classification, if the pertinent characteristics for such naming or classification can be identified.

# 6 DEVELOPMENT OF GENETIC CLASSIFICATIONS OF TILL

When the widespread occurrence of glacial sediments was first recognized during the middle of the 19th century (see Chapter 3), the deposits were not further classified into genetic subdivisions, and general terms were applied to them such as nonstratified drift, commingled drift, moraine, till, boulder clay.

## 6.1 Early period: 1840-1900

During the early studies of glaciers and their deposits, the main attention was paid to landforms. Observations were gathered mainly along modern Alpine glaciers, and they led to attempts at deciphering the origin of both the landforms and their sediments, including till or moraine.

In the Alps, superglacial till, called "moraines superficielles" (de Charpentier 1841) was distinguished from basal till or ground moraine, then called "moraine profonde" (Martins 1842:343). Similar distinction was made in the areas of continental glaciations, for instance by Goodchild (1875), Krapotkin (1876), and Torrell (1877), who distinguished a basal compact till overlain by a looser and coarser textured or laminated upper till. Several terms developed for each of these two varieties of till, such as:

1. upper till, superficial till, superglacial till, englacial till, surface moraines, or superglacial moraines, and similar terms in other languages;

2. (true) till, subglacial till, basal till, lodgement till, moraine profonde (even in the English papers), ground moraine, and equivalent terms in other languages.

T.C. Chamberlin, who was an excellent observer, was particularly interested in the genetic classification of glacial deposits, both the sediments and the landforms, and he published his classifications in two papers: 1883 and 1894. (The 1894 paper had been presented previously at the International Geological Congress, 1891).

Chamberlin (1883) considered the following three classes of till in his genetic classification of drift:

1. subglacial till;

2. upper till — englacial or superglacial till; and

3. subaqueous till — berg till, floe till.

Chamberlin (1883:297) referred to the subglacial till as "true till" and suggested that the term "upper till" used by others was unfortunate, since "...it is liable to confusion with a second true till overlying the first." Both the subglacial and superglacial tills were considered by him to be produced by the

direct action of glaciers, but in a polygenetic manner. Chamberlin (1894:518)
emphasized that glaciers were merely
assumed to be the primary and chief
agents, and that the secondary and associated agencies were very important. His
third class, subaqueous tills, are
entirely secondary products, and he later
renamed them "glacionatant tills" (ibid.
1894:536).

The terms "superglacial," "englacial,"
"subglacial," as used by Chamberlin (1883,
1894) were applied not only to deposition,
but also to glacial transport, particularly the term "englacial."

Concerning the subglacial transport of
debris, two mechanisms had been considered
by glacial geologists during the 19th century: glacial drag under the ice sole, and
transport of debris in the basal part of
glacier ice, as discussed for instance by
Penck (1882:37-39), Heim (1885:349-351),
and von Böhm (1901:232-233). Chamberlin
(1882:259) adhered to the view of subsole
transport when discussing his observations
on the recent glacial deposits in the
Alps, but later on (ibid. 1883, 1894) he
did not specify the mode of transport
prior to the deposition of subglacial
till. As for the deposition of subglacial
till, Chamberlin (1894:25) suggested that
it was lodged. However, he did not
describe the process of lodgement, and
even used the term "lodged" in reference
to the deposition of loess by wind. Though
T.C. Chamberlin is usually credited with
the introduction of the concept of lodgement for the deposition of subglacial
till, Upham (1892:136-137) also used the
term "lodged" and described the process
more or less in the same way as it is
understood now.

Chamberlin's (1894:527) descriptions of
englacial and superglacial till emphasized
that this till was let down by passive
melting. It should be mentioned here that
passive melting out of stagnant ice was
considered by Goodchild (1875) to be the
main process of deposition of all tills,
and Upham (1891, 1892) had discussed
englacial drift in great detail.

In the 1894 classification, Chamberlin
added another process in glacial deposition under the heading "Formations produced by the direct action of Pleistocene
glaciers" (Chamberlin 1894:521): "...the
mechanical action of the edge of the ice"
affecting both glacial and nonglacial
material. Two varieties of landforms were
included here — push moraines and lateral
moraines, but no name was assigned to the
material.

A classification of tills, similar to
Chamberlin's (1894) was published by

Woodworth (1899). He distinguished the
following varieties of till, with examples
of landforms or materials:
 I Intraglacial till
   - superglacial
   - englacial
   - subglacial
 II Extraglacial till
   - ice bound
        (e.g., in moraines)
   - ice-free
        (e.g., berg till)
In that same year (1899), E. Richter
organized a Glacier Conference at Gletsch,
Switzerland, attended by 17 French- and
German-speaking European scientists and
one American. This conference developed a
classification and nomenclature of
moraines (Richter 1901, discussed in von
Böhm 1901:218-268), in three languages
(German, French, and English). This
classification did not contribute much to
the existing classifications of till or
moraines as deposited material. Its emphasis was on classification of landforms
with two main groups: moving moraines, and
deposited moraines. Before the conference
the group of terms "Grundmoräne = moraines
de fond = moraines profondes = ground
moraine" had been applied both to basal
till and basal debris, but the new classification used these terms for the deposited moraines only, and proposed a new
term "Untermoräne = moraines inferieures"
for glacial debris still moving with the
glacier. Von Böhm (1901:220-268) sharply
criticized the exclusion of moving
material from the term "Grundmoräne" (=
basal till), because glacial drag of the
subsole material was a very popular concept at that time. Also, the exclusion of
surface moraines (= superglacial till)
from the deposited moraines would imply
that ground moraine (Grundmoräne) as
material consists both of subglacial and
supraglacial till.

In summary, by the turn of the century,
the twofold division of till into
superglacial and subglacial till was
generally accepted, except for the Glacier
Conference 1899, Subaquatic secondary or
extraglacial tills were also recognized.


6.2 First half of the 20th century

No significant change in genetic classification or nomenclature is noticeable in
the literature on glacial deposits during
the first half of the 20th century,
although several new terms began to
appear. Tarr (1909) proposed the term
"ablation till" which, together with its
equivalent "ablation moraine," became used

interchangeably with "superglacial till" and "superglacial moraine." The term "tillite" was first used for indurated till by Penck (1906).

Von Klebelsberg (1948:157, 158) introduced two specific derivation-related terms for the supraglacially transported debris and supraglacial till: "Hangschuttmoräne" (talus moraine, talus till) and "Bergsturzmoräne" (landslide moraine, landslide till). Both terms were used for the material and the landform.

After some 50 years, Flint (1947) still referred to the twofold genetic classification:
1. "superglacial till" in "ablation moraine"; and
2. "basal till."

It appears that the position of deposition of till (superglacial versus basal) was considered to be of particular significance, since it indirectly indicated different processes of till deposition (Flint 1947:111). More attention was paid now to the descriptive textural and lithologic characteristics of tills, and the identification of its erratic components, than to their genesis. Such descriptive lithologic characteristics were found to be useful as practical criteria for mapping tills, their stratigraphic correlation, and the deciphering of the directions of glacial movements.

The scope of this report does not permit discussion of the investigations of the physical characteristics of till, but it may be mentioned that one of them, granulometric composition, has been used continuously in Sweden as the basis for the classification of till since G. Lundqvist, Sr., proposed it in 1930 (see Lundqvist 1940). Similar trends in the investigation and classification of tills by their texture and lithology also prevailed in other countries, and this approach influenced the development of genetic classifications during the second half of the 20th century, as discussed below.

## 6.3 Revival of genetic classification, middle of 20th century

Since certain groupings of the physical characteristics of till occasionally implied genetic significance, interest in the genesis of till was renewed. This was already noticeable in the second edition of Flint's textbook on glacial and Pleistocene geology (Flint 1957:120-121). There terms denoting depositional processes are used rather than those referring merely to the position of deposition. They now dominate the still traditional twofold classification of till:
1. ablation till (=superglacial till); and
2. lodgement till (=basal till).

The lodgement process was questioned by some investigators of till, and they considered or even preferred the basal melt-out interpretation, for instance, G. Lundqvist (1940) and Harrison (1957).

In an attempt to explain various small-scale structures observed abundantly in subglacial till of eastern Finland, Virkkala (1952:107-109) proposed a combination of accretion of basal drift with subsequent shearing, sporadic erosion, and deformation underneath a moving glacier. In other words, he postulated a combination of lodgement with subsole erosion and deformation and also, that "...the difference between deposited till and basal drift in transport may meanwhile not have been distinct" (ibid.:108).

Flint (1957:122) pointed out that ablation till might derive from superglacial drift, described in detail by Sharp (1949), that was more thoroughly modified by nonglacial reworking than is basal drift, but no specific name was assigned to this modified till. Harrison (1957:298) informally called it "solifluction till" and "mudflow." Hartshorn (1958) proposed a new genetic term "flowtill" for secondary resedimented till in the superglacial ablation environment, and Boulton (1968) described it in detail from present-day glaciers in Spitsbergen.

Elson (1961) proposed the following classification of tills:
a. superglacial ablation till; b. subglacial ablation till; c. comminution till; d. deformation till.

Elson's (1961) subglacial ablation till corresponds to that which was later called "subglacial melt-out till" by Boulton (1971). Together with the "comminution till," also a new term, the new names (b) and (c) cover what was called "lodgement till" at that time. However, the process proposed by Elson (1961) for the formation of comminution till was new (see also Elson, this volume). Deformation till as proposed by Elson (1961; see also Elson this volume) was a new term, proposed for subsole material that was not only deformed by glacial drag, but also translocated by the glacier before its final deposition; it has commonly been called "soft till" by engineers (Elson 1961).

The participation of water in the formation and deposition of secondary till, proposed by Chamberlin (1883, 1894) and

Table 1.  Genetic classifications of till.
A – Derivation of till and its classification (Dreimanis 1969: Table I).
B – Proposal of genetic classification of tills by Till Work Group, November 20, 1979,
in its Circular No. 16, published in Hambrey and Harland (1981:24).
C – Classification (B) published by the Chinese Research Group of Glaciol.Sediments (1981: 79).

**A**

| GLACIAL DRIFT IN TRANSPORT | GLACIAL DRIFT DEPOSITED AS TILL | |
|---|---|---|
| | ON LAND | IN LAKE OR SEA |
| SUPERGLACIAL DRIFT → | ABLATION TILL | WATERLAID TILL |
| GLACIAL ICE — ENGLACIAL DRIFT | | |
| BASAL DRIFT | BASAL TILL — DEPOSITED BY BASAL MELTING / DEPOSITED BY LODGMENT | |
| | ══ LOCAL TILL ══ | |
| DEFORMED BEDROCK OR SEDIMENTS → | DEFORMATION TILL | |

**B**

| Glacial debris in transport | Facies of tills by position of deposition | | Terrestrial tills | Waterlain tills |
|---|---|---|---|---|
| | | | Facies of tills related to process of deposition | |
| Supraglacial debris | Proglacial | Ablation till | Flow till | Waterlain flow till |
| | Supraglacial till | | Lowered till / Flow till / Melt-out till / Sublimation till | — |
| Englacial debris | Subglacial (or basal) till | | Melt-out till / Lodgement till / Deformation till / Flow till | Waterlain melt-out till / Waterlain flow till / Iceberg till |
| Basal debris | | | | |
| Deformed bedrock or deformed sediments and/or glacially eroded surface of rocks or sediments | | | | |

**C**

表1  冰碛物的成因分类（国际第四纪委员会，1979.11）

| 搬运中的冰川岩屑 (Glacial Debris) | 冰 碛 (Till) | | | |
|---|---|---|---|---|
| | 根据沉积作用位置的冰碛相 | 陆地冰碛 | | 水域冰碛 (Waterlain Till) |
| | | 与沉积作用有关的冰碛相 | | |
| 冰面岩屑 | 前 碛 | 碛融溜 (Ablation Till) | 流 碛 | 水 成 流 碛 |
| 冰内岩屑 | 表 碛 | | 低流碛 / 融出碛 / 升华碛 | |
| 冰下岩屑 | 下碛（或底碛） | | 融出碛 / 滞 碛 / 变形碛 / 流 碛 | 水成融出碛 / 水成流碛 / 冰 出 碛 (Iceberg Till) |
| 变 形 基 岩 或 变 形 沉 积 物 和 1 或 基 岩 或 沉 积 物 的 冰 川 侵 蚀 面 | | | | |

Woodworth (1899), had been disregarded for more than half a century, with a few exceptions (Beskow 1935; Flint 1947 and 1957). A new term for subaquatic till was proposed by pedologists Odynski, Wynnyk, and Newton (1952) — "lacustro-till" — for crudely stratified till-like sediment in Alberta, Canada, but the term remained unnoticed for some time.

Möller (1960) proposed the term "lee side lens till" (described in more detail in Hillefors 1973) for till rich in lenses of sorted material and deposited on the lee side of bedrock knobs in Sweden. Some other varieties of tills containing an abundance of lenses of sorted material had been noticed previously in Sweden and named "Kalix till" (Beskow 1935; Lundqvist 1977). A close interrelationship of basal melting and the deposition of all varieties of drift subglacially was proposed by Carruthers (1939) in England, in his very extreme monoglacialistic under-melt hypothesis, which entirely disregarded glacioaquatic sedimentation outside glaciers.

Rukhina (1960) proposed a genetic classification of the "deposited moraines" (olhozhennye moreni), meaning tills, by considering the following criteria: 1. the place of the accumulation of material in relation to the body of the glacier; 2. mobility of the glacier; 3. composition of the morainic material; 4. transformation of morainic material (syndepositional and postdepositional).

The resulting classification of Rukhina (1960:21) recognizes two main groups of moraines (tills): 1. typical moraine; and 2. transformed moraines, with five subtypes.

The transformation or modification is accomplished by syndepositional and/or postdepositional mass movements and deposition in water, for instance by deposition from icebergs. The transformed moraines form a transition from true moraines to fluvioglacial, limnoglacial, glaciomarine "basin facies" and alluvial deposits. Since this classification was published in Russian only, it remained unnoticed by most Western glacial geologists.

## 6.4 The 1961-73 period of the INQUA Commission on Genesis and Lithology of Quaternary Deposits

In 1961 the Commission on Genesis and Lithology of Quaternary Deposits was established at INQUA, under the chairmanship of B. Krygowski and E.V. Shantser. One of its tasks was to develop a classi-fication of glacial deposits. Symposia and seminars were organized in Poland in 1963, 1967, 1968, and 1970-71, in the Soviet Union in 1964 and 1969, and in Paris, during the VIII INQUA Congress in 1969 (Krygowski 1971). As a result of these discussions, several papers on genetic and petrogenetic classifications of moraines or tills were published in the Soviet Union and Poland — for instance by Strelkov (1965), Shantser (1966), Goretskyi (1968, mentioned in Lavrushin 1976), Dreimanis (1969), Krygowski, Rzechowski, and Stankowski (1969), Raukas (1969), Lavrushin (1970a), and Rukhina (1973).

Because of an overall increase in interest in tills and other glacigenic deposits, several conferences independent of the INQUA Commission C-2 were held in North America and Europe: in Edmonton, Alberta, 1962; Columbus, Ohio, 1969; Guelph, Ontario, 1971; London, Ontario, 1972; Stockholm, 1969; Uppsala, 1972; Trondheim, 1973 (Dreimanis 1976). Although a genetic classification of tills was not the main topic at these conferences, they contributed very much to the development of genetic classifications.

Returning to the INQUA Commission C-2 meetings, the 1967 conference in Poland (Krygowska 1969) should be mentioned in particular, because several proposals for classifications of till were presented and discussed there. Three of them were genetic: Shantser (published in 1966), Dreimanis (1969:16), and Raukas (1969:12). The Dreimanis (1969) classification is reproduced here in Table 1-A, since it became the basis for further classifications discussed at the beginning of the Till Work Group's activities during the 1974-86 period.

Krygowski, Rzechowski, and Stankowski (1969) presented three proposals for a comprehensive genetic-lithologic classification. The application of the main one of these three Polish classifications is well illustrated in Krygowski (1974: Fig. 2). Krygowski's classification is reproduced here as Table 2 and its application as Table 3. Probably because of their complexity, none of the three above classifications have been applied outside Poland, but they have been used in Poland, particularly for stratigraphic investigations, as discussed in the review paper of Niewiarowski (1976).

Several classifications of till that considered the genetic aspects were published independently of the INQUA Commission's work during the period 1961-73.

Thus Lliboutry (1965:690-705) discussed

Table 2. B. Krygowski's (1969) classification of tills.

| Numeration of distinctions | Classificatory distinctions (units and subunits) | | | | | Criteria – basis of classificatory distinctions | |
|---|---|---|---|---|---|---|---|
| I Genetic type | morainic deposits (I) | | | | | general habitus | genetic criteria |
| II Subtype | "continental" (IIa) | | subwater (IIb) | | | general habitus, structure, erratics, a.o. | |
| III Facies | basal (IIIa) | ablational (IIIb) | ...of thalassotope* (IIIc) | ...of fluviotope* (IIId) | ...of limnotope* (IIIe) | general habitus, structure, erratics, stratigraphic and morphologic situation a.o. | |
| IV Subfacies | ...of active ice (IVa) | ...of stagnant ice (IVb) | ...of dead ice (IVc) | a.s.o. | | general habitus, structure, erratics, stratigraphic and morphologic situation a.o. | |
| IVa | morainic deposits of active ice, "continental", basal | | | | | see IV | |
| V Order | glacial tills | | glacial sands and gravels | | | as above and; mechanical composition, grain abration a.o. | physical–chemical criteria |
| VI Family | structureless (VIa) | stratified (VIb) | slaty (VIc) | disturbed (VId) | a.s.o. | as above and; structure, chaotic (structureless), slaty, disturbed a.o. | |
| VIa | structureless glacial tills | | | | | see IV | |
| VII Subfamily | ...with distinct orientation of pebbles (VIIa) | | ...with weak orientation of pebbles (VIIb) | | | as above and texture (fabric) | |
| VIII Genus | ...with dominating petrographical type d/D (VIIIa) | ...with dominating petrographical type c/D (VIIIb) | a.s.o. | | | as above and petrographic types | |
| VIIIa | glacial tills with dominating petrographical type d/D | | | | | see VIII | |
| IX Subgenus | ...with higher quantity of carbonates (IXa) | | ...with lower quantity of carbonates (IXb) | | | as above and quantity of carbonates | |
| X Species | ...with higher quantity of montmorillonite (Xa) | ...with lower quantity of montmorillonite (Xb) | a.s.o. | | | as above and quantity of montmorillonite | |
| XI Variety | grey (XIa) | brown (XIb) | | | | as above and color | |

Table 3. Application of Krygowski's (1969) classification of tills: an example of gray till from the Baltic Sea cliff at Reval, Poland (Krygowski 1974: Fig. 2).

**Classificatory units**

- I genetic type
- II genetic subtype
- III facies
- IV subfacies
- V order
- IV family
- VII subfamily
- VIII genus
- IX subgenus
- XI species
- X variety

**A — According to genetical criteria**

- morainic deposits
  - underwater morainic deposits
  - continental morainic deposits
    - ablation morainic deposits
      - basal morainic deposits of stagnant ice
    - basal morainic deposits
      - basal morainic deposits of active ice

**B — According to physical-chemical criteria**

- glacial till
  - gl. till disturbed
  - gl. till stratified
  - gl. till structureless
    - gl. till with distinct orientation of pebbles
    - gl. till with chaotic orientation of pebbles
      - gl. till with prevailing petrographic type c/D
      - gl. till with prevailing petrographic type d/D
        - gl. till without carbonates
        - gl. till with carbonates
          - gl. till with lower amount of montmorillonite <8%
          - gl. till with higher amount of montmorillonite >8%
            - gl. till brown
            - gl. till gray

a classification of moraines that applied the term "moraine" to glacial debris, tills, and glacial landforms.

Harland, Herod, and Krinsley (1966) proposed a very broad "textural-genetic classification of tills, tillites and related rocks" (Table 4). It included a wide scope of diamicts ranging from glacial ortho-till(ite)s to nonglacial till-like pseudo-till(ite)s. The subgroup of ortho-tills included all the varieties of glacial deposits that had been considered tills at that time. The next broader group, the autochthonous tills, contained even the ice-rafted para-tills, which were included with tills only in the classification of Rukhina (1960). This very broad classification was used by those who were dealing with tillites, but most glacial geologists still adhered to narrower classifications, such as that proposed by Boulton (1971; 1972):

Supraglacial tills $\left\{\begin{array}{l}\text{Flow till} \\ \\ \text{Melt-out till}\end{array}\right.$

Subglacial tills $\left\{\begin{array}{l}\text{Melt-out till} \\ \\ \text{Lodgement till}\end{array}\right.$

Waterlaid and deformation till were included in the classification of Dreimanis (1969): see Table 1-A.

Lavrushin (1970a) proposed a classification of dynamic facies and subfacies of ground moraine (in the sense of till) where several descriptive names of genetically diverse moraines (tills) appeared, such as facies of plastic flow, or monolithic moraine, facies of large rafts, facies of imbricate or scaly moraine, facies of altered moraine. This classification appeared both in English (Lavrushin 1970b) and German (Lavrushin 1971).

Towards the end of the 1961-73 period it became quite obvious that till, moraine, and other related terms have been used with very diverse meanings in different countries and by different authors. The number of scientists interested in glacial and/or glacigenic deposits and processes (see Appendix B for the usage of these terms) had been growing rapidly, but the papers published became more and more specialized, often dealing mainly with compositional aspects of the sediments investigated. The origin of the glacial or glacigenic deposits still had to be deciphered more thoroughly.

A summary of our knowledge about the origin of till was presented by Goldthwait (1971) as an introduction to the "Till: A symposium" volume. The Columbus till symposium itself, May 1969, the resulting

volume, and the following shorter till symposium at the IX INQUA Congress in Christchurch, New Zealand in 1973 (with abstracts only in Dreimanis and Goldthwait 1973) marked the transition to the next period.

## 6.5 The 1974-86 period of INQUA Commission C-2

During the reorganization of the INQUA Commission C-2 at the VIII INQUA Congress in Christchurch, New Zealand, it was agreed that Commission C-2 should deal mainly with the genesis of glacigenic deposits and their landforms, while continuing the work of the previous period. The author of this paper was appointed the President of the Commission, and he proposed as its principal project "(a) Genetic classification of tills (or moraines as materials), and criteria for recognition and differentiation of various genetic types of till." This project (a), later renumbered (1) and (2A), developed into what was called briefly the "Till Work Group." By the end of 1974, 50 participating members from 20 countries had either been invited or had volunteered to deal with this task. Unfortunately, the vice-chairman of the Commission during the previous period, Dr. B. Krygowski, was prevented from continuing his participation because of a stroke. All other members of Commission C-2 who had been dealing with till continued their participation.

The following work plan, subject to modifications from the members, was proposed in the first circular distributed on June 6, 1974.

"1. To find out the current practice on genetic classification of tills (or moraines — as materials), and on usage of the relevant terminology (such as till, moraine, drift, etc.) in various countries or areas (a questionnaire was enclosed).

2. The criteria and references presently used in recognizing and distinguishing the above types of tills, also to be reported by countries or areas.

3. Modifications in the current practice of genetic classification of tills, and critical re-evaluation of the criteria for recognition and differentiation of various types of tills.

4. An agreement on general principles of genetic classification of tills and, if possible, development of a general classification scheme; also an agreement on the main criteria for recognition of the main types of tills. (Individual variability will always exist, and some of the types

Table 4. Textural-genetic classification of tills, tillites and related rocks by Harland, Herod, and Krinsley (1966: Fig. 2).

| DIAMICT TEXTURE / GLACIAL ORIGIN | | TILLOID TEXTURES | | | OTHER DIAMICT TEXTURES |
|---|---|---|---|---|---|
| | | Unsorted e.g., Boulder clay | Part sorted e.g., Boulder sand | Stratified Boulder beds etc. | |
| Immediate (e.g., ablated) | TOTAL | Ortho–till (ite) — AUTOCHTHONOUS TILL(ITE)S | | | e.g., solifluxion deposits |
| Mediate (e.g., rafted) | | ← para – till (ite) → (POLYGENETIC) | | | |
| Derived — from associated tills | PARTIAL | TILL (ITE)S (POLYGENETIC) — ALLOCHTHONOUS TILL(ITE)S — varve like turbidites | | | e.g., fluvio-glacial deposits |
| Derived — from older tills | | DERIVED TILL(ITE)S | | | |
| Metamorphosed | | (MUDFLOWS TURBIDITES) META – TILLITES / META – TILLOIDS (TURBIDITES) | | | psephites |
| Not known | | TILLOIDS | | | diamictons / diamictites |
| No connection | | e.g., pebbly mudstone — PSEUDO – TILL(ITE)S | | | e.g., pebbly sandstones |

and criteria will change with time.)

5. The submissions of the individual members of the commission on the above items (1), (2), and (3), and the summary of the opinions on (4) will be assembled for publication as a monograph."

Completion of the entire project on genetic classification of glacigenic deposits and their landforms was planned by 1982, after two intercongress periods. During these, 1974-1982, the Till Work Group had grown to 115 members from 32 countries all over the world. It became the largest Work Group of INQUA Commission C-2, and till became the dominant subject at all Commission meetings. Since the meetings were usually combined with excursions to formerly glaciated areas or in the environs of present-day glaciers, it was possible to discuss till, its characteristics, and probable genesis right at the sections. Thus, our conclusions on the genesis of till were constantly tested or challenged.

By 1982, several hundred papers on till investigations had been presented at 16 meetings held in 13 countries during the 1974-82 period, and the Till Work Group had met for at least 7 discussion evenings to deal with the terminology and classification of till. In addition, 23 circulars were distributed among the Work Group members, some with questionnaires. The opinions of the members came back by correspondence or at the meetings, and, particularly during the period 1976-78 they were recirculated. Seventeen volumes of proceedings, guidebooks, and abstracts had been published (Dreimanis, in Schlüchter 1982:7-8).

The meetings, their publications, Till Work Group circulars, and private correspondence were all aimed at serving the exchange of opinions and experiences in till studies among all those interested in sharing their opinions. The main purpose of our project was no longer to develop a genetic classification of till that would satisfy most, but rather to enrich our knowledge of the genesis of till (see also Lundqvist 1984).

When the 1982 INQUA Congress approached, our initial deadline for the project, we already had enough material to produce a

Table 5. Genetic classification of till or moraines by Lavrushin (1980).

**Till – Moraines**

- **Deposits of ablation till**
  - Mudflow till
  - Melt-out till
  - Washed out or peruvial glacial debris (boulder belts)

- **Deposits of the complex of end moraines (glaciodynamic types of end moraines)**
  - Deposits of push moraines
  - Deposits of pressed moraines or squeezed out "moraines"
  - Deposits of mass movement or flow moraines

- **Basal till of ground moraine**
  - **Lodgement till (facies group: monolithic tills)**
    - Platy or layered till
    - Lamellar or slaty till (Foliated or fissile till)
    - Folded till
    - Massive till
    - Fluted till
    - Facies of plastic flow of ice
  - **Overthrust till (facies group: macroshear tills)**
    - Macrosheared till
    - Till of minor moraines
    - Till of "lobate ridges" (by Gravenor and Kupsch)
    - Facies of large floes of rafts
    - Till of De Geer moraines
    - Till of Labrador type minor moraine (ribbed moraines, rogen moraine)
    - Facies of shear – plane flow of ice

Groups:

30

monograph, our objective No. 5, but it did not appear that important anymore. It seemed to be more important to continue our discussions on till genesis and to be prepared to change our opinions and conclusions whenever new facts or new ideas were brought forward. Also we still had unresolved "borderline" problems — the relation of glacial to glaciolacustrine and glaciomarine depositional processes and the resulting genetic classifications. Since the Commission's work groups dealing with these aquatic deposits were not yet ready for their reports for the 1982 INQUA Congress, and since we hoped also to learn more from the newly-created glacitectonics work group on the relation between glacial tectonics and till (another "borderline" problem), we were glad to join all other Work Groups of the Commission C-2 to ask the INQUA Council during the XI INQUA Congress in Moscow to give all of us an extension for the completion of the project until 1987.

During the 1982-86 period, eight more field conferences and workshops were organized. Till was again one of the main topics, and five more Till Work Group circulars were distributed. By 1987, five volumes of conference proceedings dealing with tills and tillites as one of their main topics had been published (Evenson, Schlüchter, and Rabassa 1983; Serrat 1983; Deynoux 1985; Kujansuu and Saarnisto 1987; van der Meer 1987), and four more were in preparation for publication (see Dreimanis 1987b).

Several new names of till and changes in the terminology and genetic classifications of till or moraine were proposed or discussed during the 1974-87 period. Most active were the years 1975-1980, when several proposals of genetic classifications were submitted to the Till Work Group and/or published. Also, some changes in terminology were proposed.

1. Francis (1975:51) suggested that "waterlaid" be changed to the linguistically more correct term "waterlain" in reference to tills;

2. Lavrushin (1976) modified and expanded his (1970) proposal of dynamic facies of moraine (till), and published its condensed English version with comments (1980, reproduced here as Table 5).

3. Stankowski (1976) contained several papers and a review of panel discussions dealing with genetic classifications of till, from the INQUA Commission C-2 Symposium in Poland in 1975. Boulton's classification (1976b:69) considered sedimentary association, source of material, and deposition of till, and he introduced

a new term "supraglacial morainic till" for a supraglacially derived and supraglacially deposited till, discussed in more detail by Boulton and Eyles (1979:11-23) and later changed to "supratill" (Boulton and Deynoux 1981). Rozycki (1976:55) proposed "exaration moraine" for basal till (basal moraine) consisting mainly of local material that had usually been called "local moraine."

4. Schlüchter (1977) introduced the term "Schlammoräne = mud till" for a lodgement till consisting of incorporated pre-existing lacustrine sediments.

5. Shaw (1977:1244) proposed an expanded classification of terrestrial tills, reproduced here as Table 6. It was based upon the position of glacial debris during transport and the tectonic facies of debris bands in glacier ice, and the position and process of deposition. Two new process-related tills appeared there: "sublimation till" and "lowered till."

6. Dreimanis (1976:20-41; 1980) reviewed the genetic classifications of tills, and the terminology used in various official INQUA languages up to 1977.

7. McGown and Derbyshire (1977) proposed an extended genetic classification of till, based on the modes of formation, transportation and deposition, and suggested differentiation of the tills by their "total fabric" (summation of all the directional properties of a sediment, at all scales; see also Derbyshire, Edge, and Love 1985).

8. Grube (1979) proposed "Sohlmoräne" as a German term similar to deformation till, but broader; it consisted of two subunits: "Schermoräne" ("shear till" of Stephan, this volume), and "Stauchmoräne."

9. Dreimanis (1979:172) summarized the terms used for waterlain tills during 1950-77. A terminological change was proposed by Gravenor, von Brunn, and Dreimanis (1984) for the subaquatic meltout till, suggesting it be called "undermelt diamicton." A parallel term, "undermelt till," is used in some recent papers (see Chapter 9.2.5, this paper).

10. Four classifications presented at or prepared for the X INQUA Congress at Birmingham in 1977 and some of them revised in 1978 were published in Stankowski (1980): by Dreimanis (1980:9: Table 6), by Rukhina (1980:12: reproduced here as Table 7), by Boulton (1980b: Table 1).

11. Dreimanis (Till Work Group #1 Circular No. 16, 1979) rearranged the previous 1977 classification of tills, so that the position, processes, and environments were separated. It was published in Hambrey and Harland (1981:24) and is

Table 6. Genetic classification of terrestrial tills by Shaw (1977).

| Position of transportation | Position of deposition | Process of deposition | Tectonic facies |
|---|---|---|---|
| Supraglacial Englacial Basal | Proglacial Lateral ice-contact Supraglacial Subglacial | Lowered Flow Melt-out Sublimation Lodgement | Highly attenuated Poorly attenuated |

Table 7. Genetic subdivisions of glacial deposits by Rukhina (1980).

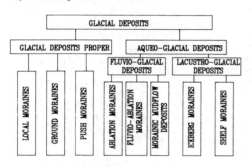

reproduced here as Table 1-B. This classification was translated into Chinese and published by the Chinese Research Group of Glaciological Sediments of the Chinese Society of Glaciology and Cryopedology (1981:79, reproduced here as Table 1-C).

12. Heuberger (1980) discussed the German nomenclature of glacial deposits in the Eastern Alps, pointing out some differences with other regions.

13. Boulton (1980a:7, Fig. 2) published a classification of till in hierarchical structure, reproduced here as Table 8.

14. Lawson (1981b: Table 2) proposed a genetic classification of primary and secondary deposits within the context of glacial environment, suggesting that only the primary deposits be called till (see

Table 8. Classification of till in hierarchical structure by Boulton (1980a).

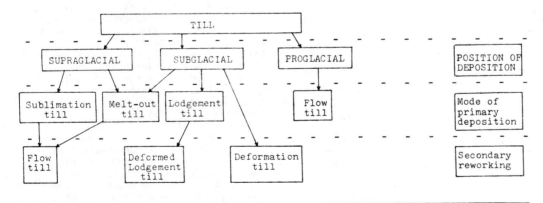

Parallel classifications:-  (1) Textural -  e.g. clast dominant
                                                matrix dominant
                                                supraglacial morainic till

                            (2) Morphological - e.g. fluted lodgement till

Lithified sediments:-       Add suffix - ite

also Lawson 1979). Boulton and Deynoux (1981), in a somewhat similar way, proposed to differentiate tills into primary and secondary tills, and Dreimanis (1982) into ortho-tills and allo-tills (see also Chapter 8.3). The classification of Boulton and Deynoux (1981) is reproduced here as Table 9.

15. Dreimanis (1982b:31) constructed a table (reproduced here as Table 10) showing the factors that influence the formation and deposition of till as a background for the expanded comprehensive classification of till. However, a depositional classification only is given in this table, with a tentative proposal to replace the term "flow till" by "mass movement till."

Table 9. Genetic classifications of tills and tillites by Boulton and Deynoux (1981).

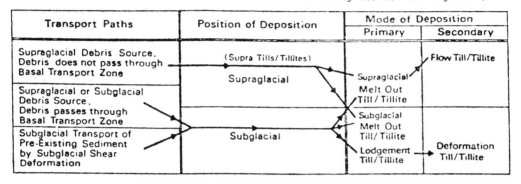

Table 10. Factors that influence the formation and deposition of tills (upper half) and tentative depositional genetic classification of tills (lower half) in Till Work Group's 1977-82 report by Dreimanis (1982a: Table 1).

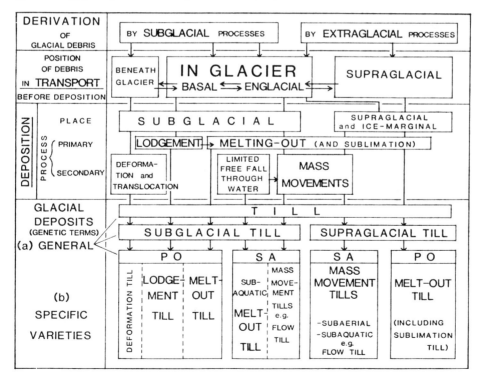

16. Lundqvist (1984:12-13, Table 3) proposed an expanded genetic classification of till based upon five parameters of transport, release, and deposition. For a similar, more comprehensive classification see Lundqvist (this volume: Table 1).

17. Warren (1987:114) proposed a term "diagenetic till" to be "...restricted to sediments that have first been deposited (glacially or otherwise) and then had their fabric and compaction altered by stresses applied by overriding ice."

18. Morawski (this volume) proposes to use the terms "primary" and "secondary" for glacial sediment in the sedimentologic sense, and not in relation to glacier ice. He (ibid. and 1985) also proposes a term "watermorainic facies" in a genetic classification of glacigenic sediment, as an intermediate group between morainic and glacioaquatic facies.

19. Stephan (this volume) proposes a term "shear till" for diamictons originating in the subsole zone of intensive shearing.

20. Warren (this volume) proposes a term "protalus till" for a specific variety of ice-marginal till.

The proposals for deletion of any specific terms are not listed above, except for Lawson's (1981b) proposal to consider only the primary tills as true tills. Because of uncertainties in distinguishing primary from secondary tills, some students of glacial deposits became reluctant to apply the term "till" at all, and began to use "diamicton" or "diamict" instead.

# 7 GENETIC DEFINITIONS OF TILL AND THEIR IMPLICATIONS

The definition of till by Geikie (1863:185; see Chapter 3 in this report), which is probably the oldest in scientific literature, combines both genetic and lithologic aspects. Most geologists mean both aspects when they use the term "till." More than one hundred years after Geikie's definition was published, Flint (1971:148) wrote that till "...is commonly defined as nonstratified sediment carried or deposited by a glacier, hence the term is both sedimentologic and genetic." Harland, Herod, and Krinsley (1966:228) also emphasize that "till and tillite have two essential elements in their definition: (a) petrographic and (b) petrogenetic," but they added another ingredient, the polygenetic nature of till, which is either not considered in some narrow definitions (e.g., Lawson 1979a:28), or is overextended (Harland, Herod, and Krinsley 1966:232).

## 7.1 Till Work Group's definition of till

The task of INQUA Commission C-2, or more specifically, its Till Work Group of 1974 was to develop a genetic classification of tills (see Chapter 6.5). A genetic classification has to be related to a genetic definition. Since every definition of till emphasizes its glacial origin, the next steps were to specify how narrow or broad this genetic glacial aspect should be, and to subordinate the lithologic components in the definition. Still, the lithologic aspects had to be considered as part of the descriptive background (Chapter 5) in order to identify and recognize till and its various specific varieties.

Before deciding on a preferred genetic definition of till, the Till Work Group first proceeded to find out (a) how the term "till" was used in the past, (b) what its present usage is throughout the world, and (c) how existing definitions have developed and how well they have served. This was done during the period 1974-1980. Some of the background information on (a) and (b) was published in Dreimanis (1978, 1980). By 1980, several Till Work Group members suggested that the time had come to decide on updated definitions of till and related terms.

In 1981, by correspondence among the most active Till Work Group members and finally, by lengthy discussions at the Commission's field conference in Wyoming and Idaho, we first established the criteria for a genetic definition of till, and then discussed various wordings of the definition. A general consensus developed that the definition should be not only (a) theoretically sound, but also (b) based upon multiple situations of the formation and deposition of till as observed at modern glaciers and concluded from thoroughly investigated ancient till sections, and (c) applicable to field mapping. The definition was finally worded as given in the next paragraph, and distributed, accompanied by eleven other published definitions of till and several definitions of other related terms, in Circular No. 23, to all members of the Till Work Group in December, 1981.

By the time the Till Work Group's preliminary report (Dreimanis 1982b) had to be written for the XI INQUA Congress, close to 90% of all those who had replied to the questionnaire of Circular No. 23, or had participated at the preceding discussions, had agreed with the following definition: "Till is a sediment that has been transported and deposited by or from glacier ice, with little or no sorting by water" (Dreimanis 1982a:21). A very similar definition was proposed earlier by Francis (1975:47): "Till may be defined as sediment deposited by or from glacier ice without the intervention of running water."

The mechanisms of the formation and deposition of till are not specified in the definition, since we still may not be aware of all of them. Further informal discussions of the above definition had resulted in the addition of the word "subsequently" before "deposited" (Dreimanis and Lundqvist 1984:9). This addition does not change the principal meaning of the first definition — it merely points out that deposition eventually follows glacial transport.

Some comments on the definition arrived after the XI INQUA Congress, and 90 members of Commission C-2 Till Work Group have now expressed their opinions on the definition. By adding four more members who did not participate in the discussions but who have published their views (two negative, two positive), we now have the opinions of 94 members from 25 countries. Only 5 out of 94 expressed preference for an entirely different definition; 76 (81%) agree with the definition as it is; 13 (14%) suggested some minor changes in the wording. Most of the latter 13 proposed an additional reference to the diamictic character of till, in various expressions. Since our definition was meant to be strictly genetic, this singular lithologic characteristic, although typical for most tills, has been left out. Among other suggestions for the modification of the definition, the exclusion of "from" has been urged. This suggestion applies particularly to "flow tills" and will be discussed further in this section.

The above definition stipulates that both glacial transport and deposition are involved in the genesis of till. The possible varieties of glacial transport will be further considered in Chapter 13.

When is till considered to be deposited? This question was also discussed at the Wyoming-Idaho 1981 field meeting, in conjunction with the definition of till, and participants agreed on the following statement:

"A till is deposited when the glacial debris that forms it undergoes no further lateral movement and the debris particles forming the till framework are in contact. The only internal vertical movement is due to reduction of pore space by removal of the remaining ice or water."

This statement was approved by more than 90% of the respondents to the questionnaire of Circular No. 23 (Dreimanis 1982a:23).

Some disagreement may exist concerning whether a supraglacial deposit of glacial debris should be called till, even though it fulfills all the above criteria but is still supported by glacier ice. This problem has been discussed in some detail by Boulton (1980a:8-9), and the reply is positive, considering particularly the experience of students of present-day glaciers. It is often hardly possible to determine whether there is still any glacier ice underneath such a supraglacial till.

The presence of glacier ice under till is relevant particularly if glacier ice remains buried for a lengthy period of time, even for thousands of years. Such a possibility was suggested by von Toll (1892) for the Novosiberian Islands. It was subsequently rejected, but has been under serious consideration for West Siberian lowlands again (Kapilanskaya and Tanogradskiy 1977).

The deposition of till may be by or from glacier ice. "By glacier ice" means deposition directly by moving glacier ice, for instance by lodgement (for details on lodgement see Chapter 9.2.1). If the terms "deformation till" or "shear till" (for details see Chapter 9.2.4) are used, they also may be considered as deposited by glacier ice.

Deposition "from glacier ice" may be understood in various ways, and therefore permits the definition of till either in a narrow or a broad sense. The narrow meaning of "from glacier ice" would be a deposition of till in direct contact with the ice only, by passive melting-out or sublimation. The resulting tills are essentially primary tills: melt-out till and sublimation till. Still, melt-out till may contain some resedimented material, as pointed out by Dreimanis (1983) when discussing the texture of Lawson's (1979a, 1981a) melt-out till at the Matanuska Glacier (see further in Chapter 8.3).

The broad meaning "from glacier ice" would include some resedimentation, after the glacial debris has been released from glacier ice, but "with little or no sorting by water," as specified in the definition of till. This specification excludes water transport that produces sorting, but it permits other transports "from glacier ice" such as squeeze flow, gravity flow, sliding, and slumping (further discussed in Chapter 8.3). However, it was agreed during the discussion of the definition of till that in order to consider such resedimented material to be till, the resedimentation

Table 11. Depositional genetic classification of till.

| RELEASE OF GLACIAL DEBRIS AND ITS DEPOSITION OR REDEPOSITION | | | DEPOSITIONAL GENETIC VARIETIES OF TILL | | |
|---|---|---|---|---|---|
| I. ENVIRONMENT | II. POSITION | III. PROCESS | IV. BY ENVIRONMENT | V. BY POSITION | VI. BY PROCESS |
| GLACIO-TERRESTRIAL | ICE-MARGINAL:<br>  -FRONTAL<br>  -LATERAL | A. PRIMARY<br>  MELTING OUT<br>  SUBLIMATION<br>  LODGEMENT<br>  SQUEEZE FLOW<br>  SUBSOLE DRAG | TERRESTIAL NONAQUATIC TILL | ICE-MARGINAL TILL | A. PRIMARY TILL<br><br>MELT-OUT TILL<br>-SUBLIMATION TILL<br><br>LODGEMENT TILL<br><br>DEFORMATION TILL<br>OR<br>GLACITECTONITE<br>SQUEEZE FLOWTILL |
| | SUPRAGLACIAL | B. SECONDARY:<br>  -GRAVITY FLOW<br>  SLUMPING<br>  SLIDING AND ROLLING<br>  FREE FALL | | SUPRAGLACIAL TILL | |
| GLACIOAQUATIC | SUBGLACIAL | | SUBAQUATIC OR WATERLAIN TILL | SUBGLACIAL TILL | B. SECONDARY TILL<br><br>FLOWTILL<br>-GRAVITY FLOWTILL |
| | SUBSTRATUM | | | | |

NOTE: EACH VERTICAL COLUMN IS INDEPENDENT FROM THE OTHER FIVE, AND NO CORRELATION HORIZONTALLY IS IMPLIED. ONLY SOME COMBINATIONS ARE FEASIBLE.

of glacial debris shall take place at its source, the glacier ice, and penecontemporaneously with the melting of ice. In other words, there should be a restriction in space and in time to the glacial environment in which the glacial debris was released.

During the discussions of the Till Work Group's definition of till and shortly afterwards, a clear majority of the Work Group members and others working with till favoured the broad application of the wording "from glacier ice"; however, the number is increasing who would rather apply it in a narrower sense, thus considering only the primary tills or orthotills as unquestionable tills. Such views are expressed, for instance by Lawson (1979, 1982), Lutenegger, Kemmis, and Hallberg (1983), Stephan and Ehlers (1983:244), De Jong and Rappol (1983). Indeed, when dealing with the end members of resedimented glacial debris deposits and such occurrences where they are easily distinguishable from primary tills, this approach is very understandable. For instance, at Matanuska Glacier Alaska, Lawson (1979a:40) estimated that only 5% of glacial sediments were melt-out till, and that all others had been resedimented, mainly by sediment flows. He could recognize and map them as extensive recognizable sedimentary units.

However, the question of the exclusion of resedimented glacial debris from tills becomes more complex, if the "true tills" are scrutinized more closely, as will be discussed in Chapter 8.3. Till is a polygenetic glacial sediment. Glacial ice is the principal agent of its deposition, but gravity and water (except for sublimation till) also participate in the formation and deposition of till. The polygenetic character of till was noted already about a century ago, for instance by Upham (1891:378) and Chamberlin (1894:518; see Chapter 6.1 of this paper).

Dreimanis and Lundqvist (1984:9) proposed that the following three conditions must be met for a sediment to be called "till": 1. it consists of debris that has been transported by a glacier; 2. close spatial relationship exists to a glacier: deposition by or from a glacier; 3. sorting by water is absent or minimal.

## 7.2 Other genetic definitions of till

We shall also discuss briefly other published genetic definitions of till. Those with a lithologic component will not be included.

Some of the definitions are very general, for instance Goldthwait's (1971:3): "Till is the only sediment stemming directly and solely from glacier ice," or Aario's (1977:99): "If the material is dominantly characterized by deposition from ice, it is till." These definitions are similar to that of the INQUA Till Work Group (Chapter 7.1), but their wording is less specific.

The most restrictive definition has been proposed by Lawson (1979a:28) based upon the sediment he studied at the Matanuska Glacier, Alaska: "Till is defined as sediment deposited directly from glacier ice that has not undergone subsequent disaggregation and resedimentation." In Lawson (1981b:2) the phrase "from glacier ice" is changed to "by the glacier ice," and emphasis is placed on a suggestion (ibid.) that "only sediments deposited by primary processes should be classified genetically as tills." These are "...lowered till, melt-out till, ...sublimation till, traction lodgement till, regelation lodgement till" (ibid.: Table 2). However, melt-out and lodgement tills may contain parts that have been disaggregated and resedimented during their formation and deposition as till, as discussed by Dreimanis (1983). Lawson's definition therefore appears to be too strict.

Another definition of till that includes some restrictions but permits consideration of most glacially related sediment flows as tills has been proposed by Boulton (1980a): "Till is a sediment whose components are brought into contact by the direct agency of glacier ice, and which, although it may suffer subsequent glacially induced flow, is not disaggregated." The first part of Boulton's definition, however, specifying that a till's "components are brought into contact by the direct agency of glacier ice," does not consider the subsidiary participation of water in the formation of till. Also, some disaggregation may occur in any genetic variety of till, as has already been mentioned above.

## 8 FACTORS TO BE CONSIDERED IN THE GENETIC DEPOSITIONAL CLASSIFICATION OF TILL

Although formational and depositional processes are most important for a genetic classification of till, the general environment of deposition and the position in relation to glacier ice must also be considered (Tables 11 and 12, columns I-III).

Table 12. Depositional genetic classification of till in the French language. It is not a direct translation of Table 11, but an adaptation by Serge Occhietti (personal communication, September 1987).

| MODE DE LÂCHAGE ET DE DÉPÔT DES DÉBRIS GLACIAIRES | | | CLASSIFICATION GÉNÉTIQUE DES TILLS, EN FONCTION DE: | | |
|---|---|---|---|---|---|
| I. ENVIRONNEMENT | II. POSITION | III. PROCESSUS | IV. L'ENVIRONNEMENT | V. LA POSITION | VI. LES PROCESSUS |
| DIRECTEMENT GLACIAIRE | MARGE GLACIAIRE: <br> -JUXTAGLACIAIRE <br> -LATÉRALE | A. PRIMAIRE: <br> FUSION SUR PLACE <br> SUBLIMATION <br> ACCRÉTION <br> FLUAGE PAR INJECTION | TILL (SENSU STRICTO) | TILL DE MARGE GLACIAIRE: <br> -TILL FRONTAL <br> -TILL LATÉRAL | A. TILL PRIMAIRE <br> -TILL DE FUSION <br> -TILL DE SUBLIMATION <br> -TILL DE FOND <br> -TILL D'ENTRAÎNEMENT <br> -TILL INJECTÉ |
| | SUPRAGLACIAIRE | TRANSLOCATION BASALE | | TILL SUPRAGLACIAIRE | |
| GLACIOAQUATIQUE | SOUS-GLACIAIRE | B. SECONDAIRE: <br> FLUAGE PAR GRAVITÉ <br> GLISSEMENT | AQUATILL (TILL SOUS-AQUATIQUE) | TILL SOUS-GLACIAIRE | B. TILL SECONDAIRE <br> -TILL FLUÉ |
| | SUBSTRATUM | ÉBOULEMENT <br> SOLIFLUXION | | | |

NOTE: LES COLONNES SONT INDÉPENDANTES LES UNES DES AUTRES. CERTAINES RELATIONS SONT POSSIBLES ENTRE LE CONTENU DES COLONNES. CE TABLEAU N'EST PAS LA TRADUCTION LITTÉRALE DU TABLEAU EN ANGLAIS (11), MAIS UNE ADAPTATION AUSSI FIDÈLE QUE POSSIBLE. EN PARTICULIER, LES TERMES DE PROCESSUS DE DÉPÔT SECONDAIRE NE SONT PAS DIRECTEMENT ÉQUIVALENTS.

## 8.1 Environment of deposition

The term environment has been used with a variety of meanings, but in sedimentology it usually means "...a geographically restricted complex where sediment accumulates, described in geomorphic terms and characterized by physical, chemical and biological conditions, influences, or forces" (Bates and Jackson 1980:205).

Usually, two large groups of environments are distinguished: terrestrial and marine. The deposition of till is commonly associated with the terrestrial, more specifically a glacier-related terrestrial, or glacioterrestrial environment. Thus, Boulton and Deynoux (1981) consider till and also glaciofluvial and glaciolacustrine sediments as products of the glacioterrestrial environment and exclude their association with the glaciomarine environment. However, many similarities exist among glaciomarine and glaciolacustrine depositional environments. Water dominates in both, and both are similarly related to glaciers: some glaciers terminate in them.

From the sedimentologic and hydrologic viewpoint, glaciomarine, glaciolacustrine, and glaciofluvial environments including subglacial water bodies may be grouped together as glacioaquatic. Terms similar to glacioaquatic, referring to all glacigenic meltwater deposits, have been used before, for instance "glacial-aqueous sediments" (Thwaites 1957:31), and "aqueous-glacial deposits" (Rukhina 1980:12). If we are transferring some of the conventional "terrestrial" environments (glaciolacustrine, glaciofluvial) to the glacioaquatic, a question arises — by what name to call the remaining glacial environment where tills are deposited from and by grounded glaciers. Is it "terrestrial" (Shaw 1977), "land-based" (Gibbard 1980:71), "continental" (Raukas 1978:189; Anderson et al. 1980:407), "glacioterrestrial" (Boulton and Deynoux 1981), "subaerial" or "supra-aquatic" (Lundqvist, this volume, 6.4), or "nonaquatic"?

According to Bates and Jackson (1980:645: terrestrial deposit) the most commonly used designation, "terrestrial," either means "on land above tidal reach," or "on land," but without the action of water. In the second meaning, the term "glacioterrestrial" would apply to most environments where the nonaquatic varieties of till are deposited. Therefore, it is tentatively proposed here to specify "nonaquatic glacioterrestrial environment" as the opposite of "glacioaquatic environment," although I do not feel happy about using three words instead of two. The two-word term "supra-aquatic environment" would not apply well to the subglacial deposits in cavities that are formed below lake level or sea level. "Continental" is similar to "terrestrial," but more restrictive, and "subaerial" would not apply to a subglacial environment.

For the time being, the following two general depositional environments may be distinguished when classifying tills: 1) glacioterrestrial and 2) glacioaquatic.

## 8.2 Position of deposition in relation to glacier ice

Glacial debris is released and deposited as till in three positions in relation to glacier ice (Tables 11 and 12; see also Lundqvist, this volume: 6.2): 1. supraglacial or superglacial; 2. ice-marginal; 3. subglacial.

These descriptive terms are self-explanatory. Supra- and superglacial mean the same, but supraglacial is currently preferred because the prefix "super-" is sometimes used as an abbreviation of "superior" in colloquial English.

The ice marginal place of deposition may be either along the front of a glacier, or along its sides (specified as "lateral") or along all sides of a mass of stagnant ice. The often used term "proglacial" may cover a wide area in front of a glacier, away from the glacier ice, and therefore should be avoided in specifying the place of deposition of till (see also Lundqvist, this volume: 6.1).

## 8.3 Processes of deposition

Two principal groups of glacial processes of deposition may be distinguished: 1) primary and 2) secondary.

"Primary processes release and deposit sediment directly from glacier ice, whereas subsequent secondary processes cause resedimentation of glacial materials" (Lawson 1981b:20). The principal primary glacial process is direct release of debris from glacier ice, either by lodgement, melt-out, or sublimation. "The secondary processes within the glacial environment are those that cause resedimentation of glacial materials. They may mobilize, rework, transport and resediment materials previously deposited by the primary till-forming processes or materials previously deposited by secondary processes which are being reworked again" (Lawson, this volume).

If the primary depositional processes would act alone, and then be followed separately by secondary processes in the glacial environment, it would be possible to separate tills, as products of primary processes, from nontills, products of secondary processes. However, one of the unique characteristics of glacial sedimentation is the very common interaction of these two groups of processes. The primary glacial depositional processes are commonly accompanied by secondary processes in such a close association that it is hard to draw a boundary between primary and secondary deposition.

The least noticeable secondary processes that commonly accompany primary sedimentation are internal mobilization, downward movement, and resedimentation of the fine particles in the voids. They are mentioned by Lawson (1981a:79) for melt-out till and further discussed by Dreimanis (1983) regarding examples of melt-out and lodgement tills observed in association with present-day glaciers. Such resedimentation normally does not affect the structural framework of primary till, but it will locally change the grain-size composition, the microfabric, and the density of till.

Locally occurring flowage, either as glacially induced squeeze flow or as gravity flow, will affect both the structure and fabric of those parts of melt-out and lodgement tills remobilized during or shortly after their primary deposition, as demonstrated by Dreimanis (1987b). Various multiphase formative processes, both primary and secondary, may be involved in the deposition of some, if not most, tills.

## 8.4 Simultaneous deposition at several levels

Glacial deposition may be in progress at several levels at the same time. Because of this, it is difficult to apply some of the laws concerning stratigraphic and facies sequences that are considered basic for most nonglacial sediments, such as the Law of Superposition and Walther's Law of the Correlation of Facies, as pointed out recently by Shaw (1985:38).

The most commonly recognized contemporaneous pair is subglacial and supraglacial sedimentation in the ice-marginal zone. In addition, sedimentation may occur at several levels at each of the subglacial or supraglacial positions, also englacially in cavities of stagnant ice, and underneath the glacier sole. Syndepositional and postdepositional glacitectonic and/or gravity flow deformation

and resedimentation may also occur.

Such complexity (Table 13) requires not only a proper recognition of all the resulting genetic varieties of the glacigenic deposits formed and deposited at these various levels and their correlation in the three dimensions of space, but also the inclusion in the interpretation of the fourth dimension — time.

## 9 GENETIC DEPOSITIONAL CLASSIFICATION OF TILL

The depositional classification proposed by the INQUA Till Work Group in this paper (Tables 11 and 12) is based upon the three factors discussed in Chapter 8 and listed here in the order of their genetic significance:
1. the process of formation and deposition;
2. the position in relation to the glacier ice; and
3. the general environment, particularly in relation to the hydrologic conditions.

## 9.1 Primary versus secondary tills

Primary and secondary processes are distinguished in every discussion of sedimentation. Because of the polygenetic nature of tills, and their deposition at several levels in relation to glacier ice, the interaction of primary and secondary processes may be very complex in the genesis of till. Still, primary glacier-related processes dominate in some instances, and secondary processes in others.

If the process of deposition is considered to be the most important criterion for genetic classification, two main groups of tills may be distinguished, depending upon their dominant processes of deposition:
1. primary till (Boulton and Deynoux 1981) or ortho-till (Harland, Herod, and Krinsley 1966) more narrowly defined by Dreimanis (1982b:23);
2. secondary till (Boulton and Deynoux 1981) or allo-till (Dreimanis 1982b).

When considering the above two pairs of terms, it may be advisable to avoid the proliferation of terms that developed unintentionally. Since ortho-till has been used with two meanings, a broad one (Harland, Herod, and Krinsley 1966) and a narrower one (Dreimanis 1982b), and since allo-till has also been used with two meanings (by Chumakov 1978 as an abbreviation of allochthonous till, and by

Table 13. Processes involved in multilevel sedimentation of till: vertical sedimentation sequence plotted versus the sequence in time.

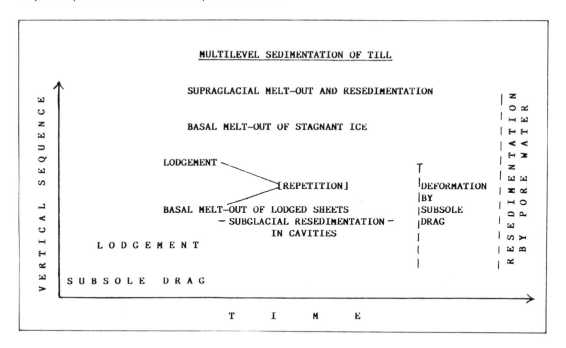

Dreimanis 1982b as a new term), it is suggested here to discontinue the use of the terms ortho-till and allo-till, as proposed by Dreimanis (1982b).

Any further hierarchical classification may proceed as outlined below, in Tables 11 and 12, or in more detail if the differentiation of very specific genetic varieties of till is desired.

The grouping of tills into primary and secondary tills as a basis for the classification of glacial sediment has some correspondence to the subdivisions of those who consider primary tills as the only true tills (e.g., Lawson 1979, 1981). However, deposits of secondary processes are also called tills in this classification because even in primary tills, secondary processes participate to various extents (Table 13).

Morawski (1984 and this volume) points out that the term "primary" applies to all sedimentary and synsedimentary structures in his "watermorainic facies" of glacigenic sediment (further discussed in Chapter 9.2.3), and "secondary" to glacitectonic structures only. Such application of the terms "primary" and "secondary" may be relevant not only to the structures of glacigenic deposits, but also to the genetic classification of till, because till

is a polygenetic sediment. Such reinterpretation of the terms "primary" and "secondary" in relation to genetic classification of till has not been discussed at the meetings of INQUA Commission C-2, because this point has been raised only recently, during the writing of this report (W. Morawski and M. Rappol, personal communications). Therefore, the terms "primary" and "secondary," in relation to the genetic classification of tills, will be used with their traditional meanings, as outlined in the first paragraphs of Chapters 9.1.1 and 9.1.2.

9.1.1 Primary tills

Primary tills are formed mainly by direct release of debris from the glacier and deposited mainly by the primary glacial processes — melt-out, sublimation, and lodgement — with subsidiary participation of secondary processes (resedimentation). Deformation till (Chapter 9.2.4) is also a primary till, since it is formed and deposited by direct action of glacier ice.

Primary tills may be divided into two end members:

1. till deposited by an actively moving glacier: lodgement and deformation till;

2. till deposited from glacier ice that is not sliding at its sole: melt-out till and sublimation till.

In the geologic literature these groups of glacial sediments are commonly listed as discrete genetic varieties of till, occurring separately as mappable units, but on closer examination it is found that they usually combine in the following pairs:

a. subglacial group: lodgement, deformation, and subglacial melt-out tills;

b. supraglacial group: supraglacial melt-out and sublimation tills, and also the lowered till of Shaw (1977) and the melt-out variety of Boulton and Deynoux (1981) — "supratill."

During subglacial deposition, lodgement and melt-out processes interfinger, particularly in those situations where basal sheets of debris-rich ice stagnate while the glacier still continues to move over them. Goldthwait (1971:16) discusses the mechanisms of primary deposition involved in such situations, and concludes that "...perhaps lodgement is the right word for these combined processes." That is the broad meaning of lodgement process and lodgement till, very common in geologic literature. The opposite view is expressed by Shaw (1985:38) who concludes a description of a similar situation by saying: "...as the dominant process is melting of nondeforming ice, they [the tills] are better described as melt-out tills." This may be considered the broad meaning of melt-out till, also common in the geologic literature. In most subglacial tills, lodgement and melt-out tills are merely the endmembers in an interdigitating continuum of components (Muller 1983). These two varieties of primary till share so many characteristics (Appendix D) that it is often difficult to differentiate them. The gradation between lodgement and melt-out tills has been noted in several detailed investigations, for instance by Haldorsen (1982), Levson and Rutter (1986), Dreimanis (1987b), and Shaw (1987). Also, lodgement till grades into deformation till or deformed till (Virkkala 1952; Ruszczyńska-Szenajch 1983; Dreimanis 1987a; Dreimanis, Hamilton, and Kelly 1987; Rappol 1987; Whiteman 1987). Unless diagnostic characteristics for one or another of these specific varieties of till dominate, it is safer to use a general term, such as "primary till" or "subglacial till."

In the supraglacial group of primary tills, melt-out till is the dominant variety. So little is known about the properties of sublimation till (Shaw, this volume), that for the time being sublimation till may be considered as a specific end member of the supraglacial melt-out till group, formed in areas of extremely cold and dry climate only.

The "lowered till" of Shaw (1977) is formed by subaerial settlement of supraglacial debris accompanying the melting of glacier ice. However, considering ice surface slopes and irregularities, and also meltwater activity, the "lowered till" will probably seldom preserve as primary till. Shaw (1985) does not use this term anymore.

9.1.2 Secondary tills

Secondary tills are products of resedimentation of glacial debris released by glacier ice, or an already deposited till in the glacial environment with little or no sorting by meltwater (see Chapter 8.3 and Lundqvist, this volume).

The main agents of resedimentation are the force of gravity and the load of glacier ice. High pore pressure of water aids the resedimentation by reducing shear strength.

If gravity had produced resedimentation from a glacier ice surface, a broad term "mass movement till" was used recently by several Till Work Group members including the present author (Table 10). Still, the already entrenched term "flowtill" proposed by Hartshorn (1958) appears to be more popular. This term is shorter, and flowage is probably the most common mass movement process anyway.

The term "flowtill" may be applied also to the products of squeeze flow, because of the similarity in the mechanisms of redeposition. If squeeze flow forming a flowtill is caused by glacier ice, the resulting squeeze flowtill is a primary till. Gravity flowtill, however, is a secondary till. Therefore, if it is possible to determine the cause of till flow, specific terms should be used: gravity flowtill, squeeze flowtill. Most tills mentioned in the geologic literature are probably gravity flowtills.

Ice-slope colluvium (Lawson 1981a:81-82, and this volume) and the deposits of sliding (Derbyshire 1984) may be included with gravity flowtills, because some flowage has also occurred in their formation, and since "it is often virtually impossible on a stratigraphic scale (meters) to separate them [the glacigenic secondary resedimentation deposits] into distinct units" (Lawson, this volume).

## 9.2 Principal depositional varieties of till

Considering the complexities involved in the deposition of till and discussed in Chapters 8.3, 8.4, and 9.1, the following principal depositional varieties may be distinguished as endmembers in an inverted pyramid or tetrahedron (Fig. 1).

The development of the terminology of the above tills, the derivation and transport of their materials prior to deposition, and the depositional processes will be discussed in some detail in the next four sections (9.2.1 to 9.2.4). Under the heading "Deposition," the part in bold print defines the respective depositional variety of till.

The descriptive characteristics of lodgement, melt-out, and gravity flowtill are summarized at the end of the discussion of each of these genetic varieties of till. They are also given in more detail in Appendix D, side-by-side, so that they may be compared. As many properties as possible should be considered to classify a till.

Only selected references on the properties of tills will be given. My apologies to those authors whose papers have not been mentioned. Also, all the opinions submitted to me by correspondence during the last 12 years could not be cited because of space restrictions.

Sections 9.2.4 and 9.2.5 will be devoted to controversial glacigenic materials that are classified as depositional varieties of till by some glacial geologists, but considered "non-tills" by others.

Several relatively recent published genetic classifications of till are presented for comparison in Tables 1B and 5-10.

### 9.2.1 Lodgement till

Terminology.

In English, the British spelling "lodgement till" is becoming more widely used than the shorter American version "lodgment till." Upham (1892:136-137) was probably the first to use the term "lodged" for the deposition of subglacial till. Chamberlin (1894) also used the term "lodged" and "lodge moraines" about the same time. Translations in other languages, some of them proposed recently, are listed in Appendix C.

Derivation, transport.

Glacial debris deposited as lodgement till is mainly subglacially derived and transported in the basal zone or the lowest englacial part of a glacier. In mountain glaciers, some of the rockfall and avalanche debris falling on the glacier surface above the firn line also may become englacial and basal because of the strongly downward moving flow lines in the glacier (Boulton 1978:774-775). Some debris is also transported as a debris-water system, between the glacier sole and the substratum (Engelhardt, Harrison, and Kamb 1978; Muller 1983a:14, 1983b:21).

Deposition.

**Lodgement till is deposited by plastering of glacial debris from the sliding base of a moving glacier by pressure melting and/or other mechanical processes.** Several of them were discussed by Goldthwait (1971:16), but there might be others not recognized yet. Basal debris may be lodged particle by particle, possibly with some size-sorting (Boulton 1975) or without such sorting (Hallet 1981). Also, sheets of debris-rich ice may become lodged if the force imparted by ice flow is insufficient to maintain their forward motion, and they become buried by the continuous lodgement over them (Boulton 1970, 1971; Lindner and Ruszczyńska-Szenajch 1979). The sheets of debris-rich ice either lose their ice by gradual negative regelation (Lavrushin 1976, 1980) or by melting due

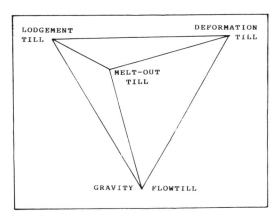

Fig. 1. Schematic genetic classification of till, with four principal endmembers at the apices of an inverted pyramid. Many combinations of the endmembers are possible, but their graphic presentation is hampered by the involvement of a fourth dimension, time.

to geothermal heat and heat produced by basal sliding (Drewry 1986). Strictly speaking, these lenses of till resulting from the negative regelation or basal melting of debris-rich ice are basal melt-out tills.

Deformation of already deposited lodgement till by glacial drag under the glacier sole may accompany the accretional lodgement process (Boulton 1979, 1982, 1987; Dreimanis 1987a), particularly if the till has been dilated. Further microstructural changes take place by the dewatering of the dilatant system (Muller 1983a).

Lodgement is not always a continuous process. It may be locally interrupted by glacial erosion, as evidenced by strong faceting of the upper surfaces of some intra-till boulder pavements along a continuous plane, or other intra-till erosional surfaces (Virkkala 1952; Dreimanis 1987b), or by meltwater erosion at the glacier sole. The latter is evidenced by downward-convex glaciofluvial channel fills of various dimensions (Eyles, Sladen, and Gilroy 1982), produced at various levels in lodgement till.

Diagnostic properties (for more details see Appendix D).

Most diagnostic properties reflect basal transport and subglacial deposition by a moving glacier, often accompanied by syn- to postdepositional deformation by glacial drag under the same glacier.

1. One set of characteristics is the laterally and generally also vertically consistent orientation of the various elements of fabric, structures, and surface marks on clasts, for instance: a) the parallel striae on the lodged clasts; b) the dominantly parallel orientation of the a-axes in macro- and micro-fabrics (Fig. 2), and commonly a gentle upglacier dip of the a-b planes of clasts; c) the soft-sediment smudges, thrust structures, shear planes, and that fissility related to lodgement or subsequent glacial drag (Fig. 3). Concentrations or pavements of clasts (Fig. 4) with striae parallel to glacial movement are common. Usually the basal contact with the substratum is sharp and erosional.

2. Comminution of debris during the basal transport and lodgement produces bimodal or multimodal particle size distribution of the various lithic components, at least one of the modes among clasts and another in the matrix, particularly in the silt size. Materials of local derivation increase in abundance towards the base of lodgement till.

Fig. 2. Lodgement till, recently deposited in front of the Solheimajøkull, Iceland, and exposed in a creek cut about 1 m below the surface. Note the fissility of the till, the faceted and striated surface of the cobble, and the orientation of most elongate clasts parallel to the glacial advance that was from the left to the right. The coin under the cobble is 2.5 cm in diameter. Photo taken in 1977.

Particle sorting is usually very poor, unless pre-sorted sediments have been incorporated, and most lodgement tills are diamictons.

3. Most lodgement tills are over-consolidated if the subglacial drainage was adequate during lodgement. Their bulk density, penetration resistance, and seismic velocity are usually higher than in other varieties of tills of the same region or similar composition.

Lodgement till occurs commonly in ground moraine, flutings, over drumlins, and on the proximal sides of some end moraines. The main summaries of diagnostic properties are in Goldthwait (1971:16-17), Boulton (1976b: Table 2), Dreimanis (1976:37), Boulton and Deynoux (1981: Table 1), McGown and Derbyshire (1977), Eyles, Sladen, and Gilroy (1982: Table 1), and Shaw (1985:30-38). Geotechnical properties are summarized in Boulton and Paul (1976) and Sladen and Wrigley (1983). Structures and/or fabric are discussed by Boulton (1970, 1971), Krüger (1979, 1984), Derbyshire (1980), Derbyshire and Jones (1980), Olszewski and Szupryczynski (1980), Derbyshire, Edge, and Love (1985), Rappol (1985), and Dowdeswell and Sharp (1986). The development of particle-size distribution by abrasion during lodgement is discussed by Boulton (1978) and Haldorsen (1981), the clast shapes by Boulton (1978) and Krüger (1984).

Fig. 3. Lodgement till of Wisconsinan age, exposed in a creek cut south of Buffalo, N.Y., U.S.A. Note the strong parallel fabric of clasts, and their orientation at right angles to the partly eroded upglacier limb of folded varved clays, visible closed to the creek level. The local glacial movement was away from the viewer. The handle of the pick is 40 cm long. Photo taken in 1958.

## 9.2.2 Melt-out till

Terminology and concept.

The term "melt-out till" was proposed by Boulton (1970), who also investigated and described the melt-out process in detail in Spitsbergen. The melt-out process has been considered by various authors, the earliest probably being Goodchild (1875) and Brückner (1886:11). Haldorsen and Shaw (1982) discussed with many references the further development of the melt-out concept and also the problem of recognizing melt-out till. The melt-out concept is now widely accepted, and many publications refer to melt-out till. However, opinions still differ on diagnostic characteristics of melt-out till. Sublimation till (Shaw 1977 and this volume) may be considered a variety of supraglacial melt-out till (see below).

Derivation, transport.

The derivation and transport of glacial debris that produce most melt-out till are similar to those of lodgement till (9.2.1). In addition, some melt-out till, particularly supraglacial melt-out till in mountain areas, derives from englacially transported debris that has never been in the basal zone of traction, and therefore has not been subjected to intensive glacial comminution and abrasion.

Deposition.

**Melt-out till is deposited by a slow release of glacial debris from ice that is not sliding or deforming internally.**
Therefore it may retain some of the structures of the debris-rich ice from which it is derived.
Melt-out till is deposited (1) as subglacial melt-out till at the base of stagnant ice — a stagnant glacier, or a stagnant zone underneath a moving glacier — by slow melting or negative regelation, or (2) as supraglacial melt-out till on the surface of a glacier by melting or in a very cold dry environment by sublimation also.
The sublimation process, being very slow, probably interacts at times with melt-out (Heuberger, personal communication, 1981). However, if it dominates the release of glacial debris, the resulting till may be called sublimation till (Shaw, this volume).

Diagnostic properties. (For more details see Appendix D).

Some properties of melt-out till are inherited from transport in glacier ice, while others derive from the melt-out process.
Properties inherited from basal and englacial transport of basal debris:
1. Debris banding in basal debris-rich ice may be preserved, but with gradational contacts and flattening resulting from the melting out of ice. In some melt-out tills the debris banding, if present, is destroyed by the melt-out process or subsequent deformation, and the till appears massive.
2. Clasts of unlithified sediments may be present; in some areas megablocks (rafts, floes) of bedrock or unlithified sediments and large-scale shear stacking of such megablocks are reported.
3. Clasts preserve the general orientation inherited from the glacial stress field, and are generally aligned parallel to glacier flow, but may also be oriented

Fig. 4. Catfish Creek till, exposed 1-3 m above Lake Erie at Bradtville, Ontario, Canada. This photo, taken in 1985, shows part of a silty and sandy subglacial till unit of complex origin, described by Dreimanis, Hamilton, and Kelly (1987). Local glacial movement was from right to left.

The intratill clast pavement, near the base of the photo, was probably formed by a combination of lodgement and glacial erosion. Sets of striae along the lower and upper surfaces of most clasts in the pavement and the long axes of clasts in the pavement, and also in the till 0.1-0.2 m above it, are parallel to the local glacial movement. The hookfolds 0.3 m above the pavement and the flat asymmetric folds along the base of the overlying large sand lens strike transverse to the direction of glacial movement. The folds were probably formed by subsole drag of dilated lodgement till. The large sand lens was deposited by a meltwater stream in a subglacial cavity; it also contains lenses of flowtill. Several planar sand-coated shear planes rising gently towards the left, are visible in the massive till overlying the large sand lens; they probably represent the former position of glacier sole. The handle of pick is 80 cm long.

transverse to flow where compressive stress conditions have occurred. The dip of a-b planes is generally reduced during melt-out to subhorizontal low angles.

4. Lithology and particle size composition are similar to that of lodgement till in the same area, but distantly-derived debris may be more abundant in melt-out till, if transported englacially. Grain-size composition is more variable than in lodgement till; according to Haldorsen (1981), melt-out till is usually coarser.

5. Clasts bear abrasion marks, but faceted and friction-cracked clasts are rare.

Properties inherited from englacial passive transport of supraglacial debris in mountain glaciers only:

1. Coarser than related melt-out till derived from subglacial debris.

2. Most clasts are angular, with no glacial abrasion marks.

3. Clast orientation is weak or random, except in some lateral moraines.

4. Usually normally consolidated.

Properties derived during the melting-out process of basally transported debris-rich glacier ice:

1. Symmetrical draping of the palimpsest banding and syndepositional layers and lenses of stratified sediments over large

clasts, or their truncation against the clasts (Fig. 5).

2. In macro-fabric: reduction of the inclination of a-b planes; micro-fabric modes weakened or destroyed.

3. Winnowing of fine particle from some parts of melting-out ice or melt-out till and redeposition in others, for instance as coatings around clasts.

4. Lenses and veins filled with sorted sediments become deposited by meltwater in cavities and crevasses of glacier ice during the melt-out process.

5. Consolidation usually normal, in the engineering sense, in supraglacial melt-out tills; overconsolidation may develop in subglacial melt-out tills if there is adequate drainage of meltwater.

Melt-out tills occur particularly in landforms related to stagnant glacier ice.

The main summaries of diagnostic properties are in Boulton (1976b: Table 2), Dreimanis (1976:26-27, 37), McGown and Derbyshire (1977), Lawson (1979a: Table XIII; 1981b:5 and Table 1), Boulton and Deynoux (1981: Table 1), and Shaw (1985:38-47). For more specific details see Boulton (1970), Ruszczyńska-Szenajch and Lindner (1976), Lawson (1979), Shaw (1979, 1982, 1983), Haldorsen and Shaw (1982), and Amark (1986).

Fig. 5. Sveg till, in the Overberg pit northwest of Sveg, central Sweden. It is a stratified silty sand diamicton. Lundqvist (1969) thought that this till "...was probably formed under active ice at sub-glacial outlets from the ice-lake region in the upper Ljusnan valley." Shaw (1979) interpreted it as basal melt-out till; Muller (1983a) as lodgement till. The handle of the shovel is 50 cm long. Photo taken in 1976.

### 9.2.3 Flowtill or flow till (broad meaning), flow till complex

Terminology and development of the concept.

Descriptions of glacigenic sediment flows have appeared in the geologic literature from time to time, for instance in Gripp (1929) and Harrison (1957). Harrison applied the terms "solifluction till" and "mudflow." The term "flowtill" was proposed first by Hartshorn (1958). He applied it to "...ablation moraine that moves as a mudflow off a glacier onto adjacent lower surfaces" (ibid.:481), and gave examples from Late Pleistocene sections in the eastern United States and recent sediments at the Malaspina Glacier in Alaska.

Detailed investigations of flowtill along present-day glaciers have been done particularly by Boulton (1968, 1971, 1972, 1973) who spelled the term as two words (flow till) and Lawson (1979, 1981, 1982). Lawson argued that the correct term should be "sediment flow deposit," since he did not consider this resedimented material to be true till, according to his restricted definition of till.

To emphasize that the flowage has occurred subsequent to the deposition of till, such terms as "flowed till," "slumped till," and "till flows" were suggested by several members of the Till Work Group during the discussions around 1980. These terms would not mean a genetic variety of till, but merely a recognition of a postdepositional resedimentation of till by flowage, including solifluction.

Even when dealing with glacigenic flow deposits there is a growing reluctance to use the term "flowtill." As already mentioned, Lawson (1979, 1981) used the term "sediment flow deposit" and "ice-slope colluvium"; Gravenor, von Brunn, and Dreimanis (1984) "glaciogenic sediment gravity flows"; Shaw (1985) "glacigenic mass-flow deposits" and "debris flow deposits." Morawski (1984, 1985, and this volume) proposed calling the intercalated flowtill and glaciofluvial sediment complexes "watermorainic sediments." He considered the watermorainic facies to be genetically intermediate between morainic (till) and fluvioglacial sediments. Lutenegger, Kemmis, and Hallberg (1983) applied the positional term "supraglacial deposits" to such complexes, avoiding the term "flowtill." On the other hand, Boulton (1980a:9) and Boulton and Deynoux (1981) proposed a more restrictive use of the term "flow till," applying it only to those "...viscous masses with a relatively low water content which flow slowly downslope" (Boulton and Deynoux 1981:400). The flows with a higher water content and independent grain movement would be called by them ordinary "mud flows" (ibid.). The latter ones could be identified by their "...well-defined sedimentary structure" (ibid.). In the outline "Flowtill" (Till Work Group Circular No. 22, 1981) Boulton was more specific:

"...a massive poorly sorted bed would represent the flowtill, a sorted bed with recognizable sedimentary structures the non-flowtill. Thus a flowtill is identified by its association with sorted non-flowtills. When working on a very small scale, those elements can be distinguished. On a larger scale, this is not convenient and the term flowtill complex should be used, as non-tills occur within a predominantly till mass."

While the application of the term "flowtill complex" appears to be very logical, the differentiation of its components into "true flowtills" and "non-flowtills" may become just a theoretical exercise, particularly if the sorted components are in the minority. We need terms that are sufficiently broad and applicable in geological mapping (see also Lundqvist, this volume). Time will reveal whether the glacially related resedimented diamictons will be given the short, simple designation "flowtills" (broad meaning), or the longer "flowtill complexes," or "mass movement tills," or "watermorainic facies," or they will not be called tills or moraines at all. In this review, the term "flowtill" will still be used in its broadest meaning as discussed below.

Derivation, transport, deposition.

**Flowtills (broad meaning) may derive from any glacial debris upon its release from glacier ice or from a freshly deposited till, in direct association with glacier ice (Fig. 6). The redeposition is accomplished by gravitational slope process, mainly by gravity flow, or by squeeze flow, and it may take place ice-marginally, supraglacially, or subglacially, and subaerially or subaquatically.**

Diagnostic properties (for more details see Appendix D and Lawson, this volume).

Some properties of flowtill are inherited from its primary source — glacier ice where the debris is released, or an already deposited till that was resedimented shortly after its initial deposition. Others are produced by flow and other related mass movements.

The lithologic composition of flowtill is usually inherited, except for that part of the matrix incorporated by the flow or altered by sorting due to flow.

One of the main characteristics produced by the resedimentation process is the facies association with water-sorted sediments at various levels, as well as intraformationally, particularly in waterlain flowtill deposits (Fig. 7). Therefore, most waterlain flowtill deposits, if consisting of many flows, look stratified.

Resedimentation by flow may produce some sorting, inverse or normal grading, and settling of rock clasts towards the base in matrix-dominated flows. The fabric is determined by local stresses, and its strength differs in different zones of flow and in different positions laterally. Therefore, random to strong fabric with parallel and transverse maxima has been reported by various authors (Boulton 1971; Rose 1974; Lawson 1979b; Derbyshire 1980; Morawski 1984; Rappol 1985; Dowdeswell and Sharp 1986). Flow structures are most conspicuous in waterlain flowtills.

The basal contact is usually sharp, depending upon the type of flow and the substratum material. It may be conformable with the substratum, or erosional, channel-shaped.

Fig. 6. Ice-marginal resedimentation of dilated water-saturated sandy subglacial till, in the lower half of the photo, at the terminus of the Burroughs Glacier, Alaska. Photographed in 1986.
The boulder in the centre, measuring 1 x 1.5 m, fell out of the ice shortly before photographing. The impact created very fluid flows (lighter shades) downslope of the boulder, and to the left of it. The till surface on the right of the boulder is gravelly, with the fines washed out. The till resulting from the resedimentation is interpreted as ice-marginal gravity flowtill.

Fig. 7. Catfish Creek Drift section in the Lake Erie bluff on the west side of Plum Point, 1-2.5 m above the lake. The dark layers are silty diamictons interpreted as subaquatic flowtill. They are interbedded with the lighter coloured glaciolacustrine silt and fine sand. The flow was from the right to the left. Subglacial till (Dreimanis 1982c:Fig. 1, S.W. end) was exposed about 1-2 m above this subaquatic flowtill-glaciolacustrine sediment complex, to the right of the photo taken in 1986. Local glacial movement was from the right to the left.

The main summaries of diagnostic properties and further references may be found in Boulton (1976b: Table 2), Lawson (1979, 1982, and this volume), Boulton and Deynoux (1981: Table 1), Lutenegger, Kemmis, and Hallberg (1983), DeJong and Rappol (1983), Morawski (1984:70-71), Gravenor, von Brunn, and Dreimanis (1984: Table 1), Rappol (1986), Drewry (1986:134-135). Subaquatic flow tills have been described by Marcussen (1973), Evenson, Dreimanis, and Newsome (1977), Kurtz and Anderson (1979), Hicock, Dreimanis, and Broster (1981), Morawski (1984, 1985, and this volume), Anderson (1985: Table 1), and Broster and Hicock (1986).

Further detailed classification of flowtills may be developed in hierarchical order, if diagnostic criteria can be identified to differentiate them. First, two large subgroups can be distinguished:
1. gravity flowtills,
2. squeeze flowtills.
The gravity flowtills can be further divided according to the dominant slope processes, for instance those discussed by Lawson (1979a, and this volume). No further classification of gravity flowtills will be attempted here, considering their very common interstratification, and the similarity in the characteristics of their varieties. For some descriptions relating to different slope processes, see Boulton (1971), Lawson (1979, and this volume), Drewry (1986), and Warren (this volume).

### 9.2.4 Squeeze flowtill, shear till, deformation till, glacitectonite

The concept of resedimentation by subglacial or ice-marginal squeeze flow or similar mechanisms has been mentioned from time to time in the geologic literature (see below). A question arises: shall the product of squeeze flow be considered a variety of flowtill, or shall it be referred to as part of deformation till, or merely as a deformed till or other subsole sediment (glacitectonite)?

The presence of plastically deforming water-saturated diamicton under the sole of glacier ice has been observed for more than a hundred years. References to the early reports may be found in von Böhm (1901). More recently squeeze-ups of basal till into radial and transversal crevasses forming minor till ridges have been reported by Okko (1955: Plates 4 and 6) from Iceland and by Mickelson (1971:50-61), Goldthwait (1971:16; 1974: Fig. 9; 1986), and Haselton (1979: Fig. 7a) from Alaskan glaciers. A thin "active subsole drift" was recorded by Engelhardt, Harrison, and Kamb (1978) under Blue Glacier, Washington, by bore-hole photography. Also, a similar deformed debris-rich subsole zone, a few centimetres thick, was observed by Schlüchter (1983) under the edge of Findelenglacier in the Alps, but it was already frozen when exposed in February, 1981. A much thicker (5-6 m) layer of deforming very porous and water-saturated sediment with high pore water pressure was postulated by Blankenship et al. (1986) from seismic measurements under an Antarctic ice stream.

Boulton (1979:28-30; 1987:31-32) reported experiments on subsole deformation of viscous dilatant lodgement till under Breidamerkurjökul in Iceland, and named it (1987:32) a "rapidly deforming dilatant A horizon."

Squeezing or pressing of till by the weight or movement of glacier ice has been proposed for the formation of a variety of ridges of till since the beginning of this century. Thus till ridges in the cores of some eskers have been explained by this mechanism by Korn (1910, mentioned in Woldstedt 1954), and Kraus (1928). The formation of rim ridges was suggested by ice-pressing (Hoppe 1952), and a similar process was proposed for the formation of DeGeer moraines (Hoppe 1959) and cross-valley moraines (Andrews 1963). A similar mechanism of subglacial deformation has been suggested for the formation of till flutes (Dyson 1952; Mickelson 1971; Boulton 1976a) and drumlins (Smalley and Unwin 1968; Evenson 1971; Boulton 1982, 1987). Till dykes, filling cracks in bedrock or sediments, over or underneath basal tills, have been described from Pleistocene sections by Dreimanis (1936: Figs. 3 and 8), Lavrushin (1976), Broster, Dreimanis, and White (1979), Levkov (1980), Hicock and Dreimanis (1985), Seret (1985), Amark (1986), and many others. Seret (1985:42) calls the glacial sediment injected into the crevasses of substratum bedrock "moraine de dislocation (dislodgement till)." Subhorizontal ice-marginal extrusions of dilated subglacial till are proposed as a mechanism by Dreimanis (1987a), in a discussion of a multigenetic Late Pleistocene till tongue.

The above ridges, dykes, tongues, or layers lithologically are diamictons that consist either of (1) resedimented subglacial till deposited by the same glacier that subsequently deformed it (Okko 1955; Boulton 1976a; Broster, Dreimanis, and White 1979; Dreimanis 1987a), or (2) basal glacial debris released at the glacier sole but not deposited yet as till (Engelhardt, Harrison, and Kamb 1978; Muller 1983b), or (3) any subsole material related or unrelated to the glacier that had squeezed it (Barnett 1987; Dreimanis 1987a: related waterlain stratified drift; McClintock and Dreimanis 1964; Ramsden and Westgage 1971; Stanford and Nickelson 1985; van der Wateren 1987: an older till and other older sediments). In case (1) Boulton's (1980a: Fig. 1) term "deformed lodgement till" may be applied to the diamicton. In case (2) the "active subsole drift" is glacial debris still in transit; when deposited, the final product will be either "deformation till" (Elson 1961) or "lodgement till," if plastered as a very thin layer (Fig. 8; Muller 1983a). In case (3) the final product may be called "deformed older sediment" (specified) or "glacitectonite" (Banham 1977; Pedersen, this volume), if the primary structures can be recognized after deformation by squeeze-flow. However, if the final product is so homogenized that the primary structures have been eradicated, then Elson's (1961) term "deformation till" will be appropriate also in case (3). Goldthwait's (1980: personal communication cited in Till Work Group's Circular No. 17:12) term "squeeze till = squeeze-out till," abbreviated to "squeeze till" or

Fig. 8. Catfish Creek till exposed in the Catfish Creek valley 0.4 km east of New Sarum, Ontario, Canada. The lower part of the photo up to the cluster of clasts at the top of the pick shows a massive silty and sandy diamicton containing many deformed sand lenses, partly washed out by the creek waters. Subhorizontal, very thin, sand-coated shear planes were visible just above the sand lenses. The sand lenses and shear planes strike N.-S. and rise westward. This diamicton is interpreted as deformation till. The handle of pick is 80 cm long.
The overlying massive diamicton, above the cluster of clasts is of similar composition, but without sand lenses. It contains several intratill clast pavements, above the photo, with sets of parallel E.-W. striae, E. being the stoss side. This diamicton is interpreted as lodgement till, deposited by the Erie lobe, locally moving from the E. towards the W. Photographed in 1985.

specified as "squeeze flowtill" may be applied in all three cases.

In Boulton's (1980a: Fig. 1) classification of tills by process x environmental matrix, both "deformed lodgement till" and "deformation till" are called "subglacial flow tills." Indeed, squeeze flow is the main process of the formation of the above discussed subsole sediments in all three cases.

Flowtills are usually considered to be secondary tills (Tables 8 and 10). Such designation applies to gravity flowtill. The squeeze flowtill is more probably a primary till, since it is formed and deposited by direct action of glacier ice, the same as other deformation tills. However, the mechanism of the formation of squeeze flowtill is similar to that of gravity flowtill. Therefore its properties are expected to be similar to those listed for gravity flowtill in Appendix D, with the exclusion of supraglacial debris, supraglacial sediment associations, and supraglacial landforms.

A question arises whether the squeeze flowtill is the only glacial subsole sediment that deserves to be called deformation till, as implied by Boulton's (1987) discussion of deformation of sediments beneath glacier ice. Elson (this volume) defines deformation till as "...weak rock or unconsolidated sediment that has been detached from its source, the primary sedimentary structures distorted or destroyed, and some foreign material admixed." By this definition, both Boulton's deformed subsole horizons, A and $B_1$, detached from their source and redeposited by glacier, comprise deformation till. Boulton's (ibid.) allochthonous A-horizon is the already discussed squeeze-flowtill. The underlying $B_1$-horizon consists of parautochthonous sheared and deformed rigid substratum; it is called "Schermoräne" by Grube (1979), "shear till" by Stephan (this volume), or merely "shear zone" by Rappol (1983, 1987). Since shear banding develops also in flowtill (Boulton 1976b:74) "...due to streaking out during the flow," some shear till or shear zone diamictons may be formed by viscous response to stress, just as in the A-horizon or a squeeze flowtill. Both Boulton's (1987) A- and $B_1$-horizons (or their equivalents by any other names mentioned in this paragraph) have gone through the subsole glacial deformation and redeposition process called "morainization" by Lundqvist (1985:389), by which "...shearing, brecciation and flow as a result of glacier action deformed and mixed...sediments to become diamictons."

In some instances (Ruszczyńska-Szenajch 1983; Stephan and Ehlers 1983; van der Meer, Rappol, and Semeijn 1985), a sharp unconformity separates this strongly sheared diamicton that looks like local till from the overlying lodgement till. In other cases, a transitional interdigitating contact has been noted with the overlying basal or lodgement till (Grube 1979; Rappol 1987). The transitional contacts make it hard to separate the lodgement till from deformation till.

In all the above examples the deformation and translocation by flow and shear are described as having occurred under the glacier sole, but some geologists express strong doubts whether it is always possible to differentiate material transported under the glacier sole from that which was transported frozen to the glacier sole and thus became part of the glacier. Also, evidence of subsole deformation (overthrusting, shearing, folding, brecciation) has been noticed at various levels in tills interpreted as lodgement tills (Ruszczyńska-Szenajch 1983; Dreimanis, Hamilton, and Kelly 1987; see also Table 13 in this paper). In such cases, the subsole deformation may be considered a part of the lodgement process, as suggested by Ruszczyńska-Szenajch (1983:116). Some authors who have paid particular attention to the dynamic facies of till, such as Lavrushin (1976 to 1980) and Aboltinsh (1986), would go even further and call the entire basal or subglacial till a glacitectonite.

Most subsole-deformation sediments discussed above have been reported from glaciers where the base of ice is at pressure melting point. However, Echelmeyer and Zhongxiang (1987:83) have observed "...enhanced deformation of the frozen and ice-laden subglacial drift" under the subpolar Urumqi Glacier No. 1 in China. The shear deformation of the ice-laden drift was possible because the sediment had "...an effective viscosity of more than one hundred times less than that measured in the overlying ice."

All the above references suggest that subsole deformation and material transport by subsole drag is a common process, and the resulting sediment, deformation till or glacitectonite, should be quite common, though not necessarily as abundant as suggested by Boulton (1987:42): "...most fine-grained Quaternary tills found in lowland areas have this origin." Depending upon a combination of a variety of factors, the presence and absence of a deformed substratum may change rapidly even over short distances (Dreimanis, Hamilton, and Kelly 1987:81-85).

Warren (1987) describes overconsolidated till from several sections in southern Ireland, interpreted by him to be gelifluctued diamicton (head deposit) that has been "...restructured and compacted rather than redeposited by overriding ice. It has undergone glaciodiagenesis" (ibid.:105). Therefore Warren proposes to call it "diagenetic till," rather than "deformation till." Since the term "diagenesis" has a wide range of meaning (Bates and Jackson 1980:171), and since this till "...has been, in part at least, redeposited by glacial ice" (Warren 1987:114), it might be called "deformation till"; it was redeposited for a very short distance.

From all the above examples of till, it may be concluded that subglacial till, if formed underneath a moving glacier, may consist of the entire range from glacitectonically deformed substratum that is not yet till, to deformation till and lodgement till. Therefore, depending upon the dominant features visible in such a subglacial material and the dominant field of interest of the observer (sedimentation, stratigraphy, structural geology, etc.), different terms have been applied, ranging from glacitectonite to various varieties of till or diamicton (see, for instance Elson, Pedersen, and Stephan, this volume). Whatever term is used, the lithology, structures, and fabric of the material have to be objectively described, and the orientations measured, before assigning any specific name.

Criteria for identification of deformation till are not given in Appendix D, because they have not been discussed sufficiently by the Till Work Group. A tentative outline of the criteria was proposed by J.A. Elson in the Till Work Group Circular 20 in 1982 (reproduced in Elson, this volume), but most comments received dealt with the term itself, rather than the criteria. The discussions at the INQUA Commission C-2 field conferences in 1986 also focussed more on the two main terminologic alternatives, "deformation till" and "glacitectonite," than on their properties. The majority of discussants favoured "glacitectonite" at that time. However, several publications that appeared since then (see references of 1986 and 1987 in this Chapter) suggest that "deformation till" is a reality, although more complex than originally envisaged. We have to learn more about the processes of its formation and the resulting descriptive criteria. Some of them are similar, as already mentioned in this chapter, to those of flowtill (Appendix D).

### 9.2.5 Undermelt diamicton or undermelt till?

Another much disputed variety of secondary glacigenic deposit has been called "subaquatic melt-out till" (Dreimanis 1969), "undermelt diamicton" (Gravenor, von Brunn, and Dreimanis 1984) or "undermelt till" (Drozdowski 1986; Parkin and Hicock, this volume). These names have been proposed and used for waterlain stratified to massive diamictons deposited under the glacier sole. The glacial debris is produced by melt-out at the glacier sole (Gibbard 1980). If the debris falls a short distance through standing water,

very little sorting of sand-size and coarser particles takes place, but some clay-size particles may be transferred laterally. Strictly speaking, this sedimentation process is more aquatic than glacial. Therefore, I am hesitant now to call the resulting "undermelt diamicton" a true till. If the water layer under the glacier sole is so thin (a few mm to a few cm) that the clasts retain their glacial orientation, it will be hard to separate the resulting sediment from melt-out till. However, in such a case it will probably be considered part of melt-out till, with some participation of secondary redeposition of the matrix components of the till.

## 10 GENETIC CLASSIFICATION OF TILL BY THE POSITION OF ITS FORMATION AND DEPOSITION

The positions of the formation and deposition of till in relation to glacier ice were briefly discussed in Chapter 8.2. The supraglacial and ice-marginal positions are generally considered together.

### 10.1 Subglacial versus supraglacial tills

Classification by the place of deposition in relation to glacier ice has been most popular in the past, both in English and other languages. Prior to 1980, the following genetic terms have been most commonly used in Europe and North America (Dreimanis 1980: Table 4):

1. for supraglacial till: ablation till, ablation moraine;
2. for subglacial till: basal till, ground moraine (literal translation from the French "moraine profonde," the German "Grundmoräne," and from similar terms used in other languages).

Ablation may also occur subglacially and englacially. Therefore, the usage of the term "ablation till" or "ablation moraine" for supraglacial till has been discouraged by the Till Work Group, but it still appears in the literature.

### 10.1.1 Subglacial or basal till

Terminology.

Both the terms "basal till" and "subglacial till" are commonly used. The main objection to the use of "basal till" is that it may be confused with "basal debris," the material transported in the basal traction zone of the glacier. Confusion is most likely to occur in those languages where "moraine" is used rather than "till" because "moraine" in many cases applies to both the material in glacial transport, and the material deposited by the glacier. For instance, in Russian, "donnaya morena" (ground moraine) and "osnovnaya morena" (basal moraine) are used interchangeably to denote either basal debris or subglacial till (Schukin 1980; Timofeev and Makkaveev 1986). Therefore, a direct translation of the English term subglacial till — podlednikovaya morena — has been proposed recently by several members of the INQUA Till Work Group as a less confusing alternative (Appendix C).

Subglacial till comprises all those tills that are deposited subglacially: primary lodgement, deformation and basal melt-out tills, and secondary gravity flowtill.

If the glacitectonites, deformation till, and shear till are to be classified as tills, they are subglacial because of their position (Chapter 9.2.4). However, Grube (1979) and Grube and Vollmer (1985) place shear till (Schermoräne) underneath basal or subglacial till (Grundmoräne), and classify it as the uppermost unit of a special genetic group "Sohlmoräne." Those who consider any subglacially deformed substatum to be merely a glacitectonite or a shear zone and not a till would also not include it with subglacial till.

"Subglacial till" as a broad useful term.

(The terms "subglacial till" and "basal till" will be used interchangeably, with the same meaning, in the next three paragraphs, depending upon which of them was used by the author mentioned).

Boulton (1976b:72) thought that the term "basal till" was "...redundant in view of the much more specific genetic terms which can now be deployed." That was the dominant opinion at the INQUA Till Work Group during the late 1970's. However, more recent detailed investigations of subglacial tills, for instance by Muller (1983b), Saarnisto and Peltoniemi (1984), and Dreimanis, Hamilton, and Kelly (1987), have increased the awareness of their genetic complexity. Even Boulton (1982:21), discussing the interaction of subglacial depositional and deformational processes forming streamlined landforms at present-day glaciers, concluded that "...the simple modelling...in which all [processes] are considered separately may not be a good guide in the real world in which we expect complex interactions to take place."

Certainly, the specific process-related varieties, such as lodgement, melt-out and flow tills, deformation till, glacitectonites and undermelt diamictons, can be identified in small-scale studies by applying the multiple criteria approach found in Parkin and Hicock (this volume), Shaw (1987), Whiteman (1987), and Dreimanis, Hamilton, and Kelly (1987). However, in a 12-m thick Catfish Creek till section, investigated by May, Dreimanis, and Stankowski (1980) adjoining to the site of Dreimanis, Hamilton, and Kelly (1987), characteristics of lodgement, melt-out, and flow-tills deposited in a subglacial environment were noted, but their separation into subunits was hardly possible.

In many recent regional or other large-scale investigations, the broader terms "basal till" or "subglacial till" are preferred, for instance by Anderson et al. (1980) and Anderson (1985) in the Antarctic; Kemmis, Hallberg, and Lutenegger (1981), Dreimanis (1983), Anderson, Goldthwait, and McKenzie (1986), Rappol (1986), and Hansel and Johnson (1987) in North America; Bergersen and Garnes (1983), Lundqvist (1983), Stephan and Ehlers (1983), Ringberg, Holland, and Miller (1984), Serebryanny et al. (1984), and van der Meer, Rappol, and Semeijn (1985) in Europe; and Ono (1984) in Asia. Also, in many discussions of diagnostic properties of tills reference is made to "subglacial till," "basal till," or merely "subglacial deposition," without distinguishing among the subglacial genetic varieties of tills. Thus, Anderson (1985: Table 1) compares basal till with glaciomarine sediments, and Lutenegger, Kemmis, and Hallberg (1983) with supraglacial sediments; Olszewski and Szupryczyński (1980) discuss the structures and fabric of recent basal till in Spitsbergen; Lavrushin (1980) the dynamic facies of basal till; Serebryanny and Orlov (1982) use the term "basal till" when discussing the origin of present-day marginal moraines of the Caucasus; and Evenson and Clinch (1987) in Alaska; Mills (1977) and Vorren (1977) present regional reviews of the grain-size characteristics of basal and other tills.

10.1.2 Supraglacial till

Terminology

"Supraglacial till" is a British term, equivalent to the American "superglacial till," and now preferred in English.

Depositional varieties.

Supraglacial till includes the following supraglacially derived, and supraglacially and ice-marginally deposited genetic varieties of till: primary supraglacial melt-out and sublimation tills, and secondary flowtill, in its broad meaning.

Considering the sloping surface of melting glacier ice and the changes in the slopes of melting ice surfaces over time during deglaciation, most supraglacial melt-out till and other supraglacial debris may become resedimented and transformed into flowtill. "The transition from melt-out till to a slowly flowing mass is difficult to identify" state Boulton and Deynoux (1981:407). Therefore, it is often safer to apply the term "supraglacial till" in such situations rather than try to draw the boundary between a supraglacial melt-out till and the initial phases of flow tills unless a special small-scale investigation is done. This differentiation is particularly difficult where the supraglacial till forms a thin surface mantle over subglacial till, and where this surface has been exposed to a variety of nonglacial agents since its deposition, for instance in many Pleistocene ground moraine areas. Even in some thicker occurrences of surface tills, it is difficult to differentiate between melt-out till and flowtill by very detailed investigations, as shown by Haldorsen (1982) and Haldorsen and Shaw (1982:273-276) in a study of a till and gravel ridge in Astedalen, Norway.

The "supra-till" of Boulton and Deynoux (1981), formerly proposed by Boulton (1976) as "supraglacial morainic till" is an endmember of that supraglacial till of valley glaciers where no basal material has been admixed.

# 11 GENETIC CLASSIFICATION OF TILLS BY ENVIRONMENT OF DEPOSITION

As discussed in Chapter 8.1, tills may be deposited in the following two environments: 1) glacioterrestrial and 2) glacioaquatic.

## 11.1 Glacioterrestrial tills

Though meltwater participates to some extent in the formation and deposition of till (see Chapter 7.1), most tills are usually classified as glacioterrestrial deposits, because they are not deposited into a body of water. This applies to typical lodgement, deformation, and melt-

out tills or their combinations in subglacial tills, to supraglacial flowtills, to sublimation till, and all those ice-marginal and subglacial flowtills that are not deposited subaquatically.

## 11.2 Glacioaquatic or subaquatic tills, waterlain tills

A common denominator for the tills called glacioaquatic, subaquatic, or waterlain is that they are deposited in water bodies, but without much if any sorting. It does not matter whether they are deposited above or below sea level, in fresh water, or in brackish water. Their depositional environment is similar: subaquatic and glacially related (see 8.1). The following depositional varieties of tills are included in this group: subaquatic ice-marginal or subglacial flowtills deposited by debris flow or squeeze flows that ended their flow in water.

When I proposed the term "waterlaid tills" in 1967 (Dreimanis 1969), later changing it to "waterlain till" following the suggestion of Francis (1975), I applied this term also to diamicton deposited by free fall in water from the base of melting glacier ice (see Chapter 9.2.5) and called it waterlain melt-out till. Since the free fall may cause some sorting of the debris, it may be better to discontinue calling this diamicton a till.

The probable processes of formation of waterlain flow tills and their lithologic structural and textural characteristics are described in Evenson, Dreimanis, and Newsome (1977), Kurtz and Anderson (1979), Hicock, Dreimanis, and Broster (1981), Morawski (1984, 1985, and this volume), Broster and Hicock (1986), Parkin and Hicock (this volume), and Dreimanis (1987a). Morawski (this volume) classifies them and the associated meltwater sediments as "water-morainic sediments."

# 12 GENETIC NAMING OF TILLS BY COMBINING PROCESS, PLACE OF DEPOSITION, AND DEPOSITIONAL ENVIRONMENT

## 12.1 Sequence of naming (using Table 11 or 12)

Generally, the genetic naming of till begins with the assignment of the position of its deposition in relation to glacier ice, for the obvious reason that at present-day glaciers the place of deposition is seen first. When studying past glacial deposits, the position of sedimentation of till may be judged at first from the regional relationships, landforms, and other gross characteristics of the till being investigated.

In the hierarchical order the process of deposition should be the second designation. Its identification requires more detailed investigation, for instance, to determine whether a subglacial till has been deposited by lodgement, basal meltout, subsole drag, or any secondary subglacial depositional process. The possibility of postdepositional or syn-depositional deformation must also be investigated.

While doing this investigation, the third factor, depositional environment, comes into question — was the till deposited subaquatically or not? If the designation of subaquatic deposition — waterlain — is not added, this usually means a nonaquatic "glacioterrestrial" environment. During detailed investigations to decipher the deposition process of the sediment in question and the position of deposition, the entire classification of the sediment as a till has to be reconsidered, particularly if the initial assignment was only preliminary.

## 12.2 Examples

1. "Supraglacial flowtill": flowtill (most probably gravity flowtill) deposited on glacier ice in a nonaquatic glacio-terrestrial environment.
2. "Ice-marginal waterlain flowtill": flowtill deposited subaquatically at the margin of a glacier.
3. "Subglacial melt-out till": subglacial till deposited by the melt-out process at close contact with glacier ice and substratum.
4. "Subglacial flowtill": subglacially deposited flowtill in a cavity that probably was not filled with water.
5. "Subglacial waterlain flowtill": subglacially deposited flowtill in a cavity that was filled with water, or deposited under a partly floating glacier at its grounding line; in the latter case a more specific name may be applied: "subglacial ice-marginal waterlain flow-till."
6. "Subglacial till": subglacially deposited till, without specifying the process of deposition.

A similar sequence in naming tills is also applied in French, by using terms listed in Table 12, for instance "till supraglaciaire flué" for "supraglacial flow till."

## 13 COMPREHENSIVE GENETIC CLASSIFICATION OF TILL

Most genetic classifications of till are based upon depositional criteria only, including those discussed in Chapters 9-12. However, in the descriptions of the main depositional varieties of tills (in Chapters 9.2.1-9.2.3) the release of debris from glacier ice and the final transport just before deposition is given more as background information.

Several previous schemes of genetic depositional classification of tills have also pointed out the derivation and transport of glacial debris prior to their deposition as background information, not as a basis for the classification of tills; for instance the till classification charts of Dreimanis (1969: Table 1; 1976: Fig. 5; 1982: Table 1, reproduced here as Table 10); Francis (1975), Boulton (1980: Fig. 2, reproduced here as Table 8).

The importance of recognizing the mode of glacial transport in classifying tills was realized by about a hundred years ago by Chamberlin (1883), Upham (1891), and Crosby (1896), for instance. These early investigators used the term "englacial till" with an emphasis on the mode of transport, not deposition.

The difference in the derivation of glacial debris before transport was also recognized at least a century ago. Thus Chamberlin (1883:301) wrote: "the material borne by glaciers belongs to two diverse classes: 1st that which falls upon the ice from cliffs and towering peaks; and 2nd, that which is abraded from the rocks over or against which it moves." The source and processes of derivation were still not emphasized in the genetic classification of till at that time, however.

As pointed out in Chapters 6.2, 6.3, and 6.5, predepositional processes have been considered to some extent in the genetic classification of till, but little was done to include them in genetic classifications before 1976. They were included, however, in several classifications published between 1976 and 1984 (see Chapter 6.5). In the Till Work Group's preliminary report (Dreimanis 1982b:27-28), the development of a comprehensive or expanded genetic classification of till was proposed as one of the objectives for the 1982-87 period.

Tables of criteria for such an expanded classification were distributed with the Till Work Group Circulars Nos. 26 and 27 and discussed at the INQUA Till Work Group meetings in 1985 and 1986. However, most

Table 14. Selected parameters for expanded genetic classifications of till.

| RELEASE OF GLACIAL DEBRIS AND ITS DEPOSITION | | | | |
|---|---|---|---|---|
| I. ENVIRONMENT | II. POSITION | III. PROCESS | IV. TRANSPORT | V. DERIVATION |
| GLACIOTERRESTRIAL | ICE MARGINAL | LOWERING OF SUPRAGLACIAL DEBRIS | SUPRAGLACIAL | SUPRAGLACIAL-EXOGENOUS |
| | SUPRAGLACIAL | MELTING OUT | ENGLACIAL | |
| | | SUBLIMATION | | |
| | SUBGLACIAL | PRESSURE MELTING AND REGELATION | BASAL (IN ICE) | SUBGLACIAL |
| GLACIOAQUATIC | | LODGEMENT | | |
| | | SUBSOLE DRAG & SHEAR | SUBSOLE (TRANSLOCATION BY GLACIAL DRAG) | |
| | | FLOWAGE: <br> -GRAVITY FLOW <br> -SQUEEZE FLOW | | |
| | | OTHER MASS MOVEMENTS: <br> -SLIDING <br> -FALLING | | |

NOTE: EACH VERTICAL COLUMN IS INDEPENDENT FROM THE OTHER FOUR, AND NO CORRELATION HORIZONTALLY IS IMPLIED. ONLY SOME COMBINATIONS ARE FEASIBLE. THE PROCESS OF DEPOSITION (COLUMN III) INCLUDES ALSO SOME TRANSPORT IMMEDIATELY PRIOR TO THE DEPOSITION.

discussants considered such a detailed classification to be merely a theoretical exercise (see also Lundqvist, this volume) without much applicability. Still, even Lundqvist, who is quite skeptical about the usefulness of such an expanded classification, has presented a table consisting of four columns of predepositional parameters and four columns of depositional parameters.

A similar provisional table of selected parameters for an expanded genetic classification of till is presented here as Table 14. For practical reasons, the columns in this table are arranged in an order opposite to the derivation-transport-deposition of till. If a till is named and classified according to this table, then its naming should begin by the use of the depositional parameters followed by the designation of the mode of transport and the derivation. The word "till" should be inserted between the depositional and transport designations. Several examples are given below, with the column numbers (I-V) listed after each designation, to assist the reader in relating the naming to Table 14. Normally, the column numbers should not be included.

1. Subaquatic (I) ice-marginal (II) gravity flow (III) till, of englacial transport (IV) and subglacial derivation (V).

2. Glacioterrestrial (I) subglacial (II), melt-out (III) till, of basal transport and subglacial derivation ("Glacioterrestrial" may be left out, since most tills belong to this group).

3. (Glacioterrestrial, I, left out), supraglacial (II), lowered and gravity flow (III) till of supraglacial transport (IV) and exogenous derivation (V).

Admittedly, the above names are long and they appear cumbersome. Because of their length, they will not be used repeatedly in a discussion. However, they may be useful in theoretical studies of till. In applied studies, particularly when using indicator tracing for mineral exploration, the most important nondepositional parameter is the mode of glacial transport (Virkkala 1974; Stephens et al. 1983; Geddes 1984). It may be necessary to use the full length of classification only once, and then use an abbreviation. The entire application of an expanded comprehensive classification of till is considered as a tentative attempt to classify till as fully as our knowledge permits.

More parameters may be included in a comprehensive expanded classification. Thus, deformation structures were added to the original proposal of the expanded

classification circulated among the Till Work Group members. However, this was found to be too complex. Deformation structures could have been formed within the glacier ice, by glacial transport (Shaw 1977), or subglacially by the load of glacier ice or by subsole drag, syndepositionally or postdepositionally, or by gravity flows. Because of the nonsynchroneity of the formation of such structures in relation to glacial transport and deposition, and the different processes involved, the products of deformation require placement in several columns, making an already complex classification proposal even more so. The exclusion of deformation structures from the present comprehensive till classification proposal does not mean that deformation is not worth considering in such a classification. The opposite is more true, and the inclusion of deformation structures may be worked out during the next intercongress period, since the Glaciotectonics Work Group will still continue its activities.

14 CONCLUSIONS

Glacial deposits have been studied for about 150 years: first mainly along present-day mountain glaciers. Later, investigations expanded to areas of former Pleistocene and pre-Pleistocene glaciations. During the last few decades, more attention has again been paid to present-day glaciers, particularly in the Arctic and Antarctic regions.

We are learning more and more about glacial processes of sedimentation, both from field and laboratory studies and from theoretical considerations, but opinions differ about what is meant by the glacial sedimentation process.

Glacial sedimentation is polygenetic. Since glacier ice is the primary and main agent in this process, some would like to exclude the participation of the secondary agents — gravity and water. Such a separation is possible outside a glacier. However, in the domain of glacier ice, underneath or over it, and often also ice-marginally, the primary glacial sedimentation and secondary resedimentation are so intimately related to the primary process in both time and space that their separation is often impossible. This complexity leads to adherence to a broad definition of glacial sediment, generally called "till" in English and "moraine" in most other languages.

Another typical complexity for till is

its genetic variability within a continuum, both vertically and laterally, and the synchronism of its deposition at various levels, for instance subglacially and supraglacially in the ice-marginal zone. Therefore, the distinction of genetically different till units is often difficult, particularly at mappable scales. To accommodate this complexity, the genetic classification of till must begin with the distinction of broad genetic groups of till such as (1) primary versus secondary tills or (2) subglacial, supraglacial, and ice-marginal till, the latter two often being combined under the term "supraglacial."

Furthermore, more detailed genetic hierarchical classification is possible, by doing thorough multicriterial investigations. For instance, in subglacial tills, the endmembers in their sedimentologic pyramid (Fig. 1) — lodgement, basal melt-out, deformation and gravity flowtills — may be distinguished locally. Opinions still differ as to whether to include the subsole material deformed and translocated by glacial drag, with subglacial till, particularly with lodgement till, or to distinguish it as a separate variety of till (deformation till, shear till), or to consider it merely as a glacitectonite.

The criteria for the distinction of more detailed genetic varieties of till are not discussed here, but the parameters for such further hierarchical classification are proposed here in Tables 10, 11, 12, and 14 and Table 1 of Lundqvist (this volume). Some controversial varieties are discussed in Chapters 9.2.4 and 9.2.5 or reference is made to discussions about them elsewhere in this volume to stimulate more investigations of the till-forming and related glacigenic processes and their products.

As already mentioned, this is not a final, conclusive report on genetic classifications of till. It is merely another progress report on the till-like continuum of accumulation of opinions about the genetic varieties of till.

## REFERENCES

Aario, R. 1977. Classification and terminology of morainic landforms in Finland. Boreas 6:87-100.

Aboltinsh, O.P. 1986. Analiz trekhosnykh lineinykh strukturnykh elementov moren i interpretatsiya ego rezultatov. In O.P. Aboltinsh, G. Ya. Eberhards, V. Ya. Klane (eds.), Morenogenez reliefa i paleogeografiya Latvii: 19-35. Riga: Latv. gosud. universit. P. Stuchki.

Agassiz, L. 1838. Des glaciers, des moraines, et des blocs eratiques. Actes Soc. Helv. Sc. Nat. 22:V-XXXII.

Agassiz, L. 1842. On glaciers and the evidence of their having once existed in Scotland, Ireland, and England. Proc. Geol. Soc. London 3:327-332.

Åmark, M. 1986. Glacial tectonics and deposition of stratified drift during formation of tills beneath an active glacier: examples from Skåne, southern Sweden. Boreas 15:155-171.

Anderson, J.B. 1985. Antarctic glacial marine sedimentation: A core workshop. Houston, Rice University.

Anderson, J.B., D.D. Kurtz, E.W. Domack, and K.M. Balshaw 1980. Glacial and glacial marine sediments of the Antarctic continental shelf. J. Geol. 88:399-414.

Anderson, P.J., R.P. Goldthwait, and G.D. McKenzie (eds.) 1986. Observed processes of glacial deposition in Glacier Bay, Alaska. Columbus, Instit. Polar Studies, Ohio State Univ.

Andrews, J.T. 1963. The cross-valley moraines of north-central Baffin Island: A quantitative analysis. Geogr. Bull. 20:82-129.

Banham, P.H. 1977. Glacitectonites in till stratigraphy. Boreas 6:101-105.

Barnett, P. 1987. Quaternary stratigraphy and sedimentology, north-central shore, Lake Erie, Ontario, Canada. Unpublished Ph.D. thesis, Univ. of Waterloo.

Bates, R.L. and J.A. Jackson 1980. Glossary of geology, 2nd edition. Falls Church, Virginia, American Geol. Inst.

Bergersen, O.F. and K. Garnes 1983. Glacial deposits in the culmination zone of the Scandinavian ice sheet. In J. Ehlers (ed.), Glacial deposits in northwest Europe: 29-40. Rotterdam, Balkema.

Beskow, G. 1935. Praktiska och kvartargeologiska resultat av grusinventeringen i Norrbottens lan. Foredragsref. Geol. Fören. Stockh. Förh. 57:120-123.

Blankenship, D.D., C.R. Bentley, S.T. Rooney, and R.B. Alley 1986. Seismic measurements reveal a saturated porous layer beneath an active Antarctic ice stream. Nature 322:54-57.

Böhm, A. von 1901. Geschichte der Moränenkunde. Abhandl. K.K. Geogr. Gesellsch. Wien, III-4. Vienna, R. Lechner.

Boulton, G.S. 1968. Flow tills and related deposits on some Vestspitsbergen glaciers. J. Glaciol. 7:391-412.

Boulton, G.S. 1970. On the deposition of subglacial and meltout tills of the margin of certain Svalbard glaciers. J.

Glaciol. 9:231-245.

Boulton, G.S. 1971. Till genesis and fabric in Svalbard, Spitsbergen. In R.P. Goldthwait (ed.), Till: A symposium: 41-72. Columbus, Ohio State Univ. Press.

Boulton, G.S. 1972. Modern arctic glaciers as depositional models for former ice sheets. J. Geol. Soc. 128:361-393.

Boulton, G.S. 1975. Processes and patterns of subglacial sedimentation: A theoretical approach. In A.E. Wright and F. Moseley (eds.), Ice ages: Ancient and modern. Geol. J. Spec. Issue 6:7-42.

Boulton, G.S. 1976a. The origin of glacially fluted surfaces: Observations and theory. J. Glaciol. 17:287-309.

Boulton, G.S. 1976b. A genetic classification of tills and criteria for distinguishing tills of different origin. In W. Stankowski (ed.), Till, its genesis and diagenesis. Univ. A. Mickiewicza w Poznaniu. Ser. Geogr. 12:65-80.

Boulton, G.S. 1978. Boulder shapes and grain-size distributions of debris as indicators of transport paths through a glacier and till genesis. Sedimentology 25:773-799.

Boulton, G.S. 1979. Processes of glacier erosion on different substrata. J. Glaciology 23:15-38.

Boulton, G.S. 1980a. Classification of till. Quaternary Newsletter 31:1-12.

Boulton, G.S. 1980b. Genesis and classification of glacial sediments. In W. Stankowski (ed.), Tills and glacigene deposits. Univ. A Mickiewicza w Poznaniu. Ser. Geogr. 20:15-17.

Boulton, G.S. 1982. Subglacial processes and the development of glacial bedforms. In R. Davidson-Arnott, W. Nickling, and B.D. Fahey (eds.), Research in glacial, glaciofluvial and glacio-lacustrine systems, p. 1-31. Proc. 6th Guelph Symposium on Geomorphology, 1980. Norwich, Geo Books.

Boulton, G.S. 1987. A theory of drumlin formation by subglacial sediment deformation. In J. Menzies and J. Rose (eds.), Drumlin Symposium, p. 25-80. Rotterdam, Balkema.

Boulton, G.S. and M. Deynoux 1981. Sedimentation in glacial environments and the identification of tills and tillites in ancient sedimentary sequences. Precambrian Research 15:397-420.

Boulton, G.S. and N. Eyles 1979. Sedimentation by valley glaciers: A model and genetic classification. In Ch. Schlüchter (ed.), Moraines and varves, p. 11-23. Rotterdam, Balkema.

Boulton, G.S. and M.A. Paul 1976. The influence of genetic processes on some geotechnical properties of glacial tills. J. Engin. Geol. 9:159-194.

Broster, B.E., A. Dreimanis, and J.C. White 1979. A sequence of glacial deformation, erosion, and deposition at the ice-rock interface during the last glaciation, Cranbrook, B.C., Canada. J. Glaciol. 23:283-295.

Broster, B.E. and S.R. Hicock 1986. Multiple flow and support mechanisms and the development of inverse grading in a subaquatic glacigenic debris flow. Sedimentology 32:645-657.

Brückner, E. 1886. Die Vergletscherung des Salzachgebietes, nebst Beobachtungen über die Eiszeit in der Schweiz. Penck's Geogr. Abh. I-1.

Carruthers, R.G. 1939. On northern glacial drifts: Some peculiarities and their significance. Geol. Soc. London, Quart. Jour. 95:299-333.

Chamberlin, T.C. 1882. Observations on the recent glacial drift of the Alps. Trans. Wisconsin Acad. Sci., Arts and Letters 5:238-270.

Chamberlin, T.C. 1883. Preliminary paper on the terminal moraine of the second glacial epoch. U.S. Geol. Survey 3rd Ann. Report: 291-402.

Chamberlin, T.C. 1894. Proposed genetic classification of Pleistocene glacial formations. J. Geol. 2:517-538.

Charlesworth, J.K. 1957. The quaternary era, with special reference to its glaciation. London, Edward Arnold.

Charpentier, J. de 1841. Essai sur les glaciers et sur le terrain eratique du bassin du Rhone. Lausanne, Marc Ducloux.

Chinese Research Group of Glaciological Sediment 1981. On the terminology of moraine terms and their concept and translation. J. Glaciol. and Cryopedol. 3(1):78-84 (in Chinese, with some terms in English).

Chumakov, N.M. 1978. Precambrian tillites and tilloids. Moscow, Nauka (Russian).

Clayton, L. and S.R. Moran 1974. A glacial process-form model. In D.R. Coates (ed.) Glacial geomorphology, p. 89-119. Binghamton, N.Y., State Univ. New York.

Crosby, W.O. 1896. Englacial drift. American Geologist 17:203-234.

Derbyshire, E. 1980. The relationship between depositional mode and fabric strength in tills: schema and test from the temperate glaciers. In W. Stankowski (ed.), Tills and glaciogene deposits. Univ. A. Mickiewicza w Poznaniu. Ser. Geografia 20:41-48.

Derbyshire, E. 1984. Sedimentological analysis of glacial and proglacial debris: A framework for the study of

Karakorum glaciers. In K.J. Miller
(ed.), International Karakorum Project
1:347-364. Cambridge University Press.

Derbyshire, E., M.J. Edge, and M. Love
1985. Soil fabric variability in some
glacial diamicts. In M.C. Forde (ed.),
Glacial tills 85: Proceedings Internat.
Confer. on Construct. in Glac. Tills and
Bould. Clays, 12-14 March 1985, p.
169-175. Edinburgh, Engin. Technics
Press.

Derbyshire, E. and P.F. Jones 1980.
Systematic fissuring of a matrix-
dominated lodgement till at Church
Wilne, Derbyshire, England. Geol. Mag.
117:243-254.

Deynoux, M. (ed.) 1985. Glacial record.
Proceedings of the Till Mauretania '83
Symposium, Nouakchott-Atar, 4-15 January
1983. Special issue. Paleogeogr.,
Paleoclimatol., Paleoecology 51.

Dowdeswell, J.A. and M.J. Sharp. 1986.
Characterization of pebble fabrics in
modern terrestrial glacigenic sediments.
Sedimentology 33:699-710.

Dreimanis, A. 1936. The rock-deformations
caused by the inland-ice, on the left
bank of the Daugava at Dole Island, near
Riga in Latvia (in Latvian, with English
summary). Riga, A. Gulbis.

Dreimanis, A. 1969. Selection of genetically significant parameters for
investigation of tills. In L. Krygowska
(ed.), Zesz. Nauk. Univ. A. Mickiewicza
w Poznaniu. Geografia 8:15-20.

Dreimanis, A. 1976. Tills: Their origin
and properties. In R.F. Legget (ed.),
Glacial till. Roy. Soc. Can. Spec. Publ.
12:11-49.

Dreimanis, A. 1978. Terminology and genetic classifications of tills or moraines
currently used in Europe and North
America. In E.V. Shantser and Yu. A.
Lavrushin (eds.), Ground moraines of
continental deposits (Materials of
Internat. Sympos.): 12-27. Moscow: Geol.
Inst. Acad. Sci. (Russian with English
tables).

Dreimanis, A. 1979. The problems of
waterlain tills. In Ch. Schlüchter
(ed.), Moraines and varves, p. 167-177.
Rotterdam, Balkema.

Dreimanis, A. 1980. Terminology and development of genetic classifications of
materials transported and deposited by
glaciers. In W. Stankowski (ed.), Tills
and glacigene deposits. Univ. A.
Mickiewicza w Poznaniu. Ser. Geogr.
20:5-10.

Dreimanis, A. 1982a. Work Group
(I)-Genetic classification of tills and
criteria for their differentiation:
Progress report on activities 1977-1982,
and definitions of glacigenetic terms.
In Ch. Schlüchter (ed.), INQUA
Commission on genesis and lithology of
Quaternary deposits, Report on activi-
ties 1977-1982, p. 12-31. Zürich.

Dreimanis, A. 1982b. Quaternary glacial
deposits: Implications for the interpretation of Proterozoic glacial deposits.
Geol. Soc. Amer. Memoir 161:299-307.

Dreimanis, A. 1982c. Two origins of the
stratified Catfish Creek Till at Plum
Point, Ontario. Boreas 11:173-180.

Dreimanis, A. 1983. Penecontemporaneous
partial disaggregation and/or resedimentation during the formation and deposition of subglacial till. Acta Geol.
Hispanica 18:153-160.

Dreimanis, A. 1987a. Genetic complexity of
a subaquatic till tongue at Port Talbot,
Ontario, Canada. In R. Kujansuu and M.
Saarnisto (eds.), INQUA Till Symposium,
Finland 1985. Geol. Survey Finland,
Spec. Paper 3:23-38.

Dreimanis, A. 1987b. Commission on genesis
and lithology of glacial deposits (C-2).
Report for the period 1982-1987. In Ch.
Schlüchter (ed.), INQUA Report on activities 1982-1987, p. 68-78. Zürich, ETH.

Dreimanis, A. and R.P. Goldthwait 1973.
Till. IX INQUA Congress Symposium 10,
Abstracts. Christchurch.

Dreimanis, A., J.P. Hamilton, and P.E.
Kelly 1987. Complex subglacial sedimentation of Catfish Creek till at
Bradtville, Ontario, Canada. In J.J.M.
van der Meer (ed.), Tills and glaciotectonics, p. 73-87. Rotterdam, Balkema.

Dreimanis, A. and J. Lundqvist 1984. What
should be called till? In L.K. Königsson
(ed.), Ten years of Nordic till
research. Striae 20:5-10.

Dreimanis, A. and Ch. Schlüchter 1985.
Field criteria for the recognition of
till or tillite. Paleogeogr.,
Paleoclimat., Paleoecol. 51:7-14.

Drewry, D. 1986. Glacial geologic
processes. Edward Arnold, London.

Drozdowski, E. 1986. Stratigraphy and origin of Vistulian Glaciation deposits in
northern part of the Lower Vistula
Region. Polska Akad. Nauk., Prace Geogr.
146 (Polish with Russian and English
summaries).

Dyson, J.L. 1952. Ice-ridged moraines and
their relation to glaciers. Amer. J.
Sci. 250:204-211.

Echelmeyer, K. and W. Zhongxiang 1987.
Direct observation of basal sliding and
deformation of basal drift at sub-
freezing temperatures. J. Glaciol.
33:83-98.

Elson, J.A. 1961. The geology of tills. In
E. Penner and J. Butler (eds.), Proceed.

14th Canad. Soil Mechanics Confer., N.R.C. Canada, Assoc. Comm. Soil and Snow Mechanics. Techn. Memor. 69:5-36.

Engelhardt, J.F., W.D. Harrison, and B. Kamb 1978. Basal sliding and conditions at the glacier bed as revealed by borehole photography. J. Glaciol. 20:469-508.

Evenson, E.B. 1971. The relationship of macro- and microfabric of till and the genesis of glacial landforms in Jefferson County, Wisconsin. In R.P. Goldthwait (ed.), Till: A symposium, p. 345-364. Columbus, Ohio State Univ. Press.

Evenson, E.B. and J.M. Clinch 1987. Debris transport mechanisms at active alpine glacier margins: Alaskan case studies. In R. Kujansuu and M. Saarnisto (eds.), INQUA Till Symposium, Finland 1985. Geol. Survey Finland, Spec. Paper 3:111-136.

Evenson, E.B., A. Dreimanis, and J.W. Newsome 1977. Subaquatic flow till: A new interpretation for the genesis of some laminated till deposits. Boreas 6:115-130.

Evenson, E.B., Ch. Schlüchter, and J. Rabassa (eds.) 1983. Tills and related deposits. Rotterdam, Balkema.

Eyles, N. (ed.) 1983. Glacial geology. Oxford, Pergamon Press.

Eyles, N., J.A. Sladen, and S. Gilroy 1982. A depositional model for stratigraphic complexes and facies superimposition in lodgement tills. Boreas 11:317-333.

Flint, R.F. 1947. Glacial geology and the Pleistocene Epoch. New York, Wiley and Sons.

Flint, R.F. 1957. Glacial and Pleistocene geology. New York, Wiley and Sons.

Flint, R.F. 1971. Glacial and Quaternary geology. New York, Wiley and Sons.

Flint, R.F., J.E. Sanders, and J. Rodgers 1960. Diamictite: A substitute term for symmictite. Geol. Soc. Amer. Bull. 71:1809-1810.

Francis, E.A. 1975. Glacial sediments: A selective review. In A.E. Wright and F. Moseley (eds.), Ice ages: Ancient and modern. Geol. J. Spec. Issue 6:43-68.

Francis, E.A. 1984. Glaciofluvial etc., or What is in a name? Quatern. Newsletter 44:32-37.

Geddes, R.S. 1984. Long distance glacial dispersion and mineral exploration. Till tomorrow '84. Kirland Lake, Ontario, May 8-12, 1984: Can. Inst. Mining and Metallurgy, Ont. Geol. Survey: 35.

Geikie, A. 1863. On the phenomena of the glacial drift of Scotland. Geol. Soc. Glasgow Trans. 1.

Gibbard, P. 1980. The origin of stratified Catfish Creek Till by basal melting. Boreas 9:71-85.

Goldthwait, R.P. 1971. Introduction to till, today. In R.P. Goldthwait (ed.), Till: A symposium, p. 3-26. Columbus, Ohio State Univ. Press.

Goldthwait, R.P. 1974. Rates of formation of glacial features in Glacier Bay, Alaska. In D.R. Coates (ed.), Glacial geomorphology, p. 163-185. Binghamton, State Univ. New York.

Goldthwait, R.P. 1986. Glacial history of Glacier Bay park area. In P.J. Anderson, R.P. Goldthwait, and G.D. McKenzie (eds.), Observed processes of glacial deposition in Glacier Bay, Alaska, p. 5-16. Columbus, Instit. Polar Studies, Ohio State Univ.

Goodchild, J.G. 1875. The glacial phenomena of the Eden valley etc. Geol. Soc. London Quart. J. 31:55-99.

Gravenor, C.P., V. von Brunn, and A. Dreimanis 1984. Nature and classification of waterlain glaciogenic sediments, exemplified by Pleistocene, Late Paleozoic and Late Precambrian deposits. Earth Sci. Reviews 20:105-166.

Gripp, K. 1929. Glaziologische und geologische Ergebnisse der Hamburgischen Spitzbergen-Expedition 1927. Naturwiss. Ver. Hamburg, Abhandl. a.d. Gebiet d. Naturwiss. 22:146-249.

Grisak, G.E., J.A. Cherry, J.A. Vonhof, and J.P. Blumele 1976. Hydrogeologic and hydrochemical properties of fractured till in the Interior Plains Region. In R.P. Legget (ed.), Glacial till. Roy. Soc. Canada Spec. Publ. 12:304-335.

Grube, F. 1979. Zur Morphogenese und Sedimentation im quartären Vereisungsgebiet Nordwestdeutschland. Verh. naturwiss. Ver. Hamburg (NF) 23:69-80.

Grube, F. and Th. Vollmer 1985. Der geologische Bau pleistozäner Inlandgletschersedimente Norddeutschlands. Bull. Geol. Soc. Denmark 34:13-25.

Haldorsen, S. 1981. Grain-size distribution of subglacial till and its relation to glacial crushing and abrasion. Boreas 10:91-105.

Haldorsen, S. 1982. The genesis of tills from Åstadalen, southeastern Norway. Norsk Geol. Tidskr. 62:11-32.

Haldorsen, S. and J. Shaw 1982. The problems of recognizing melt-out tills. Boreas 11:261-277.

Hallet, B. 1981. Glacial abrasion and sliding: Their dependence on the debris concentration in basal ice. Ann. Glaciol. 2:23-28.

Hambrey, M.J. and W.B. Harland (eds.) 1981. Earth's pre-Pleistocene glacial record. Cambridge, Cambridge Univ. Press.

Hansel, A.K. and W.H. Johnson 1987. Ice marginal sedimentation in a late Wisconsinan end moraine complex, northeastern Illinois, U.S.A. In J.J.M. van der Meer (ed.), Tills and glaciotectonics, p. 97-104. Rotterdam, Balkema.

Harland, W.B., K.N. Herod, and D.H. Krinsley 1966. The definition and identification of tills and tillites. Earth Sci. Reviews 2:225-256.

Harrison, P.W. 1957. A clay-till fabric: Its character and origin. J. Geol. 65:275-308.

Hartshorn, J.H. 1958. Flowtill in southeastern Massachusetts. Geol. Soc. Amer. Bull. 69:477-482.

Haselton, G.M. 1979. Some glaciogenic landforms in Glacier Bay National Monument, Southeastern Alaska. In Ch. Schlüchter (ed.), Moraines and varves, p. 197-205. Rotterdam, Balkema.

Hedberg, E.D. (ed.) 1976. International stratigraphic guide. New York, Wiley and Sons.

Heim, A. 1885. Handbuch der Gletscherkunde. Stuttgart.

Heuberger, H. 1980. Zur Nomenklatur der Glazialablagerungen aus ostalpiner Sicht. Verh. naturwissensch. Ver. Hamburg (NF)23:93-100.

Hicock, S.R., A. Dreimanis, and B.E. Broster 1981. Submarine flow tills at Victoria, British Columbia. Can. J. Earth Sci. 18:71-80.

Hicock, S.R. and A. Dreimanis 1985. Glaciotectonic structures as useful ice-movement indicators in glacial deposits: Four Canadian case studies. Can. J. Earth Sci. 22:339-346.

Hillefors, A. 1973. The stratigraphy and genesis of stoss- and lee-side moraines. Bull. Geol. Instit. Univ. Uppsala, N.S. 5:139-154.

Hoppe, G. 1952. Hummocky moraine regions, with special reference to the interior of Norbotten. Geogr. Annaler 34:1-71.

Hoppe, G. 1959. Glacial morphology and inland ice recession in northern Sweden. Geogr. Annaler 41:193-212.

Jong, M.G.G. De and M. Rappol 1983. Ice-marginal debris-flow deposits in western Allgäu, southern West Germany. Boreas 12:57-70.

Kapilanskaya, F.A. and V.D. Tarnogradskiy 1977. On the problem of formation of relict glacier ice deposits and preservation of primordially frozen tills. Isvestiya Vsesayuz. Geogr. Obschestva 109:314-319 (Russian).

Kemmis, T.J., G.R. Hallberg, and A.J. Lutenegger 1981. Depositional environments of glacial sediments and landforms on the Des Moines lobe, Iowa. Iowa Geol. Surv. Guidebook Series 6.

Klebelsberg, R. von 1948. Handbuch der Gletscherkunde und Glazial-geologie. Vienna, Springer-Verlag.

Krapotkin, P.N. 1876. Issledovaniia o lednikovom periode. Zap. Russk. geogr. ob-va 7:1.

Kraus, E. 1928. Tertiär und Quartär des Ostbaltikums Die Kriegschauplatze 1914-1918 geologisch dargestellt. H.10, T.1:142.

Krüger, J. 1979. Structures and textures in till indicating subglacial deposition. Boreas 8:323-340.

Krüger, J. 1984. Clasts with stoss-lee form in lodgement tills: A discussion. J. Glaciol. 30:241-243.

Krygowska, L. (ed.) 1969. International Symposium of the INQUA Commission on Origin and Lithology of Quaternary Deposits, 4-8 Sept. in Poznań and 9-10 Sept. 1967 in Warsaw, Zeszyty Nauk. Univ. A. Mickiewicza w Poznaniu. Geografia Nr 8.

Krygowski, B. 1971. Classification project of glacial tills. VIII Congres INQUA Paris 1969. Etudes sur le Quaternaire dans le monde 2:781-786.

Krygowski, B. (ed.) 1974. The papers of till work staff of Commission on Genesis and Lithology of Quaternary Deposits (INQUA), Sesz. Nauk. Univ. A. Mickiewicza w Poznaniu. Geografia Nr. 10.

Krygowski, B. 1974. Problematyka glin morenowych na kongresie INQUA w Paryzu (1969r.). In B. Krygowski (ed.), The papers of till work staff of Commission on genesis and lithology of Quaternary deposits. Zesz. Nauk. Univ. A. Mickiewcza w Poznaniu. Geografia 10:81-86 (Polish with German summary and English tables).

Krygowski, B., J. Rzechowski, and W. Stankowski 1969. Project of classifying glacial tills. Bull. Soc. Amis Sci. et Lettres Poznan, Ser. V, 31:141-154.

Kujansuu, R. and M. Saarnisto (eds.) 1987. INQUA till symposium, Finland 1985. Geol. Survey Finland, Spec. Paper 3.

Kurtz, D.D. and J.B. Anderson 1979. Recognition and sedimentologic description of recent debris flow deposits from the Ross and Weddel Seas, Antarctica. J. Sed. Petrol. 49:1159-1170.

Lavrushin, Yu.A. 1970a. Opyt videleniya dinamicheskikh facii i subfacii v donnoi morene materikovykh oledenenii. Litolog. i polezn. iskop. 6:38-49.

Lavrushin, Yu.A. 1970b. Recognition of

facies and subfacies in ground moraine
of continental glaciations. Lithology
and economic deposits 6:684-692.

Lavrushin, Yu.A. 1971. Dynamische Fazies
und Subfazies der Grundmoräne. Zeitschr.
Angew. Geol. 17:337-343.

Lavrushin, Yu.A. 1976. Structure and
development of ground moraines of continental glaciation, Nauka, Moscow
(Russian).

Lavrushin, Yu.A. 1980. Vital problems of
till sedimentogenesis. In W. Stankowski
(ed.), Tills and glacigene deposits.
Univ. A. Mickiewicza w Poznaniu. Ser.
Geogr. 20:19-40.

Lawson, D.E. 1979a. Sedimentological analysis of the western terminus region of
the Matanuska Glacier, Alaska. Cold Reg.
Research and Engin. Lab. Rept. 79-9.

Lawson, D.E. 1979b. A comparison of the
pebble orientations in ice and deposits
of the Matanuska glacier, Alaska. J.
Geol. 87:629-645.

Lawson, D.E. 1981a. Distinguishing characteristics of diamictons at the margin of
Matanuska Glacier, Alaska. Ann. Glaciol.
2:78-84.

Lawson, D.E. 1981b. Sedimentologic characteristics and classification of depositional processes and deposits in glacial
environment. Cold Reg. Res. and Engin.
Lab. Rept. 81-27.

Lawson, D.E. 1982. Mobilization, movement
and deposition of active subaerial sediment flows, Matanuska Glacier, Alaska.
J. Geol. 90:279-300.

Levkov, E.A. 1980. Gliatsiotektonika.
Minsk; Nauka i tehnika (Russian).

Levson, V. and N.W. Rutter 1986. A facies
approach to the stratigraphic analysis
of Late Wisconsinan sediments in the
Portal Creek area, Jaspar National Park,
Alberta. Geogr. phys. et Quatern.
40:129-144.

Lindner, L. and H. Ruszczyńska-Szenajch
1979. Changing conditions of glacial
erosion and deposition reflected by differentiation of glacial deposits at
Rozwady (Swietokryskie Mountains). In
Ch. Schlüchter (ed.), Moraines and
varves, p. 249-255. Rotterdam, Balkema.

Lliboutry, L. 1965. Traité de glaciologie,
Tome II, Glaciers, variations du climat,
sols gelés. Paris, Masson and Cie.

Lundqvist, G. 1940. Bergslagens minerogena
jordarter. Sver. Geol. Unders. Ser. C
433.

Lundqvist, J. 1977. Till in Sweden. Boreas
6:73-85.

Lundqvist, J. 1983. The glacial history of
Sweden. In J. Ehlers (ed.), Glacial
deposits in north-west Europe, p. 77-82.
Rotterdam, Balkema.

Lundqvist, J. 1984. INQUA Commission on
genesis and lithology of Quaternary
deposits. In L.-K. Königsson (ed.), Ten
years of Nordic till research. Striae
20:11-20.

Lundqvist, J. 1985. Glaciotectonic and
till or tillite genesis: Examples from
Pleistocene glacial drift in central
Sweden. In M. Deynoux (ed.), Glacial
Record. Paleogeogr. Paleoclimatol.,
Paleoecol. 51:389-395.

Lutenegger, A.J., T.J. Kemmis, and G.R.
Hallberg 1983. Origin and properties of
glacial till and diamictons. Spec-Publ.
Geol. Environm. and Soil Properties,
ASCE Geotech. Engin. Div., ASCE Conv.
Houston 1983:310-331.

Marcussen, I. 1973. Studies on flow till
in Denmark. Boreas 2:213-231.

Martin, H., H. Porada, and O.H. Walliser
1985. Mixtite deposits of the Damara
sequence, Namibia, problems of
interpretation. In M. Deynoux (ed.),
Glacial record. Paleogeogr.,
Paleoclimat., Paleoecol. 51:159-196.

Martins, Ch. 1842. Sur le formes regulieres du terrain de transport des
vallees du Rhin anterieur et du Rhin
posterieur. Bull. Soc. Geol. France
13:322-345.

May, R.W., A. Dreimanis, and W. Stankowski
1980. Quantitative evaluation of clast
fabrics within the Catfish Creek Till,
Bradtville, Ontario. Can. J. Earth Sci.
17:1064-1074.

McClintock, P. and A. Dreimanis 1964.
Reorientation of till fabric by
overriding glacier in the St. Lawrence
valley. Amer. J. Sci. 262:133-142.

McGown, A. and E. Derbyshire 1977. Genetic
influence on the properties of till.
Quat. J. Engin. Geol. 10:389-410.

Meer, J.J.M. van der (ed.) 1987. Tills and
glaciotectonics. Rotterdam, Balkema.

Meer, J.J.M. van der, M. Rappol, and J.
Semeijn 1985. Sedimentology and genesis
of glacial deposits in the Goudsberg,
central Netherlands. Mededl. Rijks Geol.
Dienst. 39-2.

Miall, A.D. 1985. Architectural-element
analysis: A new method of facies analysis applied to fluvial deposits.
Earth-Sci. Reviews 22:261-308.

Mickelson, D.M. 1971. Glacial geology of
the Burroughts Glacier area,
southeastern Alaska. Ohio State Univ.
Inst. Polar Studies Rept. 40.

Miller, D.J. 1953. Late Cenozoic marine
glacial sediments and marine terraces of
Middleton Island, Alaska. J. Geol.
61:17-40.

Milligan, V. 1976. Geotechnical aspects of
glacial tills. In R.F. Legget (ed.),

Glacial till. Roy. Soc. Canada Spec. Publ. 12:269-291.

Mills, H.H. 1977. Textural characteristics of drift from some representative Cordilleran glaciers. Geol. Soc. Amer. Bull. 88:1135-1143.

Möller, H. 1960. Moranavlagringar med linser av sorterat material i Stockholmstrakten. Geol. Fören. Stockh. Förh. 82:169-202.

Morawski, W. 1984. Watermorainic sediments. Prace Inst. Geol. CVIII (Polish, with Russian and English summaries).

Morawski, W. 1985. Pleistocene glaciogenic sediments of the watermorainic facies. Quatern. Studies in Poland 6:99-115.

Muller, E. 1983a. Dewatering during lodgement of till. In E.B. Evenson, Ch. Schlüchter and J. Rabassa (eds.), Tills and related deposits, p. 13-18. Rotterdam, Balkema.

Muller, E. 1983b. Till genesis and the glacier sole. In E.B. Evenson, Ch. Schlüchter and J. Rabassa (eds.), Tills and related deposits, p. 19-22. Rotterdam, Balkema.

Niewiarowski, W. 1976. The main directions of till studies in Poland. In W. Stankowski (ed.), Till, its genesis and diagenesis. Univ. A. Mickiewicza w Poznaniu. Ser. Geografia 12:13-32.

Odynski, W., A. Wynnyk, and J.D. Newton 1952. Soil survey of the High Prairie and McLennan sheets. Alberta Soil Surv. Rept. 17.

Okko, V. 1955. Glacial drift in Iceland: Its origin and morphology. Comm. Geol. de Finlande Bull. 170.

Olszewski, A. and J. Szupryczyński 1980. Texture of recent morainic deposits of a terminal zone of the Werenskiöld Glacier (Spitsbergen). Polish Polar Research 1, 2-3: 45-79.

Ono, Y. 1984. Annual moraine ridges and recent fluctuation of Yala (Dakpatsen) Glacier, Langtang Himal. Data Center for Glac. Res., Japan. Soc. Snow and Ice, Pub. 2:73-83.

Penck, A. 1882. Die grosse Eiszeit, Himmel u. Erde IV, Berlin.

Penck, A. 1906. Süd-Afrika und Sambesifalle. Geogr. Zeitschr. 12:600-611.

Pettijohn, F.J. 1949. Sedimentary rocks. New York, Harper and Brothers.

Pettijohn, F.J. 1957. Sedimentary rocks. 2nd ed. New York, Harper.

Ramsden, J. and J.A. Westgate 1971. Evidence for the reorientation of till fabric in the Edmonton area, Alberta. In R.P. Goldthwait (ed.), Till: A symposium, p. 335-344. Columbus, Ohio State

Univ. Press.

Rappol, M. 1983. Glacigenic properties of till. Studies in glacial sedimentology from the Allgäu Alps and the Netherlands. Publ. Fys.-Geogr. Bodemk. Lab. Univ. Amsterdam 34.

Rappol, M. 1985. Clast-fabric strength in tills and debris flows compared for different environments. Geologie en Mijnbouw 64:327-332.

Rappol, M. 1986. Aspects of ice flow patterns, glacial sediments, and stratigraphy in northwest New Brunswick. In Current Research, Part B, Geological Survey of Canada. Paper 86-1B:223-237.

Rappol, M. 1987. Saalian till in the Netherlands: A review. In J.J.M. van der Meer (ed.), Tills and glaciotectonics, p. 3-21. Rotterdam, Balkema.

Raukas, A. 1969. Composition and genesis of Estonian tills. In L. Krygowska (ed.), Internat. Symposium, INQUA Com. on origin and lithol. of Quatern. deposits, 4-8 Sept. in Poznan and 9-10 Sept. 1967 in Warsaw. Zeszyty Nauk. Univ. A. Mickiewicza w Poznaniu. Ser. Geografia 8:167-176.

Raukas, A. 1978. Pleistotsenovye otlozheniya Estonskoi SSR. Tallin, Valgus (Russian).

Richter, E. 1899. Die Gletscherkonferenz im August 1899. Verh. VII. Intern. Geol. Congr. Berlin 1899, 2:279-288.

Ringberg, B., B. Holland, and U. Miller 1984. Till stratigraphy and provenance of the chalk rafts at Kvarnby and Angdala, southern Sweden. In L.-K. Konigsson (ed.), Ten years of Nordic till research. Striae 20:79-90.

Rose, J. 1974. Small-scale spatial variability of some sedimentary properties of lodgement tills and slumped till. Proceed. Geol. Asso. 85, Pt. 2:239-258.

Rózycki, S.Z. 1976. Basal moraines composed of local material ("exaration moraines") in Malopolska Upland. In W. Stankowski (ed.), Till, its genesis and diagenesis. Univ. A. Mickiewicza w Poznaniu. Ser. Geografia 12:55-63.

Rukhina, E.V. 1960. Litologiya morennykh otlozhenii. Leningrad Univ.

Rukhina, E.V. 1973. Litologiya lednikovykh otlozhenii. Leningrad, Nedra.

Rukhina, E.V. 1980. Genesis and subdivisions of glacial deposits. In W. Stankowski (ed.), Tills and glacigene deposits. Univ. A. Mickiewicza w Poznaniu. Ser. Geografia 20:11-14.

Ruszczyńska-Szenajch, H. 1983. Lodgement tills and syndepositional glacitectonic processes related to subglacial thermal and hydrologic conditions. In E.B.

Evenson, Ch. Schlüchter and J. Rabassa (eds.), Tills and related deposits, p. 113-117. Rotterdam, Balkema.

Ruszczyńska-Szenajch, H. and L. Lindner 1976. Pleistocene melt-out till at Nowe Miasto on the Pilica River (Middle Poland). In W. Stankowski (ed.), Till, its genesis and diagenesis. Univ. A. Mickiewicza w Poznaniu. Ser. Geografia 12:149-153.

Saarnisto, M. and H. Peltoniemi 1984. Glacial stratigraphy and compositional properties of till in Kainuu, eastern Finland. Fennia 162:163-199.

Saussure, H.B. de 1779. Voyages dans les Alpes, précédés d'un essai sur l'histoire naturelle des environs de Genève, T.1. Neuchâtel.

Schermerhorn, L.J.G., 1966. Terminology of mixed coarse-fine sediments. J. Sediment. Petrol. 36:831-835.

Schlüchter, Ch. 1977. Grundmoräne versus Schlammoräne - two types of lodgement till in the Alpine Foreland of Switzerland. Boreas 6:181-188.

Schlüchter, Ch. (ed.) 1982. INQUA Commission on genesis and lithology of Quaternary deposits, Report on activities 1977-82. Zürich, ETH.

Schlüchter, Ch. 1983. The readvance of the Findelengletscher and its sedimentological implications. In E.B. Evenson, Ch. Schlüchter and J. Rabassa (eds.), Tills and related deposits, p. 95-109. Rotterdam, Balkema.

Schukin, I.S. 1980. Chetyrekhiazichnyi entsiklopedicheskii slovar terminov po fizicheskoi geogrrafiim, russko-anglo-nemetsko-frantsuzskii. Moskow, Sovetskaia entsiklopediia.

Serebryanny, L.R. and A.V. Orlov 1982. Genesis of marginal moraines in the Caucasus. Boreas 11:279-289.

Serebryanny, L.R., N.A. Golodkovskaya, A.V. Orlov, E.S. Malyasova, and E.O. Ilvesi 1984. Glacier variations and moraine accumulation processes in the central Caucasus. Moscow, Nauka (Russian with English summary).

Seret, G. 1985. Classification des sediments morainiques. Bull. l'Assoc. francaise pour l'etude du Quatern. I:41-43.

Serrat, D. (ed.) 1983. INQUA Commission Genesis and Lithology of Quaternary Deposits (Barcelona 1983). Acta Geol. Hispanica 18(3/4):151-248.

Shantser, E.V. 1966. Ocherki ucheniya o geneticheskikh tipakh kontinentalnykh osadochnykh obrazovanii. Moscow, Nauka.

Sharp, R.P. 1949. Studies of superglacial debris on valley glaciers. Amer. J. Sci. 247:289-315.

Shaw, J. 1977. Tills deposited in arid polar environments. Can. J. Earth Sci. 14:1239-1245.

Shaw, J. 1979. Genesis of the Sveg tills and Rogen moraines of central Sweden: A model of basal melt out. Boreas 8:409-426.

Shaw, J. 1982. Melt-out till in the Edmonton area, Alberta, Canada. Can. J. Earth Sci. 19:1548-1569.

Shaw, J. 1983. Forms associated with boulders in melt-out till. In E.B. Evenson, Ch. Schlüchter and J. Rabassa (eds.), Tills and related deposits, p. 3-12. Rotterdam, Balkema.

Shaw, J. 1985. Subglacial and ice marginal environments. In G.M. Ashley, J. Shaw, and H.D. Smith (eds.), Glacial Sedimentary Environments. SEPM Short Course 16. Tulsa, Soc. Econ. Paleont. and Mineral.: 7-84.

Shaw, J. 1987. Glacial sedimentary processes and environmental reconstruction based on lithofacies. Sedimentology 34:103-116.

Sladen, J.A. and W. Wrigley 1983. Geotechnical properties of lodgement till. In N. Eyles (ed.), Glacial geology, p. 184-212. Oxford, Pergamon Press.

Smalley, J.J. and D.J. Unwin 1968. The formation and shape of drumlins and their distribution and orientation in drumlin fields. J. Glaciol. 7:377-390.

Spencer, A.M. 1985. Mechanisms and environments of deposition of late Precambrian geosynclinal tillites. In M. Deynoux (ed.), Glacial record. Paleogeogr., Paleoclimat., Paleoecol. 51:143-157.

Stanford, S.D. and D.M. Mickelson 1985. Till fabric and deformational structures in drumlins near Waukesha, Wisconsin, U.S.A. J. Glaciol. 31:220-228.

Stankowski, W. (ed.) 1976. Till-its genesis and diagenesis. Univ. A. Mickiewicza w Poznaniu. Ser. Geogr. 12.

Stankowski, W. (ed.) 1980. Tills and glacigene deposits. Univ. A. Mickiewicza w Poznaniu. Ser. Geogr. 20.

St. Arnaud, R.J. 1976. Pedological aspects of glacial till. In R.F. Legget (ed.), Glacial till. Roy. Soc. Can. Spec. Publ. 12:133-155.

Stephan, H.-J. and J. Ehlers 1983. North German till types. In J. Ehlers (ed.), Glacial deposits in north-west Europe, p. 239-247. Rotterdam, Balkema.

Stephens, G.C., E.B. Evenson, R.B. Tripp, and D. Detra 1983. Active alpine glaciers as a tool for bedrock mapping and mineral exploration: A case study from Trident Glacier, Alaska. In E.B. Evenson, Ch. Schlüchter and J. Rabassa (eds.), Tills and related deposits, p. 195-204. Rotterdam, Balkema.

Strelkov, S.A. 1965. Geneticheskaya klassifikatsiya otlozhenyi materikovogo oledeneniya v sviazi s obschimi zakonomernostiami razvitiya lednikovikh pokrovov. Osnovn. problemi izuch. chetvert. perioda: 151-156. Moscow, Nauka.

Sugden, D.E. and B.S. John 1979. Glaciers and landscape. London, Edward Arnold.

Tarr, R.S. 1909. The Yakutat Bay region, Alaska. U.S. Geol. Surv. Prof. Paper 64.

Teodorovich, G.I. 1939. Ky obschei klassi-fikatsyi oblomochnikh valunov. Materiali gliatsiologicheskikh issledovanyi 21. Moscow.

Thwaites, F.T. 1957. Outline of glacial geology. Madison.

Timofeev, D.A. and A.N. Makkaveev 1986. Terminologiya glyatsialnoi geomor-fologii, Moscow, Nauka.

Toll, E. von, 1892. Die fossilen Eislager und ihre Beziehungen zu den Mammutleichen. Mem. Acad. Imp. du Sciences St. Petersburg, Ser. 7, 42:13.

Torrell, O. 1877. On the glacial phenomena of North America. Amer. J. Sci. 13:76-79.

Upham, W. 1891. Criteria of englacial and subglacial drift. American Geologist 8:376-385.

Upham, W.I. 1892. Inequality of distribu-tion of the englacial drift. Bull. Geol. Soc. Amer. 3:134-148.

Virkkala, K. 1952. On the bed structures of till in eastern Finland. Compt. Rendus de la Soc. geol. de Finlande XXV:98-109.

Virkkala, K. 1974. On the Wurmian till deposits in Finland. In B. Krygowski (ed.), The papers of Till Work Staff of Commission on genesis and lithology of Quaternary deposits (INQUA). Zeszyty Nauk. Univ. A. Mickiewicza w Poznaniu. Ser. Geografia 10:59-80.

Vorren, T.O. 1977. Grain-size distribution and grain-size parameters of different till types on Hardangervidda, south Norway. Boreas 6:219-227.

Warren, W.P. 1987. Glaciodiagenesis in geliflucted deposits on the south coast of County Cork, Ireland. In J.J.M. van der Meer (ed.), Tills and glaciotec-tonics, p. 105-115. Rotterdam, Balkema.

Wateren, D. van der 1987. Structural geology and sedimentology of the Dammer Berge push moraine, FRG. In J.J.M. van der Meer (ed.), Tills and glaciotec-tonics, p. 157-182. Rotterdam, Balkema.

Wayne, W.J. 1963. Pleistocene formations in Indiana. Indiana Geol. Surv. Bull. 25:85.

Whiteman, C.A. 1987. Till lithology and genesis near the southern margin of the Anglian ice sheet in Essex, England. In J.J.M. van der Meer (ed.), Tills and glaciotectonics, p. 55-66. Rotterdam, Balkema.

Woldstedt, P. 1954. Das Eiszeitalter. 2nd ed., vol. 1. Stuttgart, Ferdinand Enke Verlag.

Woodworth, J.B. 1899. The ice-contact in the classification of glacial deposits. Amer. Geol. 23:80-86.

Wright, G.F. 1889. The Ice Age in North America, and its bearings upon the anti-quity of man. New York.

## APPENDIX A

**Till Work Group members who have contributed to the preparation of this report during the period 1973-1987. The names of most active participants are in bold print.**

R. Aario (Finland), O. Āboltiņš (USSR), N. Ahmad (India), M. Åmark (Sweden), J.T. Andrews (USA), J.W. Attig (USA), A. Banham (UK), P.J. Barnett (Canada), P. Barrett (New Zealand), A. Ber (Poland), O.F. Bergersen (Norway), B. Bergstrøm (Norway), A. Berthelsen (Denmark), P. Blystad (Norway), **G.S. Boulton** (UK), J. Bru (Spain), D.W. Burbank (USA), P.E. Calkin (USA), A. Cailleux (France), A.C. Rocha Campos (Brazil), A.G. Cepek (GDR), J.J. Clague (Canada), **L. Clayton** (USA), J. Cohen (Ireland), R. Connell (UK), D.G. Croot (UK), J. Danilāns (USSR), E. Derbyshire (UK), M. Deynoux (France), J.C. Dionne (Canada), J.A. Dowdeswell (UK), L.A. Dredge (Canada), E. Drozdowski (Poland), J.M. Dubois (Canada), J. Ehlers (GFR), J. Elson (Canada), K. Erikkson (Sweden), E.B. Evenson (USA), V. Evzerov (USSR), N. Eyles (Canada), M.M. Fenton (Canada), E. Francis (UK), B. Frenzel (GFR), R.J. Fulton (Canada), A. Gaigalas (USSR), K. Garnes (Norway), R. German (GFR), P.L. Gibbard (UK), G. Gillberg (Sweden), **R.P. Goldthwait** (USA), P.F. Gozhik (USSR), D.R. Grant (Canada), C.P. Gravenor (Canada), K. Gripp (GFR), F. Grube (GFR), S. Haldorsen (Norway), M.J. Hambrey (UK), A. Hansel (USA), W.V. Harland (UK), G.M. Haselton (USA), H. Heuberger (Austria), **S.R. Hicock** (Canada), A Hillefors (Sweden), H. Hirvas (Finland), M. Houmark- Nielsen (Denmark), D. van Husen (Austria), D. Ingolfsson (Iceland), H.G. Johansson (Sweden), W.H. Johnson (USA), A. Karczewski (Poland), P.F. Karrow (Canada), A. Karukäpp (USSR), L.K. Kauranne (Finland), **T.J. Kemmis** (USA), R.W. Klassen (Canada), J. Krüger (Denmark), R. Kujansuu (Finland), E. Lagerlund (Sweden), P. La Salle (Canada), **Y.A. Lavrushin** (USSR), **D.E. Lawson** (USA), E.A. Levkov (USSR), V.K. Lukashev (USSR), **J. Lundqvist** (Sweden), I. Marcussen (Denmark), M.A. Marqués (Spain), **C.L. Matsch** (USA), W.H. Mathews (Canada), A.V. Matveyew (USSR), R.W. May (Canada), J.J.M. van der Meer (Netherlands), J. Menzies (Canada), K.D. Meyer (GFR), D.M. Mickelson (USA), H.H. Mills (USA), S.R. Moran (Canada), Ch. Mougeot (Canada), E.H. Muller (USA), A.M. McCabe (N. Ireland), A. McGown (UK), I.C. McKellar (New Zealand), W. Niewiarowski (Poland), P. Nieminen (Finland), S. Occhietti (Canada), A. Olszewski (Poland), J. Ono (Japan), G. Orombelli (Italy), G. Ostrem (Norway), M. Parent (Canada), M.A. Paul (UK), S.Sh. Pedersen (Denmark), M. Perttunen (Finland), R.D. Powell (USA), G. Prichonnet (Canada), D. Proudfoot (Canada), K.S. Petersen (Denmark), J. Rabassa (Argentina), M. Rappol (Netherlands), A. Rasmussen (Denmark), A. Raukas (USSR), B. Ringberg (Sweden), S. Rubulis (Argentina), G.H.J. Ruegg (Netherlands), L. Rudmark (Sweden), E.V. Rukhina (USSR), **H. Ruszczyńska-Szenajch** (Poland), M. Ruzicka (Czechoslovakia), J. Rzechowski (Poland), M. Saarnisto (Finland), R. Sauchez (Belgium), D. St. Onge (Canada), **Ch. Schlüchter** (Switzerland), C. Schubert (Venezuela), L.R. Serebryanny (USSR), G. Seret (Belgium), D. Serrat (Spain), M.J. Sharp (UK), D.R. Sharpe (Canada), E.V. Shantser (USSR), **J. Shaw** (Canada), Shih Ya-feng (China), W.W. Shilts (Canada), S. Sjoring (Denmark), J.L. Sollid (Norway), W. Stankowski (Poland), H.J. Stephan (GFR), R. Sutinen (Finland), V.D. Tarnogradskiy (USSR), D.G. Vanderveer (Canada), P. Vernon (Ireland), J.M. Vilaplana (Spain), J.-S. Vincent (Canada), K. Virkkala (Finland), T.O. Vorren (Norway), W.P. Warren (Ireland), C. Whiteman (UK).

## APPENDIX B
## Definitions of some terms used in discussions of till.

**Diamicton** - any non-sorted or poorly sorted unconsolidated sediment that contains a wide range of particle sizes. (Flint et al. 1960, expanded in Flint 1971; see Harland, Herod, and Krinsley 1966 re: diamict-diamicton-diamictite.)

**Glacial debris** - material being transported by a glacier in contact with glacier ice. In most cases it is disaggregated, except for clasts of various sizes, including large rafts. (Discussed by Dreimanis 1982a:15-16.)

**Glacial drift** - all rock material in transport by glacier ice, all deposits made by glacier ice, and all deposits predominantly of glacial origin made in the sea or in bodies of glacial meltwater whether rafted in icebergs or transported in the water itself. It includes till, stratified drift, and scattered clasts that lack an enclosing matrix (Flint 1979:147).

**Glacigene, glacigenic, glaciogenic,** versus **glacial** - of glacial origin (Bowes 1970, in Wyatt 1986:139). It is used here with a broad meaning - "originating from glacier", and "glacial" is used in a narrower meaning when dealing with glacial erosion and sedimentation, except for the already entrenched term "glacial drift." In the geologic literature the usage of these terms varies.

APPENDIX C. SELECTED GLACIGENIC TERMS
IN 21 LANGUAGES

Papers on glacial deposits have been published in many languages and a variety of terms has been used. Though a considerable number of scientific terms are international, there are still some differences in their spellings Also, specific linguistically different terms have developed in some languages. They also may differ in the same language from one country to another. Thus, in the column of Spanish terms, those used in Spain are indicated by (Sp), but the terms used in Argentina by (Arg).

The main purpose of this multilingual listing is to assist in deciphering the meaning of some commonly used terms when confronted with them in a foreign language paper, particularly when trying to understand the legend of an illustration, or the title of a paper.

The initial list of terms was twice as long as in the present publication, but it had to be shortened because of space restrictions. Some of the terms left out were those which were very similar in most languages, such as 'sublimation', or those which had meanings more petrographic than genetic.

In some languages the variety of terms used for any specific glacigenic sediment or their group appears to be greater than in other languages. However, this difference depends partly upon the contributors: some had selected the terms that are most commonly used or preferred, while others listed a greater variety, including even new terms, not much used yet. Several geologic dictionaries were also consulted for additional terms and their meanings. Antiquated or less commonly used terms are placed in parentheses.

A couple of comments. - Though the most common spelling of 'glaciofluvial', 'glaciolacustrine', and 'glaciomarine' in the English language is with 'o' in the middle (Francis, 1984), some authors prefer to omit the 'o'. To indicate both spellings, 'o' is placed in parentheses (o), also in the term 'glaci(o)genic'. The terms 'basal till' and 'subglacial till' usually are equivalent in their meaning; they are listed here separately, to indicate their specific equivalents in those languages, where both terms are used. The term 'moraine' is used both for glacigenic landforms and glacial sediment (= till) in many languages, and its spell-ing does not differ much from the original French 'moraine'. However, in several languages where 'till' and the specific linguistic form or equivalent of 'moraine' are listed in their column, the term 'till' tends to become more popular. In the French of Quebec 'till' has already replaced 'moraine'. In the Japanese, the usage of the 'moraine' equivalents 'taiseki' and 'hyotaiseki', and their derivatives 'shomo-taiseki' for ablation till and 'tei-taiseki' for lodgement till are presently being discouraged, preferring their usage for landforms ('tei-taiseki' = ground moraine), according to Yugo Ono (personal communication, 1987). In several languages, for instance in the German, 'Moräne' is still more commonly used than 'till'. The development of terminology differs from one language to another, and from one region to another.

The following members of the INQUA Commission C-2 have contributed to this compilation of terms. Though many of them have indicated that their submissions have been discussed with other colleagues, it is impossible to list all the names of their coworkers, since some of them have not been mentioned. The names of the direct contributors are listed by language, in the same order as the language columns are arranged on ensuing pages.
French: A. Cailleux, M. Campy, M. Deynoux, S. Occhietti, G. Seret, R. Souchez. Spanish: J. Rabassa, D. Serrat. Català D. Serrat. Italian: G. Orombelli. Gaelic: W.P. Warren. German: B. Frenzel, F. Grube, K.D. Meyer, Ch. Schlüchter, D. van Husen. Dutch: J.J.M. van der Meer, M. Rappol. Danish: K.S. Petersen, L.A. Rasmussen. Norvegian: Sylvi Haldorsen. Swedish: E. Lagerlund, J. Lundqvist. Finnish: R. Kujansuu. Estonian: A. Raukas. Latvian: A. Dreimanis. Lithuanian: A. Gaigalas, A. Jurgaitis, G. Petrauskas. Russian: E.A. Levkov, A.V. Matveyew, A. Raukas, E.V. Rukhina, L.R. Serebryanny, E.V. Shantser. Ukrainian: P.F. Gozhik. Byelorussian: E.A. Levkov, A.V. Matveyew. Polish: M. Brykczynski, A. Karczewski, W. Niewiarowski, H. Ruszczyńska-Szenajch. Japanese: Y. Ono. Chinese: Shi Yafeng. The Japanese and the Chinese scripts have been handwritten by C.T. Wu.

| ENGLISH | FRENCH | SPANISH |
|---|---|---|
| till; (boulder clay) | till; dépôt morainique; sédiment morainique;(moraine; moraine déposee) | till; morrena |
| ablation till | till d'ablation; moraine d'ablation | till de ablación |
| basal till; subglacial till | till basal; till de fond; moraine de fond; (moraine profonde) | till basal |
| deformation till | till d'entraînement; till de translocation; (moraine de remaniement) | till de deformación |
| flowtill; flow till | till flué; till d'écoulement; (moraine glisée) | till de flujo |
| lodgement till; lodgment till | till de fond; moraine de plaquage | till alojado |
| melt-out till | till de fusion (sur place); till de fonte; moraine de fonte sous pression | till de fusión |
| primary till | till primaire; moraine primaire | till primario |
| secondary till | till secondaire; moraine secondaire; (para-moraine) | till secundario |
| subglacial till | till de fond; till sous-glaciaire; moraine sous-glaciaire | till subglaciar (Sp); till subglaciario (Arg) |
| supraglacial till; superglacial till | till supraglaciaire; moraine supraglaciaire | till supraglaciar (Sp); |
| waterlain till; subaquatic till | aquatill; till aquatique; (moraine sous-aquatique) | till subacuatico |
| debris flow (deposit) | depot de coulee de debris; charge detritique | depósitos de flujo de derrubios (Sp); depósitos de flujo de detritos (Arg) |
| glacial debris | débris glaciaires; charge glaciaire; moraines mouvantes | derrubio glaciar (Sp); detrito glaciario (Arg) |
| glacial drift | dépôts glaciaires | acarreo glaciar (Sp); drift glaciario (Arg) |
| glaci(o)genic; glacigene | glaciogène; glaciogenique; d'origine glaciaire | glacigénico |
| glaci(o)fluvial | fluvioglaciaire | glaciofluvial; fluvioglaciar (Sp) |
| glaci(o)lacustrine | glaciolacustre | glaciolacustre |
| glaci(o)marine; glacial marine | glaciomarin | glaciomarino |
| glaci(o)tectonic | glaciotectonique; glacitectonique | glaciotectónico |
| megablock; floe; raft | mégabloc; glaçon; radeau | megaclastos |

| CATALÀ | ITALIAN | GAELIC (Irish) |
|---|---|---|
| till; morrena | till; (morena; deposito morenico; deposito giaciale) | till |
| till d'ablacio | till di ablazione; (morenico di ablazione; deposito glaciale di ablazione) | till tochtar-oighreach |
| till basal | till basale; (morenico di fondo; deposito glaciale di fondo) | till bharr-oigreach |
| till de deformació | till di deformazione | till an mhíchumtha |
| till de flux | till di colata; till di flusso ('flow till') | till na sní |
| till d'acreció | till di fondo; (morenico di fondo; deposito glaciale di fondo) | till an ghreamaithe |
| till de fusió | till di fusione | till an leá |
| till primari | till primario | till phríomha |
| till secundari | till secondario | fo-thill |
| till subglacial | till subglaciale; (morena di fondo) | till fho-oighreach |
| till supraglacial | till supraglaciale; (morena superficiale) | till bharr-oigreach |
| till subaquatic | till subacqueo; (morenico subacqueo) | till uisce-leagtha |
| 'debris flow' | colata detritica; flusso detritico | sil-leagan; shreabhadh smionagair |
| arrossegalls glacials | detrito glaciale; (morena) | smionagar oighreach |
| 'glacial drift' | complesso glaciale; (deposito glaciale; accumulo glaciale) | deascadh aighreach |
| glacigènic | glaciogenico | oighreach |
| glacio-fluvial | fluvioglaciale | dobharoighreach |
| glacio-lacustre | glaciolacustre | lochoighreach |
| glacio-marí | glaciomarino | muiroighreach |
| glacio-tectònic | glaciotettonico | teicteon-oighreach |
| megaclastes | | sload |

| ENGLISH | GERMAN |
|---|---|
| till | Moränenmaterial; Geschiebemergel; Moräne; Gletschersediment; Till |
| ablation till | Ablationsmoräne; Ablationsmoränematerial; Obermoräne; Ablationstill |
| basal till | Basalmoräne; Grundmoränenmaterial; Grundmoräne; Grundtill |
| deformation till | Deformationsmoräne; Verformungsmoränenmaterial; Verformungstill |
| flowtill | Fliessmoräne; verflossene Moräne |
| lodgement till | Grundmoräne (sensu stricto); Grundmoränenmaterial; Setztill; 'Lodgement till' |
| melt-out till | Ausschmelzmoräne; Ausschmelztill; 'Melt-out Till' |
| primary till | Primärmoräne; primäres Moränenmaterial; Primärtill |
| secondary till | Sekundärmorane; sekundäres Moränenmaterial; Sekundärtill |
| subglacial till | Grundmoräne; Grundmoränenmaterial; Grundtill |
| supraglacial till | Obermoräne; Moränenmaterial der Gletscheroberfläche; supraglazialer Till |
| waterlain till | subaquatische Moräne; Unterwassermoräne; subaquatischer Till; (Schlammoräne) |
| debris flow | Ablagerung des Schuttstromes; Murgang |
| glacial debris | Glazialschutt; Gletscherschutt |
| glacial drift | glaziale Ablagerungen; Gletschersedimente |
| glaci(o)genic | glazigen |
| glaci(o)fluvial | glazifluvial; glazifluviatil; fluvioglazial |
| glaci(o)lacustrine | glazilakustrisch; glazilakustrin : glazilimnisch |
| glaci(o)marine | glazimarin |
| glaci(o)tectonic | glazitektonisch; glazialtektonisch |
| megabloc; floe; raft | Scholle |

| DUTCH | DANISH | NORVEGIAN |
|---|---|---|
| till; keileem | moræne; 'till' | morene |
| ablatie till; ablatie keileem; (ablatiemorene) | ablations moræne | ablasjonsmorene |
| basale till; basale keileem; (grondmorene) | bundmoræne | basalmorene; bünnmorene |
| deformatie till; deformatie keileem | 'glacitectonite' | deformasjonsmorene |
| 'flow till' | 'flow till' | flytemorene |
| 'lodgement till' | 'lodgement till' | |
| 'melt-out till' | udsmeltnings moræne | utsmeltnings morene |
| primaire till; primaire keileem | | |
| secundaire till; secundaire keileem | | |
| subglaciale till; subglaciale keileem | bundmoræne | basalmorene |
| supraglaciale till; supraglaciale keileem | | supraglasial morene; overflatemorene |
| 'waterlain till'; subaquatische keileem | aqua till | |
| 'debris flow' afzetting | slamstrøm | |
| glaciaal puin | | |
| glaciale sedimenten; glaciale afzetting | glaciale sedimenter | |
| glacigeen | glacigene | glasigen |
| fluvioglaciaal; glaciofluviatiel | glaciofluviale | glasifluvial |
| glaciolacustrien | glaciolacustrine | glasilakustrin |
| glaciomarien | glaciomarine | glasimarin |
| glaciotektoniek | glaciotektoniske | glasitektonikk |
| megablok; schol | megaklast | |

| ENGLISH | SWEDISH | FINNISH |
|---------|---------|---------|
| till | morän | moreeni(aines) |
| ablation till | ablationsmorän | ablaatiomoreeni |
| basal till | basal morän; bottenmorän | pohjamoreeni |
| deformation till | deformationsmorän | deformaatiomoreeni |
| flowtill | 'flow till'; flytmorän | valumoreeni |
| lodgement till | 'lodgement till' | lodgementmoreeni |
| melt-out till | 'melt-out till' | sulamismoreeni; vajomoreeni |
| primary till | primär morän | primärmoreeni |
| secondary till | sekundär morän | sekundärmoreeni |
| subglacial till | subglacial morän | subglasiaalinen moreeni |
| supraglacial till | ytmorän; supraglacial morän | pintamoreeni; supraglasiaalinen moreeni |
| waterlain till | vattenavlagrad morän | |
| debris flow | flytjord; 'debris flow' | |
| glacial debris | (moränmaterial) | jäätikkössä kulkeutuva aines |
| glacial drift | glaciala avlagringar | jäätikkösyntyinen kerrostuma; glasiaalikerrostuma |
| glaci(o)genic | glacigen | jäätikkösyntyinen; glasigeeninen |
| glaci(o)fluvial | glacifluvial | jäätikköjokisyntyinen; glasifluviaalinen |
| glaci(o)lacustrine | glacilakustrin; issjö- | jääjärvi-; glasilakustrinen |
| glaci(o)marine | glacimarin | glasimariinen |
| glaci(o)tectonic | glacialtektonik | glasitektooninen |
| megablock; floe; raft | skolla; jätteblock | |

74

| ESTONIAN | LATVIAN | LITHUANIAN |
|---|---|---|
| moreen | morēna | morena |
| ablatsioonimoreen | ablācijas morēna | abliacinė morena |
| basaalmoreen; pohjamoreen | pamatmorēna; bazālā morēna | dugninė morena |
| deformatsioonimoreen; survemoreen | deformācijas morēna | deformacinė morena |
| solifluktsioonimoreen; roomemoreen | plūsmas morēna | tekėjimo morena |
| pôhjamoreen; ladumoreen | sablīvējuma morēna | pamatinė morena |
| väljasulamismoreen | izkusuma morēna | ištirpusi morena |
| primaarmoreen; esmasmoreen | primārā morēna | pirminė morena |
| sukundaarmoreen; teismoreen | sekundārā morēna | antrinė morena |
| subglatsiaalne moreen | subglaciālā morēna | poledyninė morena |
| supraglatsiaalne moreen; pinnamoreen | supraglaciālā morēna; virsas morēna | antledyninė morena |
| basseinimoreen; veeskuhjunu moreen | ūdenī nogulsnēta morēna | akvalinė morena |
| roomesete | senešu plūsmas nogulas | sanašu srautas |
| liustikupurd; liustikukores | ledāju saneši | ledyninės nuotrupos |
| liustikusete | ledāju nogulas; ledāju nogulumi | ledyno nešmenys |
| liustikutekkeline; glatsigeenne | glacigēns | glacigeninis; (ledyninis) |
| liustikuvooluveeline; glatsifluviaalne | fluvioglaciāls; glacifluviāls | fliuvioglacialinis |
| liustikujärveline; limnoglatsiaalne | limnoglaciāls | limnoglacialinis |
| liustikulis-mereline | glaciomarīns | ledyninis jūrinis |
| glatsiotektooniline | glaciotektonisks | glaciotektoninis |
| hiidrahn; pangas | blāķis | blokas |

| ENGLISH | RUSSIAN | |
|---|---|---|
| till | morena; till | морена; тилл |
| ablation till | ablyatsionnaya morena | абляционная морена |
| basal till | donnaya morena; basalnaya morena; osnovnaya morena | донная морена; базальная морена; основная морена |
| deformation till | deformatsionnaya morena; deformatsionnyi till | деформационная морена; деформационный тилл |
| flowtill | morena techeniya; morena spolzaniya | морена течения; морена сползания |
| lodgement till | morena nakopleniya | морена накопления |
| melt-out till | morena vytaivaniya | морена вытаивания |
| primary till | pervichnaya morena | первичная морена |
| secondary till | vtorichnaya morena | вторичная морена |
| subglacial till | podlednikovaya morena | подледниковая морена |
| supraglacial till | poverkhnostnaya morena, nadlednikovaya morena | поверхностная морена надледниковая морена |
| waterlain till | basseinovaya morena; vodnaya morena; podvodnyi till | бассейновая морена; водная морена; подводный тилл |
| debris flow | otlozheniya oblomochnogo potoka | отложения обломочного потока |
| glacial debris | lednikovye oblomki | ледниковые обломки |
| glacial drift | lednikovye otlozheniya; moreny | ледниковые отложения; морены |
| glaci(o)genic | gliatsigennyi; lednikovyi | гляцигенный; ледниковый |
| glaci(o)fluvial | fluviogliatsialnyi | флювиогляциальный |
| glaci(o)lacustrine | limnogliatsialnyi | лимногляциальный |
| glaci(o)marine | lednikovo-morskoy | ледниково-морской |
| glaci(o)tectonic | gliatsiotektonicheskyi | гляциотектонический |
| megablock; floe; raft | ottorzhenets; lednikovaya glyba | отторженец; ледниковая глыба |

76

| | | |
|---|---|---|
| morena | морена | glina morenowa; osad morenovy; glina zwałowa |
| ablyatsyina morena | абляційна морена | glina ablacyjna; morena ablacyjna |
| osnovna morena | основна морена | glina basalna; morena bazalna |
| deformatsyina morena | деформаційна морена | zdeformowana glina morenowa |
| fluidalna morena | флюідальна морена | glina spływowa; morena spływowa |
| donna morena | донна морена | glina denna; morena denna |
| | | glina morenowa wytopienia; morena wytopienia |
| pervinna morena | первинна морена | pierwotny osad morenowy |
| vtorinna morena | вторинна морена | wtórny osad morenowy; redeponowana glina morenowa |
| podliodovikova morena | подльодовикова морена | glina subglacjalna; morena subglacjalna |
| nadliodovikova morena | надльодовикова морена | morena powierzchniowa; osad supraglacjalny; morena supraglacjalna |
| | | osady morenowe podwodne; subakwatyczna glina morenowa |
| vidklady potoku ulamkiv | відклади потоку уламкив | osad spływu rumowiskowego |
| liodovykovi ulamki | льодовикови уламки | rumowisko lodowcowe |
| liodovykovi vidklady | льодовикови відклади | osady morenowe; morena |
| glyatsigenni; liodovykovi | гляцігенний; льодовиковий | glacigeniczny; lodowcowy |
| vodno-liodovykovi | водно-льодовиковий | fluvioglacjalny; glacifluvialny |
| ozerno-liodovykovi | озеро-льодовиковий | limnoglacjalny; glaciolimniczny |
| liodovykovo-morski | льодовиково-морський | glacjalnomorski |
| glyatsiotektonichni | гляціотектонічний | glacitektoniczny; glacjotektoniczny |
| megaklasti | метакласти | porwak |

| ENGLISH | BYELORUSSIAN | |
|---|---|---|
| till | marena | марэна |
| ablation till | ablyatsyinaya marena | абляцыйная марена |
| basal till | asnoynaya marena | асноўная марена |
| deformation till | defarmatsyinaya marena | дэфармацыйная марена |
| flowtill | marena-tsyakun | марэна-цякун |
| lodgement till | donnaya marena; (marena donnaga namnazhennya) | донная марэна; (марэна доннага намнажэння) |
| melt-out till | marena vytavannya | марэна вытавання |
| primary till | pershasnaya marena | першасная марэна |
| secondary till | drugasnaya marena | другасная марэна |
| subglacial till | padledavikovaya marena; asnoynaya marena | падледавіковая марэна; асноўная марэна |
| supraglacial till | nadledavikovaya marena | надледавіковая марэна |
| waterlain till | | |
| debris flow | adklady ablomkavaga patoka | адклады абломкавага патока |
| glacial debris | ledavikovyya ablomkavyya parody | ледавіковыя абломкавыя пароды |
| glacial drift | ledavikovyya adklady | ледавіковыя адклады |
| glaci(o)genic | glyatsygenny | глячыгенны |
| glaci(o)fluvial | fliuviyaglyatsyyalny; vodnaledavikovy | флювіягляцыяльны; водналедавіковы |
| glaci(o)lacustrine | limnaglyatsyyalny; azernaledavikovy | лімнагляцыяльны; азёрналедавіковы |
| glaci(o)marine | ledavikova-marski | ледавікова-марскі |
| glaci(o)tectonic | glatsyyatektanichny | гляцыятэктанічны |
| megablock; floe; raft | vyalizny adorven | вялізны адорвень |

| JAPANESE | JAPANESE | CHINESE |
|---|---|---|
| 'till' ('moraine') taiseki, hyotaiseki | 堆石 | 冰磧 |
| shomo-taiseki | 消耗堆石 | 消融磧 |
| (tei-taiseki) | 底堆石 | 底磧 |
| | | 变形冰磧 |
| | | 流磧 |
| (tei-taiseki) | 底堆石 | 滯磧 |
| | | 融出磧 |
| | | 原生冰磧 |
| | | 次生冰磧 |
| | | 冰下磧 |
| | | 冰面磧 |
| | | 水域冰磧 |
| doeskiryu-taisekibutsu | 土石流堆積物 | 泥石流冰磧 |
| hyoga-gansetsu | 冰河岩屑 | 冰川岩屑 |
| hyosei-taisekibutsu; hyoreki | 氷成堆積物 漂礫 | 冰川堆磧 |
| hyosei | 氷成 | 冰成 |
| yuhyo-karyu | 融氷河流 | 冰水河流 |
| nyoga-kosei | 氷河-湖成 | 冰水湖泊 |
| hyoga-kaisei | 氷河-海成 | 冰水海洋 |
| hyoga-kozosei | 氷河-構造成 | 冰成構造 |
| 'megaclast' | | |

| Criterion | Lodgement till | Melt-out till | Gravity flowtill |
|---|---|---|---|
| Position and sequence in relation to other glacigenic sediments | Under advancing glacier: lodged over older pre-advance sediments and over glacitectonites, unless they have been eroded. Under retreating glaciers: the lowermost depositional unit, if the deposits related to glacial advance have been eroded. Locally underlain by meltwater channel deposits. May be overlain by any glacigenic sediments. | Usually deposited during glacial retreat over any glacially eroded substratum, or over lodgement till. Also as lenses in lodgement till. May be interbedded with lenses of englacial meltwater deposits, and locally is underlain by syndepositional subglacial meltwater sediments and subglacial flowtill. At the surface, melt-out till grades into flowtill, and in cold arid climates into sublimation till, as a specific variety of melt-out till. | Most commonly it is the uppermost glacial sediment in a non-aquatic facies association. Associated also locally with subglacial tills, where cavities were present under glacier ice, or where the glacier had overridden the ice-marginal flowtill. May be interbedded or interdigitated with glaciofluvial, glaciolacustrine or glaciomarine sediments, particularly away from its original source at the glacier ice. |
| Basal contact | Since both lodgement and melt-out tills begin their formation and deposition at the glacier sole, their basal contact with the substratum (bedrock or unconsolidated sediments) is similar in large scale, being usually erosional and sharp. The glacial erosion-marks underneath the contact and the alignment of clasts immediately above the contact have the same orientation. Glacitectonic deformation structures formed by the till-depositing glacier may occur under both tills, and they strike transverse to the direction of local glacial stress. ⟨⟨spanning Lodgement + Melt-out⟩⟩ | | Variable; seldom planar over longer distances. The flows may fill shallow channels or depressions. The contact may be either concordant, or erosional, with sole marks parallel to the local direction of sediment flow. Loading structures may be present at the basal contact of waterlain flowtill and the underlying soft sediment. |
| | Basal contact, representing the sliding base of the glacier, is generally planar if over unconsolidated substratum, but it may be grooved. The bedrock contact is usually abraded, particularly on stoss sides of bedrock protrusions. Since the sliding base of the glacier represents a large shear plane, sheared and strongly attenuated substratum material may be deposited as a thin layer along this plane, and from place to place it is sheared up into the lodgement till. Clast pavements, both erosional and depositional, may be present along the basal contact, but they occur also higher up in lodgement till. If lodgement till becomes deformed by glacial drag shortly after its deposition, the basal contact may become involved in the deformation by tight recumbent folding, overthrusting, and shearing. | If the basal contact of glacier ice was tight with the substratum during the melting, the pre-depositional erosional marks characteristic for moving glaciers, are as well preserved as under lodgement till. However, subsole meltwater may modify the basal contact locally, and produce convex-up channel fills and various other meltwater scour features. | |

| Criterion | Lodgement till | Melt-out till | Gravity flowtill |
|---|---|---|---|
| Surface expres- sion, landforms | Mainly in ground moraines and other subglacial landforms (Goldtnwait, this volume, I-A-2, I-B). Also along the proximal side of some end moraines (ibid., I-C-1). Small-scale flutes (ibid., I-A-2c) are always associated with lodgement till. | In those ice-marginal landforms where glacier ice had stagnated (Goldthwait, this volume: I-C-1a, I-C-1e, I-C-2c, I-C-3). | Associated witn most ice marginal landforms, (Goldthwait, this volume, I-C). Also, as a thin surface layer on many other direct glacial landforms (ibid., I-A, I-B). |
| Tnickness | Typically one to a few metres; relatively constant laterally over long distances. | Single units are usually a few centimetres to a few metres thick, but they may be stacked to much greater tnickness. | Very variable.  Individual flows are usually a few decimetres to metres tnick, but tney may locally stack up to many metres, particularly in ablation moraines (Goldtnwait, this volume: I-C-3) and some lateral moraines (I-C-2). |
| Structure, folding, faulting | Usually described as massive, but on closer examination, a variety of consistently oriented macro- and micro-structures indicative of shear or thrust may be found. Folds are overturned, witn anti- clines attenuated downglacier. Deformation structures are parti- cularly noticeable, if underlying sediments are involved, or incor- porated in the till, developing smudges.  Subhorizontal jointing or fissility is common.  Vertical joint systems, bisected by the stress direction, and transverse joints steeply dipping down- glacier, may be formed by glacier deforming its own lodgement till. The orientation of all the deformation structures is related to the stress applied by the moving glacier, and tnerefore it is laterally consistent for some distance. | Either massive, or with palimpsest structures partially preserved from debris stratification in basal debris-rich ice.  Lenses, clasts, and pods of texturally different material preserve best, for instance soft-sediment inclusions of various sizes, and englacial channel-fills.  Loss of volume with melting leads to tne draping of sorted sediments over large clasts. Most large rafts or floes of sub- stratum are associated with melt- out tills, and they may be deformed by glacial transport and by dif- ferential settlement during the melting. | Tne structure depend upon the type of flow and associated otner mass move- ments, the water content and tne posi- tion in tne flow (see Lawson, this volume).  Eitner massive, or displaying a variety of flow structures, such as: (a) overturned folds witn flat-lying isoclinal anticlines, (b) slump folds or flow lobes with tneir base usually sloping downflow, (c) roll-up struc- tures, (d) stretched-out silt and sand clasts, (e) intraformationally sneared lenses of sediments incorporated from substratum, with tneir upper downflow end attenuated, if consisting of fine- grained material, or banana snaped, (f) diapirs injected by dewatering and drawn out into flame structures, (g) small to medium-scale stringers of silt displaying a variety of flow patterns (shears, folds, etc.), (h) load struc- tures, particularly in the lower part of flows.  Tne orientation of struc- tures is related to the local stress, and tnerefore the orientation varies from place to place.  Melting of under- lying ice may produce local sagging in tne structures, or gravity faulting. |
| Grain size composi- tion | Usually a diamicton, containing clasts of various sizes.  Grain-size composition depends greatly upon tne lithology and grain-size com- position of the substrata up-glacier and the distance and mode of transport (basal, englacial) from there.  Comminution during glacial transport and lodgement has produced a multimodal particle size dis- tribution.  Most resulting subglacial tills are poorly to very poorly sorted ($\sigma$ = 2-5), described also as well graded, and their skewness nas a nearly symmetrical distribution (Sk = -0.2 to 0.2), except for those tills that are rich in incorporated pre-sorted materials. | | Usually a diamicton with polymodal particle size distribution.  It is texturally similar to that primary till to which it is related, but with a greater variability in grain size com- position, due to washing out of, or enrichment in fines, or incorporation of soft substratum sediments during the flow.  Some particle size redistribu- |

81

| Criterion | Lodgement till | Melt-out till | Gravity flowtill |
|---|---|---|---|
| | The abrasion in the zone of traction during lodgement produces particularly silt-size particles typical for lodgement tills. Most lodgement tills have a relatively consistent grain-size composition, traceable laterally for kilometers, except for the lower 0.5 to 1 m that strongly reflects the local material. Clusters or pavements of clasts are common. | The winnowing of silt- and clay-size particles in the voids during the melt-out may reduce the abundance of these particle sizes in some parts of melt-out tills in comparison with their lodged equivalents. Some particle size variability is inherited from texturally different debris bands in ice. Extreme variations in grain size may occur over short distances in the vicinity of large rafts and other inclusions of soft sediment. In supra-glacial melt-out tills of mountain glaciers those melt-out tills are particularly coarse grained which derive mainly from supraglacial debris. | tion takes place during the flow. The grain size composition depends greatly upon the type of flow, and the position or zone in it (Lawson, this volume). Sorting, inverse or normal grading may develop in some zones of flows, and parts of clasts may sink to the base of flow. |
| Lithology of clasts and matrix | Lithologic composition tends to be less variable than in other genetic varieties of tills; most constant is the mineralogic and geochemical composition of the till matrix. Materials of local derivation increase in abundance towards the basal contact of the tills with the substratum. | Since glacial debris of distant derivation is more common in the englacial zone than in the basal zone of a glacier, and since the englacial zone has a greater possibility to be deposited as melt-out till, rather than by lodgement, materials of distant derivation may be more abundant in the melt-out till than in the lodgement till of the same till unit, particularly in supraglacial melt-out till. Great compositional variability occurs in the vicinity of incorporated 'mega-clasts,' 'rafts' or 'floes' of sub-till material. Soft sediment clasts, for instance consisting of sand, may be found in melt-out till, but not in typical undeformed lodgement till. | The lithologic composition is generally the same as that of the source material of flowtill – a primary till or glacial debris, plus some substratum material incorporated during the flowage. Material of distant derivation dominates in the flowtills derived from supraglacial and englacial debris, but dominance of local material indicates derivation from basal debris. Soft sediment clasts derived from the substratum, or from sediment interbeds in multiple flows, are common. |
| Clast shapes and their surface marks | Following criteria apply to lodgement till and basal melt-out till where most clasts are derived from a single cycle transport: subangular to subrounded shapes dominate, depending mainly upon the distance of transport in the basal zone of traction. Bullet-shaped ('flat-iron,' 'elongate pentagonal') clasts are more common than in other tills and nonglacial deposits, and their tapered ends usually point upglacier. Some of the elongate clasts have a keel at their base. Glacial striae are visible mainly on medium-hard fine grained rock surfaces. Elongate clasts are striated mainly parallel to their long axes, unless they have been lodged or transported by rolling. | | If present, soft sediments clasts are either rounded or deformed by shear or dewatering. The more resistant rock clasts are in the same shape as they were in the source material when resedimented by the flowage. Therefore, the relative abundance of glacially abraded subangular to subrounded clasts versus completely angular clasts in flowtills of moun- |

| Criterion | Lodgement till | Melt-out till | Gravity flowtill |
|---|---|---|---|
| | The bullet-shaped and faceted clasts, also crushed, sheared and streaked-out clasts are more common in lodgement till than in other tills. Lodged clasts are striated parallel to the direction of the lodging glacial movement, and they have impact marks on both the upper and lower surfaces, but in opposite orient-ation: on the surface the stoss end is upglacier, but on the underside--the stoss end is downglacier. Clast pavements with sets of striae parallel to the direction of the latest glacial movement over them may occur at several lodgement levels. Their top facets are either parallel with the general plane of lodgement, or they dip upglacier. | If, in an area of mountain glacia-tion, the source of supraglacial melt-out till is englacially or even supraglacially transported supraglacially derived debris, then the clasts are angular. Most commonly, supraglacial melt-out till in such areas also contains an admixture of glacially abraded basal debris, also englacially transported. | tain glaciers will indicate the approx-imate participation of basal debris versus supraglacial debris in the formation of the flowtill. Some rounded water-reworked clasts, without striations, may derive in flowtills from melt-water stream deposits. |
| Fabric: macro-fabric (orienta-tion of clasts) or micro-fabric (orienta-tion of particles in the matrix) | Strong macro-fabrics with the long axes parallel to the local direction of glacial movement are reported from diamictons identified either as lodgement or melt-out tills. Occasionally transverse maxima have developed, associated with folding and shearing. The fabric strength may vary also, depending upon the grainsize of till, the abundance of clasts, and postdepositional modification. The lodgement till fabric may be of complex origin: produced by lodgement, or by deformation of the already deposited dilated till, under the same glacier. If both stress directions coincide, a strong fabric will develop; if not - the lodgement fabric be-comes weakened. Typically, the a-b planes dip slightly upglacier, if lodgement alone is involved. The micro-fabric is usually as strong as the macro-fabric. | In melt-out tills, fabric is inherited from glacier transport, where strong fabric dominates, parallel to the direction of glacial movement, unless deforma-tion changes it to transverse fabric locally. However, the melting-out process may weaken the fabric, particularly the micro-fabric. Also, the dip of the inclination of clasts becomes reduced by the reduction of the volume of ice during melting. | Variable, and depending greatly upon the type of flow and the position in the flow. It may range from randomly oriented to strong fabric, in thin flow tills. Fabric maxima are either parallel or transverse to the local flow direction, unrelated to glacial movement; the a-b planes are either subparallel to the base of flow, or they dip up-flow. Fabric maxima may also differ laterally on short distances. |
| Consolida-tion, perme-ability, density | Most lodgement tills, particu-larly the poorly sorted matrix-supported varieties, are over-consolidated, provided there was adequate subglacial drainage. Their bulk density, penetration resistance, and seismic velocity are usually high, permeability - low, relative to other varieties of till of the region. | Supraglacially formed melt-out tills are usually less (normally to weakly) consolidated than the subglacially formed, commonly over-consolidated melt-out tills, pro-vided there was adequate drainage of meltwater. Bulk density and penetration resistance may be lower and more variable than in related lodgement till. Also, permeability is more variable. | Primarily normally consolidated and relatively permeable. If clayey, may become overconsolidated due to post-depositional desiccation. Density lower than in primary tills. |

83

*Genetic Classification of Glacigenic Deposits, Goldthwait & Matsch (eds)*
© *1988 Balkema, Rotterdam. ISBN 90 6191 694 1*

# Comment on glacitectonite, deformation till, and comminution till

John A.Elson
*Department of Geological Sciences, McGill University, Montreal, Quebec, Canada*

ABSTRACT: Deformation till comprises weak rock or unconsolidated sediment that has been detached from its source, the primary sedimentary structures distorted or destroyed, and some foreign material admixed. The grain size of the subglacially acquired debris is not changed significantly in milling. Deformation till is thought to be formed and deposited primarily beneath the glacier and generally does not become incorporated into it, except by freezing to the base.

"Comminution till" is a term applied to very dense till that appears to have formed by abrasion of bedrock and the crushing of detritus dragged along underneath the ice, accompanied by a mixing process that results in the incorporation of rock powder produced by abrasion at the till-rock interface into the overlying glacial load.

The term glacitectonite does not adequately incorporate the concepts employed in defining deformation till and comminution till because its definition is restricted to material derived from soft sedimentary rocks. The term "deformed till" totally lacks comprehension of the original concept of deformation and comminution tills.

## 1 INTRODUCTION

In 1960 I was asked to present a description of till to a group of civil engineers, and if possible to tell them why soft, low-density till was sometimes found beneath dense, compact till, thus complicating the prediction of foundation conditions (Elson 1961). At the time, my academic interest in till involved the energy represented by the surface area of the till matrix and clasts, and whether this is related to the regime of the glacier that produced it, and how. The subject was too complex to address in the time that was available to me, but the concepts are inherent in the terms "deformation till" and "comminution till."

Professor Dreimanis (personal communications) has kept me informed of the recent debate on these terms, and on the recent interest in the term "glacitectonite." He also pointed out S.A. Schack Pedersen's (this volume) perversion of my term "deformation till" into "deformed till"; I agree that the latter term is virtually meaningless, and it certainly does not represent my original concept.

Both deformation till and comminution till were conceived as being formed underneath a moving ice sheet, rather than within the basal ice, and both would qualify as being "glacitectonite"; however, neither is a consolidated rock, and the suffix "ite" is not appropriate. Pedersen's glacitectonite, according to his description, is derived from sedimentary rock, and the suffix may be applicable to it. On the other hand, deformation till is an unconsolidated sediment derived from unconsolidated sediments.

## 2 DEFORMATION TILL

The definition and characteristics of deformation till were given in an appendix to Circular 20 of the INQUA Work Group on Till, March 20, 1982. Because of the limited circulation of that document, the appendix is reproduced below, with minor revisions, in the format used in the circular:

Deformation till comprises weak rock or unconsolidated sediment that has been detached from its source, the primary sedimentary structures distorted or destroyed, and some foreign material admixed. The term was used first in a classification of tills by Elson (1961). The grain size of the subglacially acquired debris has already been determined by a previous process such as lacustrine deposition, and it is not changed significantly by glacial milling, although the primary structures may be destroyed and foreign material from upglacier may be mixed into it. The parent materials of deformation till are saturated when acquired and may have high pore pressures and impeded drainage when redeposited; hence its density is usually low.

Deformation till is thought by Elson (1961) to be formed and deposited primarily beneath the glacier and generally does not become incorporated into it, except possibly by freezing to the base in large units. Deformation till is the "deforming-bed" of Boulton and Jones (1979), described in more detail by Boulton (1979), but when it is derived from an older till recognition may be difficult or impossible.

## 2.1 Deformation till characteristics and associations

Derivation: Deformation till is derived from beneath the glacier and typically incorporates unconsolidated material such as lake sediments, or weak rocks such as some siltstones and shales. The materials usually are in an undrained state (see Position).

Position: Deformation till is formed and deposited beneath the glacier, usually in undrained basins or where the glacier is flowing up a slope.

Thickness: The thickness seems to vary with the amount of time the glacier was active, and ranges from a thin zone not much thicker than the soil profile to thicknesses of many meters.

Extent: Deformation tills may be very extensive, especially in former lake basins, or may form only small patches near the crests of scarps or valley sides where the glacier has advanced upslope over weak rocks.

Incorporated unlithified and weakly lithified stratified sediment: The range of deformation tills encompasses: 1) thoroughly homogenized, previously deposited, stratified sediment in which all primary sedimentary structures have been destroyed, with a few clasts of different origin; 2) breccia of disoriented fragments retaining some of the original sedimentary structures in a matrix of the same material in a homogenized state; 3) strongly contorted and displaced sediment

containing sparse foreign clasts that has moved only a short distance. The distinction between the last variety and glacially deformed (glacitectonic) bedrock becomes arbitrary, and should be based on the presence of foreign clasts and evidence that the material, if it is till, has no attachment to its undisturbed source.

Bedding: Bedding of a primary sedimentary origin may be present. It is usually deformed, especially in the upper part of deformation till, which may blend into other massive till.

Clast orientation: Clasts of a size appropriate for orientation studies are usually sparse or rare and such studies are apt to be impractical. No such studies are known, but any preferred orientation would reflect deformation of a shearing matrix beneath the glacier. The preferred orientation probably would be less well developed than fabrics in basal meltout or lodgement tills.

Clast shape and surface marks: The shape and surface marks of clasts are mainly inherited from previous sedimentation processes and hence are not diagnostic. In deformation till, the clasts are transported more or less isolated from each other in a passive milieu that is maintained by dilatancy and high pore pressures, and are not appreciably modified.

Grain-size distribution: Grain-size characteristics depend entirely on those of the overridden material if the deformation till is derived from unconsolidated sediments. Such deformation till is likely to be better sorted than other tills. Clay may produce pebble- to cobble-size clay clasts in a clay matrix, forming a breccia. Deformation tills derived from weak rocks are coarse-grained with pebble- to cobble-sized platy clasts separated by a minor amount of finer matrix; the sizes of these clasts reflect the thickness of the original bedding. Pebble- to cobble-size erratics are typically present but sparse.

Lithologic composition: Deformation till generally has the same lithologic composition as the underlying sediments; there are often sparse to rare erratics from upglacier, particularly in the upper part of the till.

Consolidation (compression): Since deformation till commonly occurs in situations where drainage is impeded (basins, reverse slopes), pore pressures are high and it is commonly underconsolidated, or if free drainage develops as the glacier retreats it may become normally consolidated. The low density also reflects dilatancy resulting from the continuous shear stress applied by the moving glacier. It is common to find underconsolidated deformation till underlying younger, higher density normally- or overconsolidated lodgement till derived from different source materials farther upglacier, especially in glacial lake basins.

Surface expression: The surface expression of deformation till is seldom, if ever, diagnostic. It tends to form a low-relief ground moraine topography with gentle undulations with relief of only a meter or two disrupting the surface of a lake plain. Where it is derived from rock it may form isolated low mounds several meters high that have asymmetric slopes with the gentle slope upglacier; these mounds are most common at the top of valley slopes and other scarps. There are areas of low, subparallel ridges transverse to the direction of ice flow, a species of minor moraine, that may be deformation till in some glacial lake basins.

## 3 DISCUSSION

My type "deformation till" was an occurrence west of Dauphin, Manitoba, where about 1 m or less of diamicton rested on thinly interbedded silt and sand deposited in a glacial lake. The diamicton was the same silt and sand, more or less homogenized, with common to sparse pebble clasts typical of other tills in the vicinity. There was minor disturbance of the underlying beds at the contact. The glacier appeared to have advanced over the lake beds, and the setting required that the ice must have advanced in water, disturbing the upper layers and mixing in a few exotic clasts. This diamicton represents a mobile bed.

A second impressive example was in southern Quebec, where the Lennoxville till, in a locality shown to me by B.C. McDonald, had a facies composed of clasts of brecciated varved clay derived from underlying undisturbed varved clay, grading upward with an increasing proportion of matrix into massive clay, in the upper part of which were a few clasts similar to those in adjacent tills.

In both these examples (and there are many more) there is no evidence of reduction in grain size (or increase in surface area) by glacial action. Rather, grain size is inherited from the reworked sediments. The sedimentary structures of the reworked sediments have been destroyed, although there is sometimes a transitional

phase that would conform to Pedersen's glacitecton(ite). The glacial energy represented by deformation till is very small, and one might infer that the resulting lack of resistance to glacier flow of the deformable bed was associated with relatively rapid flow of the glacier, which in some instances at least was surge flow. Deformation till has a low density because the deforming sediment, usually waterlaid and in a basin susceptible to closure by an advancing ice margin, had high pore water pressure during the deformation and there was no escape for the pore water.

## 4 COMMINUTION TILL

The first very dense till that lead to the term "comminution till" was observed in excavations for the St. Lawrence Seaway near Cornwall, Ontario. For the purpose of construction contract specifications, till was defined by granulometric characteristics, but attempts at excavation by earth-moving machinery were ineffective and litigation resulted. The till is calcareous, and about as dense as the Ordovician (Utica) shale that was also being excavated. Except for the very high density, it fitted the description of lodgement till. It seems unlikely that material melted out of glacier ice and subjected to shearing that produces some dilatancy could create such a dense deposit. However, if the detritus was not englacial but was dragged along underneath the ice with other clasts, the process of crushing and abrasion ultimately would produce a size distribution such that void space would be minimized and resistance to shear would increase until it greatly exceeded that of overlying glacier ice. At this point, motion ceases and the till is deposited. The size distribution curves of this type of till are similar to those of the products of crushing and grinding rocks in industrial operations. Glacier energy is consumed in creating a large amount of new surface area within the till matrix; one might speculate that this comminution till forms under thick, slow-moving glaciers, with surface gradients conforming to the theoretical profiles for ice, rather than for a deformable bed (Boulton and Jones 1979).

Further reinforcement for the concept of comminution till has come from observations in a denuded area near Sudbury, Ontario, a quarry near Philipsburg, Quebec, and from cursory experience in grinding glass. Near Sudbury, erosion has stripped off most surficial deposits, exposing the glaciated surfaces of meta-gabbro and quartzite. Beneath the till remaining in small depressions there is commonly a white powder adhering to the rock surface, forming a rigid film usually less than 1 mm thick. A grain-size analysis (courtesy G.S. Boulton and colleagues) shows the same distribution that Boulton found in the fine fraction of tills that he studied. Locally, the powder layer is several mm thick and small lenses of it project obliquely upward into till. A similar occurrence on a highly polished limestone was exposed at the quarry near Philipsburg; here small coherent slabs and wedges of powdered limestone occur in the till up to 20 cm above the rock surface. There appears to be a mixing process so that the powder produced by abrasion at the till-rock interface becomes incorporated into the overlying till. There is obviously some glacitectonic action along with abrasion, and the abrasion is concentrated at the bedrock surface rather than within the diamicton. People who grind glass (e.g., telescope mirrors) know well that as finer and finer abrasives are used, the wetting of the rapidly increasing surface area of the abrasive and powdered glass consumes the free water, and if more water is not added to maintain excess void space, the tool and mirror suddenly adhere and are difficult to separate without breakage. It is likely that this phenomenon plays a role in the formation of comminution till.


5 CONCLUSION

Pedersen's term glacitectonite does not adequately incorporate the concepts deformation till and comminution till and has been restricted by him to material derived from soft sedimentary rocks. I think these terms still are useful. I agree that "deformation till" is somewhat ambiguous and have tried without success to think of a better word. In a glacitectonite, relicts of the original structures and disparate lithologies are recognizable, but in deformation till they are homogenized. Pedersen's term "deformed till" totally lacks comprehension of the original concept. Comminution till can be thought of as a "glacimylonite," but why add another term?

I am not optimistic that all the different tills in the present classification can always be recognized in the field, even by meticulous descriptions and laboratory analyses. Pedersen's concept may be quite appropriate for Denmark, but it does not necessarily fit the style of

glacial deposits on the Precambrian shield or on the flat-lying Paleozoic sediments of central North America.

REFERENCES

Boulton, G.S. 1979. Processes of glacier erosion on different substrata. Journal of Glaciology 23:15-38.
Boulton, G.S. & A.S. Jones 1979. Stability of temperate ice caps and ice sheets resting on beds of deformable sediment. Journal of Glaciology 24:29-43.
Elson, J.A. 1961. The geology of tills. Proceedings of the Fourteenth Soil Mechanics Conference 13 and 14 October, 1960. National Research Council of Canada, Associate Committee on soil and snow mechanics, Technical Memorandum No. 69, Ottawa, June 1961, p. 5-17, Discussion and replies, p. 18-35.

*Genetic Classification of Glacigenic Deposits, Goldthwait & Matsch (eds)*
*© 1988 Balkema, Rotterdam. ISBN 90 6191 694 1*

# Glacitectonite: Brecciated sediments and cataclastic sedimentary rocks formed subglacially

S.A.Schack Pedersen
*Geological Survey of Denmark, Copenhagen, Denmark*

ABSTRACT: The term glacitectonite is suggested to designate pre-existing sediments and sedimentary rocks that have been subjected to cataclastic glaciotectonic deformation. Glacitectonite is characterized by an identifiable primary lithology superimposed by cataclastic brecciation and/or a glacitectonic fabric.

## 1 INTRODUCTION

Sediments related to glacial processes are commonly referred to as tills, glacial diamictons, and glaciogene sediments. However, a special group of brecciated sediments and cataclastically deformed sedimentary rocks is not included in this well-known and commonly applied terminology for glaciogene deposits. The reason for this is obvious; while the glaciogene sediments are coeval with the ice or its meltwater, the brecciated sediments and glacio-cataclastites are pre-existing sediments or sedimentary rocks subjected to glaciotectonic deformation.

The use of the term glacitectonite (Banham 1977) is here recommended for these glacio-brecciated sediments and glacio-cataclastic sedimentary rocks. The term has recently been discussed at the First Field Meeting of the INQUA Working Group on Glacial Tectonics, October 1986, on the island of Moen, Denmark, where the majority of the participants agreed on the application and range of the term glacitectonite for such glaciotectonically deformed sediments and sedimentary rocks.

## 2 DEFINITION

The term tectonite was introduced by Bruno Sander (1912) for "a rock whose fabric clearly displays coordinated geometric features related to continuous flow during its formation" (Higgins 1971). Although Sander (1912) primarily directed the definition towards metamorphic rocks, he integrated the term to cover tectonic movements as a whole -- thus glaciotectonic deformations can be considered also. A similar concept was expressed by Banham (1977), who based the introduction of the term glacitectonite on Flinn's (1965) definition of tectonite.

Hence the short definition of glacitectonite:

Glacitectonite is a brecciated sediment or a cataclastic sedimentary rock formed by glaciotectonic deformation.

The characteristic features and genetic development for glacitectonites (Fig. 1) may be summarized as follows: A glacitectonite is a sedimentary bedrock that has been subjected to increasing deformation by the load and shearing along a sole of an advancing glacier. First, initial jointing and fracturing occur; second, differential movements develop along shear planes with related rotation of the previously produced cataclasts; and third, continuous crushing and grinding of the cataclastic fragments create a more and more fine-grained matrix contemporaneously with shear folding and formation of rhomb-shaped lenses with pressure shadows around cataclastic or even shear inworked erratic grains and fragments.

## 3 CLASSIFICATION AND APPLICATION

The main geological setting of glacitectonites is in glaciated terrains with "soft" sedimentary bedrock. In Denmark the most common bedrocks to develop into glacitectonites are Tertiary clays, diatomite and lignite, Danian limestone, and chalk of Maastrichtian age. On the island of Bornholm some Jurassic sediments are also deformed into glacitectonites. According

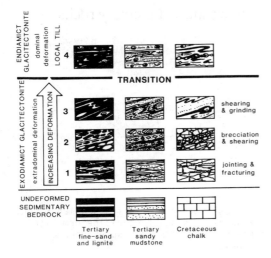

Fig. 1. Schematic development of glacitectonite exemplified for three commonly occurring sedimentary bedrock types. Distinction between exodiamict and endiamict glacitectonite was proposed by Banham (1977), whereas the concept of dominal and extradominal deformation was suggested by Berthelsen (1978).

to the main principles of classifying cataclastic rocks (Higgins 1971), the original lithology of the deformed rock should be added as a prefix to the cataclastic term. Thus a Palaeogene diatomite subjected to intense glacial deformation should be called diatomite-glacitectonite. In the same way, a deformed chalk of Maastrichtian age is classified as a chalk-glacitectonite. If relevant, a specific formation may be used as a prefix. The Palaeogene diatomite is a very characteristic unit in the western Limfjorden area, NW part of Denmark, where it is defined as the Fur Formation (Pedersen and Surlyk 1983). This area has been intensely deformed by glacial advances in the Pleistocene (Gry 1940), and it may here be convenient to use the term Fur Fm.-glacitectonite.

## 4 DISCUSSION AND CONCLUSION

The difficulty in the application of glacitectonite is the distinction between pre-existing sedimentary rock and glacitectonite, and between glacitectonite and till. The first distinction should be based on the definition of a tectonite, namely a rock in which a planar and/or linear fabric has been imposed on a pre-existing rock (Flinn 1965). Thus the sedimentary rock becomes a glacitectonite, when the primary sedimentary structures are so disturbed that they cannot be continuously traced and the planar fabric is recognized as joint fractures or shear surfaces.

The distinction between glacitectonite and till is much more difficult. In a single progressive glacial deformation event, as illustrated in Figure 2, the sedimentary bedrock is gradually deformed into a glacitectonite. The transition from glacitectonite to till may be easily defined in the field where a lodgement till with material of foreign derivation overlies the locally derived glacitectonite. In an idealized succession however, one should suspect a continuous transition from glacitectonite to local till overlain by lodgement till. The term local till refers to a till dominated by locally derived bedrock material. Locally derived is here regarded to be within a distance of one to a few kilometers.

The best criterion to differentiate between the two rock types would be the process by which erratic material is mixed. If erratic material is mixed up in the lithology by a depositional (lodgement) process (Dreimanis 1983), the sediments are classified as tills, whereas a cataclastite with erratics shear-displaced into the bedrock is classified as a glacitectonite. The most certain obvious criterion for a glacitectonite is that it

Fig. 2. Simplified sketch of the structural development of a glacitectonite. During the initial upthrusting of floes, the soft sedimentary bedrock is jointed and fractured. In the prograding glacier advance, the fractured bedrock becomes brecciated, and under the high shear strain of the glacier the brecciated glacitectonite becomes sheared and flattened. Numbers 1-4 refer to identical numbers in Fig. 1.

can be recognized as a continuous coherent but increasingly deformed bedrock (Fig. 2).

The real problem is to distinguish between glacitectonite and till when the pre-existing lithologies are glaciogene sediments. It might be possible to recognize a glacilacustrine-glacitectonite, but it is rather difficult to distinguish between a glaciofluvial-glacitectonite and a sandy till, and it is regarded as nearly impossible to distinguish between a till-glacitectonite and a till progressively deformed by the same glacial advance that initially caused the deposition of the till. Here the kineto-stratigraphical method and the dominal and extra-dominal analysis (Berthelsen 1978) are strongly recommended.

Two terms that in the writer's opinion should not be recommended are deformation till (Elson 1961, 1981) and glaciodynamic mélanges (Aber 1982). The description of deformation till by Elson (1961, 1981) is very close to the features of glacitectonite. In agreement with Ruszczyńska-Szenajch (1982) it is suggested that the term deformation till be avoided because deformation and deposition are different processes, and a deposited sediment should not be characterized by its deformation. The term mélange describes exotic blocks and minor complexes in highly deformed and metamorphic terrains (see e.g., Hsu 1974). The term is not applicable to the glaciodynamic structures described by Aber (1982), which are identical with the features characteristic of glacitectonites. Mélange is one of the adoptions from metamorphic petrology that should be avoided. Although some of the dynamic mechanisms appear to be similar, metamorphic structural geology and the structural geology in sediments are not compatible. The metamorphic facies change and the granoblastesis are so different from the water-sediment interaction that terms like mylonite, schistosity, and mélange should not be used in the terminology for deformation in sediments.

Finally, attention should be paid to the problem of sedimentary deformations in glacigene terrains, where it is questionable whether the deformation is glacitectonic, gravity gliding, or neotectonic. It should here be recommended to use the term tectonite without the prefix glaci- for brecciated and cataclastically deformed sediments.

In conclusion it is emphasized that the term glacitectonite (Banham 1977) designates glacially deformed pre-existing sediments and sedimentary rocks.

REFERENCES

Aber, I.S. 1982. Model for glaciotectonism. Bull. Geol. Soc. Denmark 30:78-90.

Banham, P.H. 1977. Glacitectonites in till stratigraphy. Boreas 6:101-105.

Berthelsen, A. 1978. The methodology of kineto-stratigraphy as applied to glacial geology. Bull. Geol. Soc. Denmark 27:25-38.

Dreimanis, A. 1983. Penecontemporaneous partial disaggregation and/or resedimentation during the formation and deposition of subglacial till. Acta Geol. Hispànica 18:153-160.

Elson, J.A. 1961. Geology of tills. Nat. Res. Council (Canada). Tech. Memorandum No. 69, p. 5-17.

Elson, J.A. 1981. Deformation till, definition given in INQUA till working group. Circular No. 20.

Flinn, D. 1965. On the symmetry principle and the deformation ellipsoid. Geol. Mag. 102:36-45.

Gry, H. 1940. De istektoniske Forhold i Moleromradet. Med Bemaerkninger om vore dislocerede Klinters Dannelse og om den negative Askeserie. Medd. dansk geol. Foren. 9:586-627.

Higgins, N.W. 1971. Cataclastic Rocks. U.S. Geol. Surv. Prof. Paper 687. Washington, United States Government Printing Office.

Hsu, K.I. 1974. Mélanges and their distinction from olistostromes. In R.H. Dott and R.H. Shaver (eds.), Modern and ancient geosynclinal sedimentation, p. 321-333. Soc. Econ. Paleontologists and Mineralogists, Spec. Publ. 19.

Pedersen, G.K. and F. Surlyk 1983. The Fur Formation, a late Paleocene ash-bearing diatomite from northern Denmark. Bull. Geol. Soc. Denmark 32:43-65.

Ruszczyńska-Szenajch, H. 1982. Lodgement tills and syndepositional glacitectonic processes related to subglacial thermal and hydrologic conditions. In E.G. Everson, J. Rabassa, and Ch. Schlüchter (eds.), Tills and related deposits, p. 113-117. Rotterdam, Balkema.

Sander, B. 1912. Uber einige Gesteinsgruppen des Tauernwestendes. Geol. Reichsanst., Wien, Jahrb. 62:219-228.

Genetic Classification of Glacigenic Deposits, Goldthwait & Matsch (eds)
© 1988 Balkema, Rotterdam. ISBN 90 6191 694 1

# Origin of a till-like diamicton by shearing

Hans-Jürgen Stephan
*Geologisches Landesamt, Kiel, FR Germany*

ABSTRACT: Till-like diamictons are produced by dragging and mixing of different grain sizes along a shear horizon in a proglacial or subglacial position with no material added by the glacier. They differ in both character and origin from "comminution till," defined as a till-like sediment resulting from the crushing and abrasion of detritus dragged along underneath a moving glacier. Therefore, the term "shear till" is proposed for those diamictons originating by pure shear.

Several times during investigations of glacially deformed sequences in gravel pits of North Germany, thin layers of till-like sediments were found in seemingly undeformed meltwater deposits. When adjacent parts were exposed by continuing excavation, it became obvious that these till-like layers actually are glacial shear zones. They usually cut the sediments at a very low angle, often almost parallel to the bedding. They are usually a few millimetres to some centimetres thick, seldom thicker. The thicker ones often appear to be faintly laminated. The main component of such a layer is a poorly sorted sand with an admixture of some silt and generally very little gravel. In a few exceptions a more silty and clayey layer was found, in which a pushed sequence consisting of glaciolimnic sediments or of interglacial or Tertiary clays or silts is cut by a shear zone.

A very special case is documented in Figure 1. The sheared layer is about 10 cm thick. This band of silty sand discordantly cuts a well-laminated meltwater sand. Subsequent to the shearing the whole complex was further thrust (overthrust and underthrust) and folded.

In these examples, a diamicton resulted from the dragging and mixing of different grain sizes along a shear horizon in a proglacial or subglacial position, with no material added by the glacier that induced the shearing. When studying small sections, such diamicton layers can easily be mistaken for a flow till or a subglacially intruded till vein.

A common process that can produce very similar diamictons is subglacial drag within the "zone of intensive shearing" (Stephan 1981, Fig. 4, caption), directly beneath the glacier sole. In an ideal case this process finally leads to a complete mixture of subsole materials in the upper part of the shear zone. Two photos with good examples have been published by Stephan and Ehlers (1983, Fig. 258, 259b). The site of 259b is illustrated with a new photo in Figure 2. It shows that more or less subhorizontal shear planes predominate in the lower part of the zone, but in the upper part -- besides a tight recumbent shear fold -- many lenticular shear bodies occur. The participants in the 1986

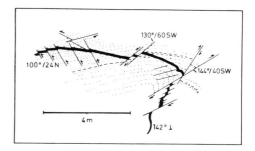

Fig. 1. WSW-ENE section in a sand pit near Wacken (Schleswig-Holstein, North-Germany) Dark band = shear zone. Sandy and silty diamicton, to the right more silty and clayey diamicton, cutting meltwater deposits, both subsequently further deformed.

Fig. 2. "Zone of intensive shearing" beneath the till of the
Weichselian Young Baltic advance in the cliff section of the
Wulfener Berg drumlin (Isle of Fehmarn). Origin of a "local till"
as "shear till" in the centimetres directly below a lodgement till
mainly composed of far transported material.

INQUA excursion (Netherlands-North Germany)
had the opportunity to study an identical
structure beneath a till sheet explained
as a glacial nappe by Dick van der Wateren
(Saalian push ridge near Damme, Lower
Saxony).

A well-developed shear zone beneath the
sole of a till can also be seen in photos
published by Dreimanis (1976, Fig. 12c)
and Rappol (1983, Fig. 55, 63, 83). Rappol
(1983:148) describes it as "zone of
penetratively sheared sediment." Banham
(1977, Fig. 2/III) depicts an example in a
simplified section and names it
"exodiamict glacitectonite." S.A. Schack
Pedersen (this volume) adopts that term
and gives examples for different fractured, brecciated, and sheared sub-till
sediments. His general conclusion that the
first phase of development of a
"glacitectonite" is initial jointing and
fracturing is not confirmed by my observations in North Germany. There, shearing
seems to be the initial phase (Fig. 3).
Subsequently, rhomb-shaped lenses develop
by alternating low-angle overthrusting and
underthrusting. The lenses become abraded
to lenticular shear bodies. Perhaps Schack
Pedersen's different view is caused by
special properties of the deformed sediments or special conditions during glaciation in the area studied by him.

Elson (this volume) defines a very dense
till-like sediment originated by dragging

beneath a glacier's sole as "comminution
till." He does not believe that the
usual grain-size distribution with minimal
void space in such till is the result of
pure mixing, but of crushing and abrasion
when the detritus was dragged along underneath the ice. In most "zones of intensive
shearing" observed by me, the density of
the material was not so extreme as
described by Elson. The process that
formed "comminution till" seems to have
been rare or absent in North Germany. It
is likely that an admixture of some abrasion powder produces a silt component in
the shear zone beneath a glacier, but in
my experience the occurrence of a more
sandy and gravelly or of a more silty and
clayey matrix in this zone strongly
depends on the composition of the bedrock
somewhat up-glacier. Mixing of the different materials overridden and dragged by
the glacier is certainly the main reason
for the origin of such diamicton. Yet in
the case of thin sandy and silty shear
layers found in pushed sand sequences as
discussed above, a production of abrasion
powder may have indeed led to the diffuse
silt component.

A subglacial zone with differential
shearing was also postulated by Boulton
and Jones (1979) for temperate glaciers.
Shearing should have occurred in a weak
horizon with high water pressure. It seems
to me that -- interpreting the kind of

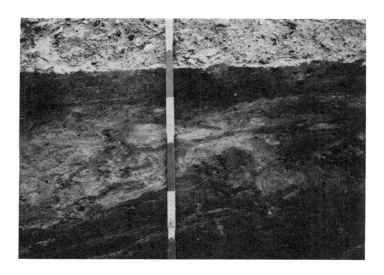

Fig. 3. Teufelsberg gravel pit (Holstein, North Germany). In the lower part an Eemian soil (Parabraunerde) is developed on a Saalian till. The disturbed zone with light sandy streaks and lenses represents a relict of the Weichselian periglacially reworked sheet. It has been sheared under the Weichselian glacier. The darker upper diamicton band is a "local till" completely consisting of reworked soil material. The very thin light streaks to the left in this "shear till" are relicts of greenish-grey soil material (Pseudogley) picked up some tenth of meters apart and dragged out in the darker (brownish) diamicton. At the top lies a bed of gravel and pebbles (washed till). Division of the scale: 10 cm.

deformation -- the observed and described examples from North Germany originated beneath cold glaciers, although the existence of a temporary thin water-saturated layer due to pressure melting cannot be excluded.

The question arises: Are these discussed diamictons "tills" or not?

The definition of the term "till" by the Till Work Group of the INQUA Commission of Genesis and Lithology of Glacial Deposits indicates that a till "is a sediment that has been transported and deposited by or from glacier ice" (Dreimanis 1982:21). I cannot apply it to those thin intercalated shear layers discussed above. The second case, a diamicton originating beneath the glacier sole, has already been discussed in the Till Work Group (cf. Dreimanis 1982: 19,20,25). Following these discussions, the subglacially sheared material could be called "deformation till." However, bearing in mind the original definition of the term by Elson (1961 and again in this volume), the "zone of intensive shearing" is not identical with

that very special till facies. It is likewise not possible to restrict the term to that part of a glacially deformed substratum that contains at least some glacial debris transported in the same glacier that caused the deformation, as proposed by Dreimanis (1982:20). Nevertheless I propose to call every diamicton that originated in the "zone of intensive shearing" a till, because it originated in closest contact with the glacier ice, excluding the less intensively sheared or deformed substratum. "Shear till" would be a good term for this diamicton. The term "Schermoräne" was used first by Gripp (1975:39) describing debris transported to a glacier's surface along overthrust planes and melting out on the surface. The material forms supraglacial till bodies or ridges, usually reworked by mass flow. The significant process of formation of those diamictons is not shearing but flowage. Grube (1979:73, Fig. 2) applied the same term to an intensively sheared basal till and thus distinguishes it from bounding, less intensively sheared till facies. In both cases the term "shear till" does not describe the genesis typifying the till reported on. However, for a till classification we need genetic definitions as a basis. Therefore I propose to apply the term "shear till" to diamictons originated by pure shearing as mentioned above.

REFERENCES

Banham, P.H. 1977. Glacitectonites in till stratigraphy. Boreas 6:101-105.

Boulton, G.S. and A.S. Jones 1979. Stability of temperate ice caps and ice sheets resting on beds of deformable sediment. Journal of Glaciology 24:29-43.

Dreimanis, A. 1976. Tills: Their origin and properties. In R.F. Legget (ed.), Glacial Till - an interdisciplinary Study. Royal Soc. Canada. Spec. Publ. No. 12:11-49.

Dreimanis, A. 1982. Work Group (1) - Genetic classification of tills and criteria for their differentiation. Progress report on activities 1977-1982, and definitions of glacigenic terms. In Ch. Schlüchter (ed.), INQUA Commission on Genesis and Lithology of Quaternary Deposits, Report on Activities 1977-1982, p. 12-31. Zurich, ETH.

Elson, J.A. 1961. The geology of tills. Proceedings of the 14th Soil Mechanics Conference, 13 and 14 October 1960. National Research Council of Canada, Associate Committee on soil and snow mechanics, Technical Memorandum No. 69: 5-17, discussion and reply, p. 18-35.

Gripp, K. 1975. 100 Jahre Untersuchungen über das Geschehen am Rande des nordeuropäischen Inlandeises. Eiszeitalter und Gegenwart 26:31-73.

Grube, F. 1979. Zur Morphogenese und Sedimentation im quartären Vereisungsgebiet Nordwestdeutschlands. Verh. naturw. Ver. Hamburg (NF) 23:69-80.

Rappol, M. 1983. Glacigenic properties of till. Studies in glacial sedimentology from the Allgäu Alps and the Netherlands. Publ. Fys.-Geogr. Bodemk. Lab. Univ. Amsterdam 34.

Stephan, H.-J. 1981. Eemzeitliche Verwitterungshorizonte im Jungmoränengebiet Schleswig-Holstein. Verh. naturw. Ver. Hamburg (NF) 24:161-175.

Stephan H.-J. and J. Ehlers 1983. North German till types. In J. Ehlers (ed.), Glacial deposits in North-West Europe, p. 239-247. Rotterdam, Balkema.

*Genetic Classification of Glacigenic Deposits, Goldthwait & Matsch (eds)*
© 1988 Balkema, Rotterdam. ISBN 90 6191 694 1

# Sedimentology of a Pleistocene glacigenic diamicton sequence near Campbell River, Vancouver Island, British Columbia

Gary W.Parkin & Stephen R.Hicock
*Department of Geology, The University of Western Ontario, London, Canada*

ABSTRACT: Multiple criteria are utilized to determine the genesis of a sequence of Late Wisconsin layered diamictons and associated sediments near Campbell River, including: descriptions of lithofacies, measurements of the orientations of bedding planes and deformation structures, stone fabric, pebble lithology, heavy mineral composition, granulometric composition of the matrix, and faunal content. The exposure is a roadcut approximately 320 m long and 20 m high. It is divided into six units (A through F) based on a combination of kineto-stratigraphy and abrupt changes in pebble lithology, heavy mineral composition, faunal content, and lithofacies. The six units comprise: glacigenic debris flows, lodgement tills, glacigenic subaquatic debris flows, subaquatic flow till, glacigenic rhythmites, subglacial melt-out till, undermelt till, glacimarine rhythmites and diamicton, iceberg-dump and iceberg-grounding tills, dropstones, and minor glaci-fluvial sediments.

The main direction of ice movement was towards the southeast as indicated by striae on the upper surface of boulders and cobbles, shear plane and extension fracture orientations, parallel stone fabric, and the dominance of Karmutsen volcanic clasts in Units A, C, D, and E. A reversal in the main ice movement direction is indicated by the orientation of shear planes in lodgement till of Unit D. Plutonic clasts derived from the Coast Plutonic Complex dominate Unit B.

The absence of extensive stratified drift deposits and weathered horizons suggests that the entire sequence may have been deposited during a single glacial event. The abrupt changes in ice flow direction and pebble lithology may be attributed to deposition by at least two competing, oscillating ice streams within the marginal zone of the Cordilleran Ice Sheet. Rapid glacier retreat from the immediate area by calving into an isostatically-depressed ice marginal zone is indicated by the thin sequence of sediment deposited between initiation of subaquatic conditions and the deposition of glacimarine diamicton in the distal zone.

## 1 INTRODUCTION

The formation of Pleistocene glacigenic layered diamictons has been mainly attributed to release of debris from glacier ice in a subaquatic environment (Dreimanis 1976, 1979; Evenson et al. 1977; Gibbard 1980; Hicock et al. 1981; Orheim and Elverhoi 1981; Eyles et al. 1983; Gravenor et al. 1984). The objectives of this study are: 1) to determine the genesis of a complex sequence of Late Wisconsin(?) layered glacial and glacimarine sediments exposed near Campbell River, British Columbia (Fig. 1); and 2) to contribute to the understanding of the behaviour of Pleistocene glaciers along the west coast of North America.

The exposure is a roadcut approximately 320 m long and 20 m high situated between 30 and 50 m above sea level. Figure 2 is a longitudinal section of the exposure divided into eight lithofacies based on a qualitative appraisal of overall texture and degree of stratification. In this paper, stratified diamicton is defined as consisting of greater than twenty percent stratified material in a given unit. All diamictons within the section are matrix-supported. The Wentworth scale is used to delimit the gravel-, sand-, silt-, and clay-sized fractions.

A description of lithofacies alone may not be sufficient to determine the genesis of glacigenic sediments, especially if definitive structures are absent (Dreimanis 1984; Karrow 1984; Kemmis and Hallberg 1984). Therefore, additional cri-

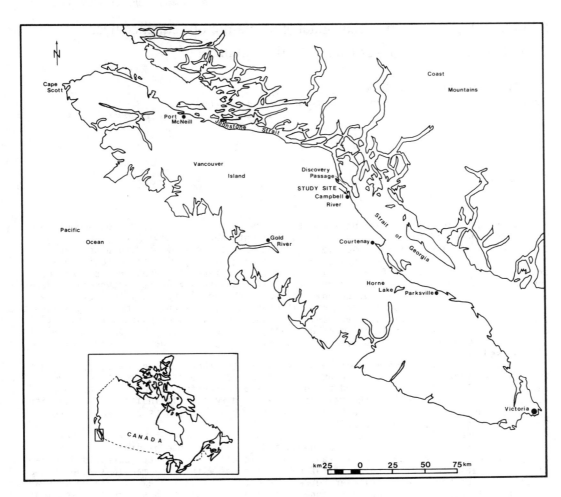

Fig. 1. Location of the exposure examined for this study near Campbell River, British Columbia.

teria were employed in this study. They include: measurements of the orientation of bedding planes and deformation structures (Fig. 2), stone fabric, pebble lithology, heavy mineral composition of the 0.250-0.125 mm size fraction, granulometric composition of the matrix, and faunal content. The technique of Mark (1973) was employed to calculate eigenvectors and eigenvalues of stone fabric measurements. A combined wet sedimentation/dry sieving technique after ASTM (1972) was used to determine the percent sand, silt, and clay in the -2.00 mm size fraction. Organic microfossils were concentrated according to the technique presented in Hicock (1980), and were then examined for the presence of dinoflagel-

late cysts, using a binocular microscope.

## 2 GENETIC TERMINOLOGY

This study uses the INQUA definition of till: "...a sediment that has been transported and is subsequently deposited by or from glacier ice, with little or no sorting by water" (Dreimanis 1982:21). The imprecise nature of the final clause is deliberate in order to allow a minor amount of disaggregation by meltwater; however, the participation of meltwater cannot be a dominant process in till formation (Dreimanis and Lundqvist 1984). The INQUA definition stresses transport by a glacier and subsequent deposition in a glacial environment, in contact or near contact with a glacier (Dreimanis 1983). Brief explanations of terms used to define the genesis of sediments in this paper are presented below.

Lodgement till forms by lodging of glacial debris from the sliding base of an active glacier by pressure melting and/or other mechanical processes (Dreimanis 1982). Deposition may occur if the frictional resistance of the substrate equals the tractional force exerted by the glacier ice (Boulton 1972). Muller (1983) has proposed lodging due to a progressive loss of pore pressure and subsequent structural collapse of a dilatant system through expulsion and escape of interstitial water. Melt-out till forms by the slow release of glacial debris by melting from ice that is not sliding or deforming internally (Dreimanis 1982; Shaw 1982). Intratill sorted sediments deposited by glacial meltwater may form contemporaneously within the melt-out till package. Melting-out can occur in both subglacial and supraglacial positions.

Glacigenic subaquatic debris flows originate directly from a glacier surface, from an inclined paleoslope, or from an unstable accumulation of debris in the ice-marginal or proglacial environment (Gravenor et al. 1984). Subaquatic flow till and submarine flow till originate directly from glacier ice in glacilacustrine and glacimarine environments, respectively (Evenson et al. 1977; Hicock et al. 1981). The less restrictive term glacigenic debris flow is applied if the environment of deposition (subaerial or subaquatic) is indeterminable.

Undermelt till is deposited from the sole of the glacier through a water column by continuous basal melting of a slowly overriding or stagnant ice shelf or floating glacier termini (Gibbard 1980; Gravenor et al. 1984). Complete disaggregation of glacial debris may occur; however, it is still considered a till as deposition takes place in the glacial environment directly from glacier ice (Dreimanis and Lundqvist 1984).

Icebergs may contribute a substantial amount of stones and diamicton to the proglacial subaquatic environment. Dropstones are released from floating ice and experience free-fall through water before deposition. Planar-based cones of diamicton, which may show a faint opposed bedding falling away from the apex of the cone, form by the rolling of debris-laden icebergs and deposition on the lake or sea bottom (Thomas and Connell 1985). Release of debris could also occur by melting on an upper surface of an iceberg and subsequent sliding or slumping of water-saturated material into the lake or sea. The resultant deposit is iceberg-dump till (Thomas and Connell 1985). Deposition of till by basal melting of a grounded iceberg may occur in the proglacial subaquatic environment (Dreimanis 1979).

Glacigenic rhythmites form in ice-contact (proglacial or subglacial) or glacier-fed glacilacustrine or glacimarine environments (Domack 1984). Sediment may be deposited by suspension from overflow-interflow currents or by traction and suspension sedimentation from underflow or turbidity currents. Glacimarine rhythmites may be distinguished by the presence of marine fauna. Varves are annual rhythmites controlled by seasonal fluctuations in meltwater discharge and suspended sediment load and may be deposited in both glacilacustrine and glacimarine environments (De Geer 1912; Stevens 1985). Glacimarine diamicton forms in a proglacial sea (Boulton and Deynoux 1981). The deposits comprise current- and suspension-derived fines, iceberg-dump and iceberg-grounding tills, dropstones and fossiliferous marine sediment (Andrews and Matsch 1983).

3 SEDIMENT CHARACTER AND GENESIS

The exposure is divided into six units (A through F) by subhorizontal contacts based on a combination of kineto-stratigraphy that utilizes directional indicators as a stratigraphic tool (Berthelsen 1978), and by abrupt changes in pebble lithology, heavy mineral composition, faunal content, and lithofacies (Fig. 3). Individual units do not represent separate geologic-event units. The purpose of the subdivision of the exposure is to allow comparison of data pertaining to possible horizontal and vertical lithofacies associations.

3.1 Unit A

Unit A, the most complex unit, is intermittently exposed at the base of the section from h0 (h = horizontal distance in metres from the southeast end of the exposure) to h109 between approximately v0.2 (v = vertical distance in metres from the base of the exposure) and v6.8 (Fig. 3).

The pebble lithology of Unit A is dominated by mafic volcanic clasts with amygdaloidal, porphyritic, breccia, and banded tuff textures (Fig. 4). Sources of the clasts include Karmutsen Formation and Bonanza Group volcanic rocks exposed on

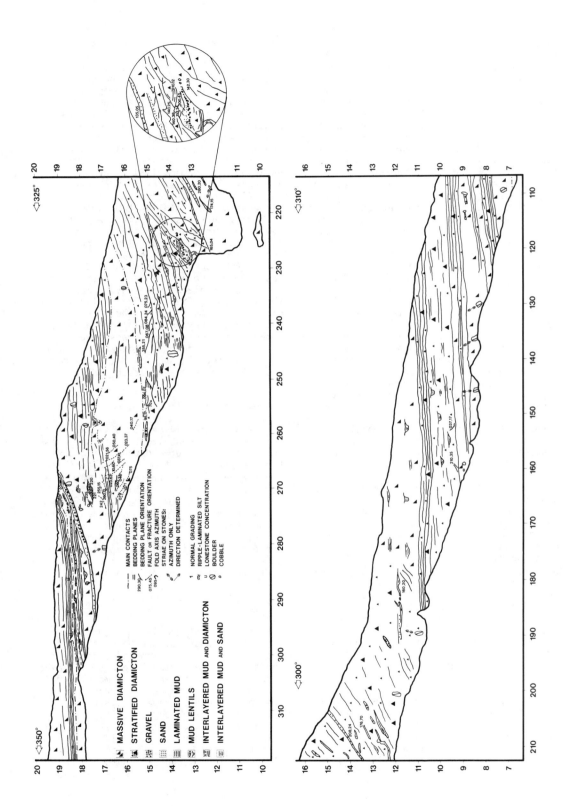

MASSIVE DIAMICTON
STRATIFIED DIAMICTON
GRAVEL
SAND
LAMINATED MUD
MUD LENTILS
INTERLAYERED MUD AND DIAMICTON
INTERLAYERED MUD AND SAND

MAIN CONTACTS
BEDDING PLANES
BEDDING PLANE ORIENTATION
FAULT OR FRACTURE ORIENTATION
FOLD AXIS AZIMUTH
STRIAE ON STONES:
    AZIMUTH ONLY
    DIRECTION DETERMINED
NORMAL GRADING
RIPPLE-LAMINATED SILT
LONESTONE CONCENTRATION
BOULDER
COBBLE

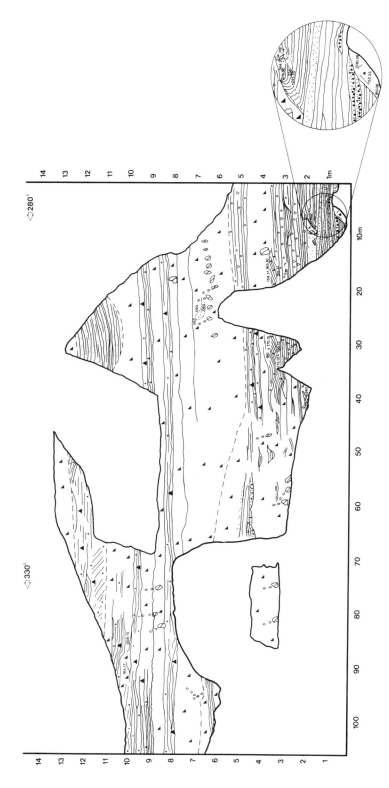

Fig. 2. Stratigraphic section of the exposure arbitrarily divided into three equal
lengths (4x vertical exaggeration).

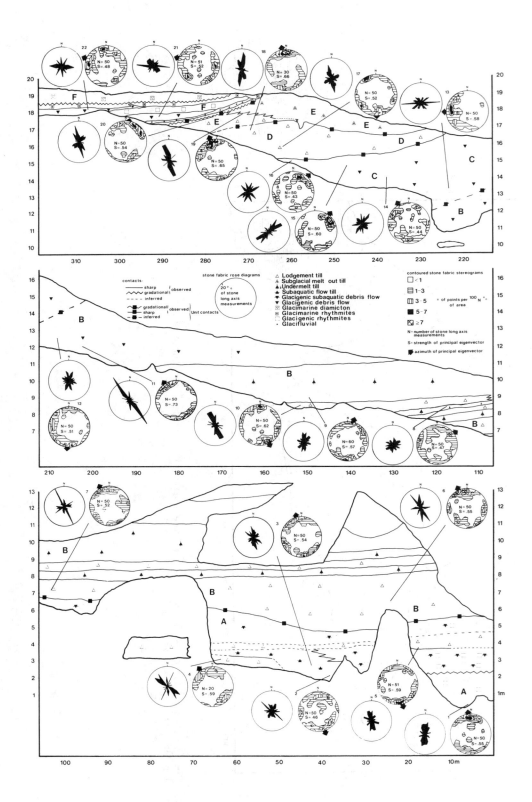

northeastern Vancouver Island (Carlisle 1972; Muller et al. 1981). The widespread distribution of petrographically similar Karmutsen volcanic rock precludes identification of precise source areas and glacier transport paths.

The plutonic clast suite in Unit A is dominated by quartz diorite and granodiorite but includes diorite, quartz monzonite, gabbro, and granite. Primary sources of these rock types include the Island Intrusions of Vancouver Island and the Coast Plutonic Complex of the British Columbia mainland. The two sources are petrographically similar and no criteria have been established to differentiate them.

A sequence of glacigenic rhythmites is exposed from h0 to h14.5 between v0.2 and v2.0 (Fig. 3). The intimate association of microlaminated mud, graded sandy gravel and sand, cross-bedded sand, lonestones, diamicton, and a lens of sandy gravel suggests that multiple depositional processes were active during the formation of this sedimentary sequence. Initial deposition of the interlayered microlaminated mud and massive sand sequence is restricted to a topographic low (Fig. 5) that suggests sedimentation by underflow or low density turbidity currents (Smith and Ashley 1985). Underflows are quasi-continuous currents generated by meltwater streams emerging from subglacial or englacial tunnels in an ice contact or glacier-fed lacustrine or marine environment (Smith and Ashley 1985). Density-driven turbidity currents occur randomly and are initiated by subaquatic slumping and debris flow, rivers in flood, and earthquakes (Walker 1984). The lack of uniform silt and clay bed thickness suggests deposition mainly by underflow currents. Normally graded gravelly sand and sand beds, such as those exposed from h0 to h10 between v0.7 and v0.9, may have been deposited by underflow or high density turbidity currents (Mackiewicz et al. 1984).

Microlaminated mud beds (less than 1 cm thick) exposed intermittently throughout the glacigenic rhythmite sequence have sharp basal contacts and show no current structures. Deposition from suspension by overflow and interflow currents is proposed. The two transport mechanisms cannot be distinguished based on the

Fig. 3. Subdivision of the section into six main units and genetic subunits. Two-dimensional rose diagrams and contoured lower hemisphere projections of stone long axes are also presented.

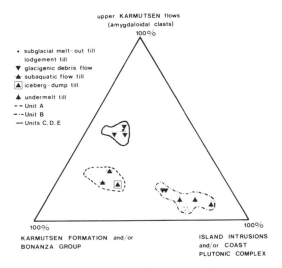

Fig. 4. Ternary comparison of rock clast proportions within five units and their genetic subunits.

Fig. 5. Depression in stratified diamicton filled with interlayered mud, sand, sandy gravel, and lonestones in Unit A. Knife 20 cm long.

character of their deposits (Smith and Ashley 1985).

Variation in the granulometric composition of the glacigenic rhythmite sequence (Fig. 6a,b) lends further support to the proposed polygenetic origin. Underflow current deposits are more coarse-grained (>60% sand) and better sorted ($\sigma_G < 2.0$) than the overflow-interflow deposits. Fine-grained, low density turbidity current deposits are also poorly sorted.

Numerous striated lonestones that disrupt bedding within the glacigenic

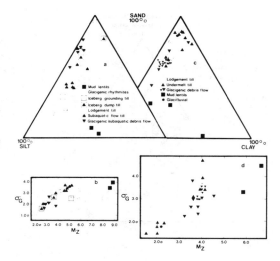

Fig. 6. Ternary and binary comparison of textural parameters in Unit A (6a,b) and Unit B (6c,d), and their genetic subunits. $M_Z$=graphic mean; $\sigma_G$=graphic standard deviation.

lake or sea floor is termed iceberg-dump till (Thomas and Connell 1985). Stone fabric 1, measured within iceberg-dump till, shows a weak cluster configuration (Fig. 3).

A zone of intense soft sediment deformation approximately 75 cm in thickness exists within the glacigenic rhythmite sequence between v1.4 and v2.1 from h4 to h15 (Fig. 2, circle inset). The deformation comprises small-scale normal and reverse faults, a sharp crested isoclinal anticline, and a series of folded, nearly overturned, and truncated sandy silt beds (Fig. 7). An isolated body of diamicton is exposed adjacent to the sharp crested isoclinal anticline.

The deformed sequence is bounded by undeformed microlaminated mud at v1.2 and v2.1, which suggests that the deformation was syngenetic. Soft-sediment deformation in the glacial environment can be attributed to glacitectonic or gravitational processes or a combination of both. The forces involved in both processes -- drag by an overriding ice mass and bed shear by flow of highly viscous sediments -- can produce almost identical structures

rhythmite sequence are "outsized"; that is, they have a diameter greater than the thickness of the host strata and could not have been transported laterally contemporaneously with the sediment (Harland et al. 1966). Lonestones penetrate or downwarp underlying strata and are draped by overlying strata. Adjacent beds onlap against most clasts. The majority of clast long axes are oriented either parallel or perpendicular to host bedding planes. Orientation of lonestones and their relationship to the host strata suggest deposition as dropstones from floating ice (Thomas and Connell 1985). Dropstones released from floating icebergs and ice shelves produce indistinguishable deposits (Shaw 1985).

A lens of sandy gravel approximately 25 cm long and a maximum of 7 cm thick disrupts bedding within the glacigenic rhythmite sequence. The lens has a convex upper surface that precludes deposition as a proglacial channel-fill (Shaw 1982). The subaquatic depositional environment represented by the host sequence and the limited extent of the lens suggest deposition by a floating iceberg as opposed to an ice shelf. Glacial debris melts out on the surface of an iceberg and is subsequently released en masse by flowing or sliding into the water or by rolling of the iceberg. The resulting deposit on the

Fig. 7. Folding and truncation of bedding within glacigenic rhythmites of Unit A. Knife 25 cm long.

(Visser et al. 1984). Deformation by a gravitational process is precluded by the absence of a significant thickness of sediment from a debris flow or slump immediately above the deformation, by the limited extent of the diamicton body, and by truncation of silt and sand beds.

The absence of major lithofacies changes generally associated with an advancing ice margin, the limited extent of the diamicton mass, and the presence of sediments of subaquatic origin immediately below and above the deformed sequence all suggest that deformation was caused by a grounded iceberg as opposed to a grounded glacier. The deformation is probably a result of compression, minor thrusting, shearing, and possibly bulldozing produced by drag at the base of the overriding grounded iceberg. Features such as the tight isoclinal fold and thrust faults, commonly produced by horizontal compressive stress, suggest that grounding was initiated as the iceberg drifted into shallow water. The isolated mass of diamicton may have been deposited by basal melting of the stranded iceberg; consequently, it is classified as iceberg-grounding till (Dreimanis 1979; Thomas and Connell 1985). Iceberg-dump till, exposed near the base of the section, is coarser-grained and better sorted than the associated iceberg-grounding till (Fig. 6a,b). Sorting is attributed to winnowing of fines by dispersion as the material fell through the water column.

The glacigenic rhythmite sequence grades vertically upwards into interlayered mud and diamicton exposed from approximately h0 to h33 between v2.2 and v3.4 (Fig. 2). Individual mud laminae are less than 2 cm thick and are disrupted by dropstones. Mud layers lack current structures and fill depressions in the upper surface of diamicton layers (Fig. 8). Deposition from suspension by overflow-interflow currents is implied by the lack of current structures and fine grain size of the mud laminae. The presence of dropstones indicates that deposition occurred in a subaquatic environment.

A lens of laminated mud in which laminae have been partially obscured is visible within a layer of diamicton (Fig. 8, above scale card). The lens is a rip-up clast eroded from the underlying sediments. The diamictic texture, deformation and incorporation of underlying fine-grained rhythmites, lack of current bedded features, and intimate association with rhythmites containing dropstones all imply deposition by cohesive debris flow in a subaquatic environment (Lowe 1982; Evenson et al.

Fig. 8. Roll-up of microlaminated mud within the interlayered mud and diamicton lithofacies of Unit A.

1977). A roll-up of laminated mud at the base of a layer of diamicton is visible in Figure 8. It is still attached to the underlying strata and fine laminae have been preserved within it. The roll-up was produced by shear at the base of the debris flow (Gustavson et al. 1975; Shaw 1985).

The direction of movement of the debris flow, visible in Figure 8, was toward the west, as indicated by the strike of the roll-up axial plane and its direction of overturning. The position of the ice-grounding line, as suggested by the presence of a series of striated boulders between h55 and h82 at v3.0 in Fig. 2, was northwest of the debris flow sequence. The direction of movement of the debris flow appears to have been toward the ice grounding line, but from a proglacial position. Consequently, the deposit is denominated glacigenic subaquatic debris flow diamicton as opposed to subaquatic flow till. Debris flow was initiated either as unconsolidated iceberg deposits or previously deposited sediments became water saturated and unstable.

Massive and stratified diamictons, exposed intermittently from approximately h36 to h67 between v1.7 and v5.8, are segregated into a series of lobate forms (less than 1 m thick) by thin layers of laminated mud, and contain zones rich in sand and gravel. Four lobes are visible in Figure 9. Cobbles are concentrated near the base of two lobes. Deformation within a mud lentil that separates two diamicton layers includes an overturned fold, a reverse fault, and a recumbent fold. A

105

Fig. 9. Interlayered massive and stratified diamicton lithofacies of Unit A. Dashed lines mark contacts between lobate flow features. Pick 45 cm long.

thin microlaminated mud lentil exposed at v3.2, h44 (Fig. 2) has been penetrated by and partially injected into an overlying diamicton layer. A micro reverse fault displaces laminae within the mud lentil. Stone fabrics 2 and 3 measured within massive diamicton show near-random horizontal girdle development (Fig. 3).

The lobate form of diamicton layers, concentration of cobbles near the base of two lobes, relatively weak stone fabrics, intimate association with mud lentils, style of deformation within mud lentils, and lateral association with sediments formed in a subaquatic environment are consistent with formation by cohesive debris flow in a subaquatic environment (Evenson et al. 1977; Hicock et al. 1981; Fraser and Cobb 1982; Lowe 1982; Broster and Hicock 1985). Deformation of mud lentils resulted from loading and horizontal compressive stress generated by drag at the base of debris flows. The direction of flow, as indicated by the orientation of the fold and thrust fault structures as seen in section and the relatively weak southeast-northwest stone fabric maxima, is approximately southeast, away from the proposed ice-grounding line position. This suggests that the debris flows were derived from glacier ice; consequently, they are denominated subaquatic flow till. Derivation directly from glacier ice is further supported by the similarity in granulometric composition of the subaquatic flow till and cogenetic lodgement till (Fig. 6a,b). The slight enrichment of silt-sized sediment in the subaquatic flow till may be the result of deposition from suspension in the turbid ice proximal zone.

Stone fabrics 4 and 5 (Fig. 3) were measured within dense, massive diamicton in Unit A. Their azimuths of maximum clustering are parallel to northwest-southeast striae on the upper surface of boulders within the same lithofacies from h47 to h80 between v3 and v4 (Fig. 2). Fabric strength values (S, Fig. 3) are greater than those measured in debris flows of Unit A. The massive diamicton truncates underlying dipping strata of the glacigenic subaquatic debris flow sequence at v3.7 (Fig. 2). These characteristics are consistent with deposition in a subglacial position by lodging (Boulton 1976; Dreimanis 1976; Kruger 1979, 1984; Boulton and Deynoux 1981; Muller 1981; Shaw 1985). A subglacial depositional environment beneath grounded, dynamically active ice is inferred.

An actively calving glacier grounded in a glacilacustrine or a brackish ice-proximal glacimarine environment is proposed for the environment of deposition of sediments within Unit A. A steep and grounded glacier margin is inferred from the occurrence of lodgement till with striated boulders at the same stratigraphic level as the proglacial subaquatic sediments and the lack of sediment deposited by basal melting beneath a floating glacier terminus or ice shelf (undermelt till). The presence of iceberg-dump and iceberg-grounding tills and deformation structures attributed to a grounded iceberg suggest that the glacier was actively calving in a proglacial water body. The alternation of underflow, turbidity, and overflow-interflow current sedimentation during the formation of the glacigenic rhythmite sequence suggests that the water was density stratified. Vertical variation in the suspended sediment concentration of the water body may be the primary cause of the stratification.

3.2 Unit B

The basal contact of Unit B is intermittently exposed from h4 to h109 between v5.0 and v6.7 (Fig. 3). The contact is sharp, probably erosional, and is accentuated by abrupt changes in diamicton colour, pebble lithology, and heavy mineral content. A substantial increase in the proportion of stones derived from the Island Intrusions and/or the Coast Plutonic Complex is evident in Fig. 4. The Coast Plutonic Complex, exposed on the British Columbia mainland and adjacent islands, is tentatively chosen as a source area for the intrusive stones as it is the most proximal and extensive concentration of the appropriate rock types. Volcanic clasts are probably derived from local exposures of Karmutsen volcanic rocks.

Three layers of massive, dense grey diamicton are exposed from h3 to h67 and h86 to h118 between v5.0 and v7.8, h13 to h128 between v8.3 and v8.8, and h106 to h155 between v8.8 and v9.2 (Fig. 2). Granulometric analyses of the diamicton matrix show a consistently poor degree of sorting and similar silty sand texture. The diamicton contains numerous striated and faceted stones, in addition to ten boulders with multiple fine striae on their upper surface with azimuths in the range of 324 to 358 degrees. A bullet-shaped boulder exposed at h95 and v6.8 shows fine striae of azimuth 340 degrees on its upper surface parallel to its long axis. The "point" of the boulder faces northwest. Three sand- and mud-filled fractures exposed above striated boulders from h22 to h26 strike approximately perpendicular to the azimuth of ice movement (indicated by northwest-southeast striae on boulders within the same lithofacies) with mean dip direction south-southeast. No net vertical displacement or deformation is associated with the fractures; consequently, they are probably extension cracks (cf. Hicock and Dreimanis 1985; Broster et al. 1979). Stone fabrics 6, 7, and 10, measured within the three massive diamicton layers, show relatively strong development of northwest-southeast fabric maxima (Fig. 3).

These characteristics suggest that the diamicton layers are lodgement till (Boulton 1976; Dreimanis 1976; Kruger 1979; Broster et al. 1979). Thus, a subglacial depositional environment beneath dynamically active ice is inferred for this lithofacies. Extension cracks are tensional fractures produced by drag beneath an active glacier (Broster et al. 1979). The fractures generally strike approximately perpendicular to ice movement direction and dip down-glacier; therefore, ice movement was toward the southeast.

A dominantly horizontal stratification is visible in a series of diamicton layers exposed from h13 to h185 between v7.7 and v11 (Figs. 2 and 10). A consistent thickness of individual beds is commonly maintained for several metres along the section. Laminated mud lentils up to 3 cm thick and less than 200 cm in length are intermittently exposed in the uppermost horizontally stratified layer. The a-axes of numerous stones are oriented horizontally and bedding beneath a stone shown near the centre of Figure 10 is bent downward. The stones are interpreted as

Fig. 10. Horizontally stratified diamicton overlain by lodgement till in Unit B. Brush 20 cm long.

dropstones deposited from floating ice. This interpretation is supported by stone fabrics 8 and 9 (Fig. 3), measured within the horizontally stratified diamicton, which have directions of maximum clustering of a-axes unrelated to the local southeast ice movement direction.

The stratified diamicton is interlayered with, and may grade laterally into, lodgement till. Basal contacts of the stratified diamicton are sharp and upper contacts are commonly gradational, which may be attributed to upshearing of substrate during lodging (Boulton 1979; Fig. 10).

The conformable association with lodgement till, as indicated by lateral gradational contacts, suggests that the horizontally stratified diamicton was deposited subglacially. The diamictic texture, lateral consistency of individual beds, presence of dropstones, near random stone fabrics with maxima unrelated to local ice movement direction, and the presence of micro-laminated mud layers and lentils are consistent with formation by basal melting through a thin water column (Dreimanis 1979; Gibbard 1980). The deposit is denominated undermelt till (Gravenor et al. 1984). Deposition of undermelt till occurs by basal melting from the underside of stagnant or advancing floating ice in glacilacustrine or glacimarine environments or by basal melting into subglacial cavities (Gibbard 1980; Gravenor et al. 1984). The intimate association with lodgement till supports deposition in the latter environment.

The undermelt till matrix is generally coarser grained and better sorted than the associated lodgement till (Fig. 6c,d). This may be a reflection of the winnowing

of fines by meltwater currents during
sedimentation through a water column;
however, minimal current activity is
suggested by the preservation of clay-
sized particles in the undermelt till
matrix (Fig. 6c) and the lack of current
bedding and internal grading. The higher
silt-sized values in the lodgement till
samples may be the result of silt production by continued abrasion during lodging
(Haldorsen 1981).

The undermelt and lodgement till succession of Unit B is overlain by glacigenic debris flows that contain intermittent
dipping mud lentils. The less restrictive
term glacigenic debris flow is preferred
as the position of the ice grounding line
is unknown. Stone fabrics 11 and 12,
measured within the glacigenic debris
flows, have directions of maximum
clustering parallel and unrelated to local
ice movement direction, respectively (Fig.
3). The strong preferred orientation of
fabric 11 may be the result of debris-
flow, reorientation by an overriding
glacier (MacClintock and Dreimanis 1964),
or the result of a late phase of ice
grounding to form a thin lodgement till
layer before deposition of the glacigenic
debris flows.


3.3 Unit C

A precise boundary between Units B and C
was not apparent in the exposure; however,
a contact is justified due to the abrupt
change in pebble lithology (Fig. 4).
Interlayered stratified and massive
diamictons with intermittent thin, microlaminated mud lentils and massive sand and
silt lenses are exposed at the base of
Unit C from approximately h205 to h228
(Fig. 2). Fabric 13 measured within stratified diamicton shows a relatively strong
a-axis girdle development dipping towards
the west, approximately parallel to the
direction of dip of the strata (Figs. 2
and 3). The direction of maximum
clustering of clast long axes is unrelated
to local ice movement direction defined
within the overlying lodgement till of
Unit D. Deformation structures include
truncation of bedding by pillars of sand
and a recumbently folded mud lentil at
h227.5, v13.0. Primary stratification is
generally preserved. The diamictic texture, similar orientation of stratification and stone fabric unrelated to local
ice flow direction, and limited deformation of underlying sediments are consistent with deposition by glacigenic
debris flow (Boulton 1968; Lawson 1982).
Truncation of bedding by two vertical sand

dikes may be the result of rapid, force-
ful, upward fluid escape initiated by
rapid deposition of the overlying debris
flow (Lowe and LoPiccolo 1974; Mills
1983).

The glacigenic debris flow sequence is
overlain by two lensoid features exposed
from h224 to h228 between v13.1 and v14.2
in Unit C. The lower lens, visible in
Figure 11, consists of a thin layer of
laminated mud overlain by stratified sandy
diamicton and in turn by a layer of normally graded gravelly sand to sand. The
form and internal composition of the two
lensoid features are consistent with deposition by glacigenic debris flow.
Accumulation of fines at the base of
debris flow lobes may be due to the
expulsion of water and elutriation of
fines from the central zone of the stratified sandy diamicton (Broster and Hicock
1985). The base of the gravelly sand layer
is marked by a series of stones that
increase in size in an upslope direction
(Fig. 11). Segregation of clasts by vertical settling may be the result of
liquefaction, waning turbulence, development of shear zones, or an increase in
water content of the debris flow (Lawson
1982). This phenomenon has been reported
to occur in both subaerial and subaquatic
environments by Lawson (1982) and Broster
and Hicock (1985), respectively.

The uppermost subunit of Unit C comprises cyclic sequences of silty sand,
gravelly sand, massive silty sand diamicton, and mud and sand lentils divided by
three main contacts. Laminated mud and
massive sand lentils commonly fill depressions in the upper surfaces of diamicton
layers. Layers of gravelly sand overlie
the lentils and accentuate contacts be-

Fig. 11. Two lensoid flow structures
exposed near the base of Unit C. Brush 20
cm long.

tween diamicton layers. Subrounded clasts of laminated mud are found near the base of silty sand layers and within diamicton matrices. Stone fabric 14, measured within a diamicton layer, shows a near random horizontal girdle configuration (Fig. 3). Stone fabric 16, measured within a lens of sandy gravel, also shows a near random girdle configuration (Fig. 3). Fabric 15, measured within diamicton, shows a stronger cluster development. The directions of maximum clustering of fabrics 15 and 16 are parallel to the depositional slope given by the attitudes of bedding planes in Unit C (Fig. 2; h227, v14.5). The sedimentologic characteristics and stone fabric of the cyclic sequences are consistent with deposition by glacigenic debris flow. A thin shear zone, commonly found at the base of debris flows, resulted in erosion and incorporation of underlying mud and silty sand. The silty sand layers show no current structures and are not internally graded. They may be the product of debris flows with high water content similar to the type 3 flows of Lawson (1982), which are erosive and mostly in shear throughout the entire thickness of the flow.

A series of shear planes and an asymmetric, overturned anticline are found in the uppermost metre of Unit C. Shear planes form by compressive thrust beneath an active glacier or by the flowing and sliding of dense masses on the substratum (Boulton 1971; Visser et al. 1984; Hicock and Dreimanis 1985). The orientation of the shear structures and the anticlinal fold axis is very similar to shear planes measured in overlying lodgement till of Unit D (Fig. 12); consequently, the former mode of genesis is preferred.

## 3.4 Unit D

Unit D consists of up to 2.0 m of massive, dense grey diamicton with a sharp, probably erosional basal contact (Fig. 13). Five sigmoidal sand-filled shear planes are found within the massive diamicton near h270 and v16.5. Fabrics 17 and 18, measured within the massive diamicton, show development of strong maxima perpendicular to the northeast-southwest strike of shear planes (Figs. 3 and 12). A secondary maximum perpendicular to the primary concentration is well developed in fabric 17.

Deposition by lodging is indicated by the uniform massive diamictic texture and well-developed stone fabric maxima perpendicular to the strikes of shear planes

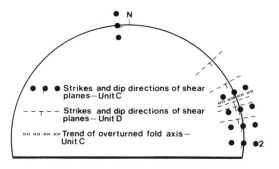

Fig. 12. Comparison of strikes and dip directions of shear planes and the trend of an overturned fold axis in Unit C, with shear planes in Unit D.

within and directly beneath the diamicton. Glacitectonic shear planes form in decelerating ice in zones of compressive thrust and beneath an active glacier in unfrozen sediments (Boulton 1971; Boulton 1979). They strike transverse to local ice movement direction and generally rise down-glacier (Dreimanis 1976). The southeast dip direction of shear planes in Units C and D (Fig. 12) suggests that local ice movement was toward the northwest.

## 3.5 Unit E

A sharp, subhorizontal contact separates Units D and E southeast of h268. The contact is obscured to the northwest by deformation and incorporation of sediment

Fig. 13. Units C (stratified), D (massive), and E (stratified) near upper limit of exposure. Chalkboard at contact of Units C and D is 23 cm wide.

from Unit D. Approximately 75 cm of
sheared, massive diamicton is exposed at
the base of Unit E northwest of h260.
Strikes and dip directions of shear planes
from Unit E and cogenetic extension fractures within massive lodgement till of
Unit D (Fig. 14) are plotted on Figure 15.
The two sets of deformation structures
strike nearly parallel and dip in opposite
directions. Stone fabric 19, measured
within the massive diamicton, reveals a
relatively strong girdle configuration
with direction of maximum clustering
approximately parallel to a striae
measurement on the surface of a boulder
and perpendicular to the strike of shear
planes and extension fractures (Figs. 3
and 15). Fabric 20 shows a polymodal
distribution with maxima both parallel and
transverse to the maximum direction of
fabric 19.

The massive diamictic texture, strong
parallel and lesser transverse stone

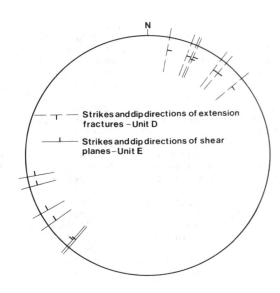

Fig. 15. Comparison of strikes and dip
directions of extension fractures in Unit
D, with shear planes in Unit E.

and does not mimic a depression in the
upper surface of Unit D (Fig. 16; above
pick). The stratified diamicton is
slightly coarser grained and better sorted
than the associated lodgement till. An
elongated clast (2.4 m by 0.5 m) of
massive diamicton is exposed within the
stratified diamicton at h255 and v17.1
southeast of a marked depression in the
lower contact of Unit E (Fig. 16). Draping
of sand layers over clasts was noted near

Fig. 14. Massive lodgement till of Unit D
(with post-depositional extension fractures) overlain by massive lodgement till
and stratified subglacial melt-out till of
Unit E. Pick 45 cm long.

fabric maxima (in relation to the orientation of shear planes), underlying extension fractures, and a striated boulder are
consistent with formation by lodging. The
direction of ice movement, as given by the
northwest dip direction of shear planes,
southeast dip direction of extension fractures, and northwest-southeast trending
striae and stone fabric, was approximately
towards the southeast.

Subhorizontally stratified diamicton is
exposed from h230 to h265 between v16.5
and v18.5 in Unit E. Stratification is
accentuated by intermittent sand lenses

Fig. 16. Massive lodgement till of Unit D
overlain by stratified diamicton of Unit
E. Note sharp contact between the two
units. Pick 45 cm long.

h272 and v18.2.

The subhorizontal stratification, relatively coarse-grained and sorted texture, preservation of an unlithified clast, draping of sand lenses over cobbles, and association with lodgement till is consistent with formation by subglacial melt-out (Boulton 1970; Shaw 1979, 1982, 1983, 1985; Lawson 1981). Draping of sand layers over two cobbles in Unit E may be the result of differential settling around the clasts (Shaw 1979). The clast of diamicton in Figure 16 was probably eroded from frozen subtill sediment by freezing to the base of the glacier and incorporated by shearing with continued forward movement before subglacial melt-out (Boulton 1979). Preservation of this clast precludes deposition under active glacier ice (Shaw 1985).

## 3.6 Unit F

The melt-out and lodgement tills of Unit E grade upwards into massive diamicton that forms the lowermost lithofacies of Unit F. A thin layer of sandy gravel overlain by approximately 15 cm of cross-bedded sand is exposed within the contact zone from h270 to h294 between v17.8 and v18.5. The cross-bedded sand appears to postdate the melt-out phase as it is not displaced by faults along 24 m of exposure. The sandy gravel layer appears to be a product of washing of fines from the underlying diamicton by current action.

The cross-bedded sand is overlain by interlayered massive diamicton and rhythmically bedded sediment. The latter lithofacies exposed from h270 to h300 between v17.9 and v18.2 consists of an alternation of massive sand, massive silt, ripple-bedded silt, and massive clayey silt beds. The sequence is denominated glacigenic rhythmites. Soft sediment deformation structures within the glacigenic rhythmites include saucer- or kidney-shaped sand lenses suspended within a clay-rich zone. The lenses may be attached to or completely isolated from the overlying silty sand. Their formation is attributed to the establishment of a reverse density gradient by deposition of silty sand over clay and subsequent sinking of the sand clasts (Mills 1983). The structures are indicative of rapid deposition and have been identified in deep and shallow water environments (Reineck and Singh 1980). Numerous "outsized" dropstones and lenses of sandy gravel disrupt bedding within the glacigenic rhythmite sequence. A lens of sandy gravel with a convex upper sur-

Fig. 17. Lens of iceberg-dump till within a glacigenic rhythmite sequence overlain by blocky glacimarine sediments in Unit F.

face is visible near the centre of Figure 17. Clasts at the base of the lens penetrate underlying sediment and a preferred horizontal a-axis orientation of elongated stones is evident. Deposition from a floating iceberg as iceberg-dump till is proposed.

Two layers of massive diamicton are interstratified with the glacigenic rhythmite sequence. Stone fabrics 21 and 22 (Fig. 3) measured within the lower and upper layers, respectively, show a similar direction of maximum clustering that is unrelated to local southeast ice movement direction as indicated by the orientation of glacitectonic structures and striae in Unit E. The stone fabrics are only slightly weaker than those measured within the underlying lodgement till of Unit E. Deposition by glacigenic subaquatic debris flow is indicated by the configuration and strength of the stone fabrics, diamictic texture, and association with glacigenic rhythmites and iceberg-dump till. The similarity of the stone fabric maxima suggests that the two flow events occurred on the same paleoslope.

The glacigenic rhythmite/subaquatic
debris flow sequence is overlain by thin
interlayers of massive sand and silt with
dropstones, and in turn by massive diamicton that shows a blocky weathering
pattern. Three mollusc shell casts in
growth position were found in the massive
diamicton, and dinoflagellate cysts were
found in both lithofacies. Preservation of
delicate processes on dinoflagellate cysts
precludes glacial transport. The layered
sequence is denominated glacimarine rhythmites and the overlying massive diamicton
is denominated glacimarine diamicton due
to the presence of marine fauna. The
matrix of the glacimarine diamicton is
generally fine-grained with sporadic zones
of coarser-grained sediment. We propose
deposition by a combination of iceberg-
dumping of coarse-grained sediment and
vertical settling of fines from turbid
overflow-interflow plumes.

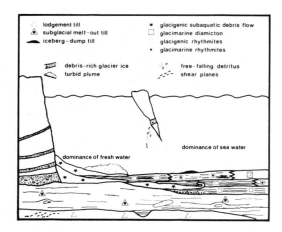

Fig. 18. Schematic model of glacier
retreat by calving of icebergs from a
tidewater glacier (not to scale).

## 4 DEPOSITIONAL MODEL OF FINAL GLACIER
RETREAT

Figure 18 is a model of glacier retreat
from the immediate study area. It is similar to the model presented by Powell
(1981) of a rapidly retreating tidewater
glacier actively calving in deep water and
to the model presented by Domack (1983) of
glacimarine deposition along a steep ice
margin. Initial glacier stagnation by in
situ down-wasting of ice resulted in the
deposition of the subglacial melt-out till
of Unit E. Initiation of a subaquatic
environment immediately following the
beginning of the stagnation phase is indicated by the narrow gradational contact
that separates the subglacial melt-out
till of Unit E and the glacigenic subaquatic debris flow diamicton/glacigenic
rhythmite sequence of Unit F (Fig. 3). The
sandy gravel layer within the contact zone
may have been produced by reworking of
sediments by stream underflows, sediment
gravity flows, and tidal currents (cf.
Powell 1981). It may mark initiation of a
proglacial sea floor. Evidence for the
release of a vast quantity of glacial
meltwater does not exist in the section.
Glacigenic subaquatic debris flow
diamicton formed in an ice marginal position from detritus melted out of basal and
englacial debris bands or by remobiliza-
tion of previously deposited sediment in a
proglacial position. Glacial debris may
have been transported upward into an
englacial (or possibly even supraglacial)
position by shear planes and/or
compressive flow planes. Evidence of
compressive glacier flow was revealed in

the lodgement till of Unit D as stone
fabrics have a significant transverse
component.
The glacigenic rhythmites formed
simultaneously with and between episodic
debris flow events by deposition from
over-, inter-, and underflow currents and
possibly low and high density turbidity
currents. The absence of marine fossils
and evidence of the hydrodynamic separation of massive clayey silt and ripple-
bedded fine silt beds indicate that
deposition occurred in an ice proximal
zone dominated by fresh water (Hicock et
al. 1981). The associated iceberg-dump
till and dropstones indicate that deposition was in a proglacial, as opposed to a
subglacial, environment.
The glacigenic rhythmites and glacigenic
subaquatic debris flows are overlain by
glacimarine rhythmites and in turn by glacimarine diamicton. The glacimarine sediments were deposited further from the ice
margin in an environment dominated by sea
water, as indicated by the presence of
marine micro- and macrofossils in growth
position, relatively poor sorting of the
fine-grained fractions, and the reduced
bed thickness within the rhythmite
sequence. Deposition by underflows is less
pronounced in the glacimarine rhythmites
due to the proposed increase in distance
from the ice margin and increase in water
density expected in a brackish to marine
environment. Poor sorting of the fine-
grained fraction is the result of aggrega-
tion of the clay- and fine silt-sized
particles by flocculation in an environment dominated by sea water (Boulton and

Deynoux 1981). Iceberg rafting contributed the coarse-grained fraction to the glacimarine sediments (cf. Domack 1983; Stoker and Bent 1985). The massive character of the glacimarine diamicton suggests that it formed beyond the influence of underflow currents in a more distal position (Domack 1983).

Rapid glacier retreat is suggested by the thin sequence of sediment deposited between initiation of subaquatic conditions (base of Unit F) and the distally deposited glacimarine diamicton. The mechanism of rapid retreat may have involved the production of large icebergs by spalling off the glacier front (Powell 1983).

## 5 CONCLUSIONS

This study demonstrates the complexity and variation of facies associations that can develop in a small geographic area of a tidewater glacial environment. The dominant ice movement direction was towards the southeast; however, a reversal in ice movement direction is suggested by the orientations of shear planes within a single layer of lodgement till and the uppermost metre of underlying glacigenic debris flows. The northwest-southeast ice flow direction may be attributed to local uplands channeling ice through Discovery Passage and down the Strait of Georgia. A similar regional direction of ice flow was determined by Fyles (1963) from tills deposited during the Fraser Glaciation near Parksville, south of the study site.

The sequence contains four lodgement tills distinguished by ice movement direction or pebble lithology, which may be attributed to deposition by at least two competing, oscillating ice streams within the Cordilleran Ice Sheet (cf. Broster and Dreimanis 1981). Ice emanating from central Vancouver Island or the Coast Mountains may have dominated the deposition of sediments within Units A, C, D, and E. The subglacial tills of Unit B, dominated by plutonic clasts, may have been deposited by ice originating in the Coast Mountains. The abrupt change in pebble lithology and heavy mineral content at the contact of Units A and B precludes deposition from different transport zones within the same ice stream; if such were the case, the change would probably be gradational.

Stone fabric maxima in lodgement tills are parallel to local ice movement direction, which suggests that glacier movement was predominantly by extending flow. A prominent component of compressive glacier flow, as indicated by the development of secondary transverse stone fabric maxima, is only evident in the uppermost lodgement till layers. Compressive flow is attributed to the gradual thinning of the ice sheet during the retreat phase, resulting in colder basal conditions.

Rapid glacier retreat from the immediate area probably occurred by calving of icebergs from a tidewater glacier. There is no evidence for the development of an extensive ice shelf during glacier retreat.

The absence of extensive stratified drift deposits and weathered horizons suggests that the entire succession was deposited during the same glacial episode. The uppermost glacimarine sediments are tentatively correlated with Late Wisconsin Capilano Sediments (12,000 years B.P.) exposed near ground surface near Parksville, south of the study site (Fyles 1963). Therefore, the bulk of the underlying glacigenic sediments could be stratigraphically equivalent to Fort Langley Formation and/or Vashon Drift (Armstrong 1981; Hicock and Armstrong 1985), deposited after the maximum of the Fraser Glaciation at the study site.

## 6 ACKNOWLEDGEMENTS

The paper is based on a Master's thesis by Parkin. Gratitude is extended to Professor A. Dreimanis for valuable advice and discussions on glacial geology. He and J.J. Clague critically reviewed the manuscript, and R.J. Fulton and R.P. Goldthwait reviewed the extended abstract. B. Hart measured stone fabrics in the field and, along with G.O. Brown, provided numerous discussions on the subject of till. I. Craig provided photographic services and J. Forth and W. Harley prepared thin sections and grain mounts. Research was funded by a Natural Sciences and Engineering Research Council of Canada grant to Hicock.

## REFERENCES

American Society for Testing and Materials 1972. Standard method for particle-size analysis of soils. Philadelphia. Annual book of ASTM standards: 112-122.

Andrews, J.T. and C.L. Matsch 1983. Glacial marine sediments and sedimentation: an annotated bibliography. Norwich, England, Geo Abstracts.

Armstrong, J.E. 1981. Post-Vashon Wisconsin glaciation, Fraser Lowland,

British Columbia. Geol. Surv. Can. Bull. 322.

Berthelsen, A. 1978. The methodology of kineto-stratigraphy as applied to glacial geology. Bull. Geol. Soc. Denmark 27:25-38.

Boulton, G.S. 1968. Flow tills and related deposits on some Vestspitsbergen glaciers. J. Glaciology 7:392-412.

Boulton, G.S. 1970. On the deposition of subglacial and melt-out tills at the margins of certain Svalbard glaciers. J. Glaciology 9:231-245.

Boulton, G.S. 1971. Till genesis and fabric in Svalbard, Spitsbergen. In R.P. Goldthwait (ed.), Till: A symposium, p. 41-72. Columbus, Ohio State Univ. Press.

Boulton, G.S. 1972. The role of thermal regime in glacial sedimentation. Inst. Brit. Geog. Spec. Pub. 4:1-20.

Boulton, G.S. 1976. A genetic classification of tills and criteria for distinguishing tills of different origin. In W. Stankowski (ed.), Till: Its genesis and diagenesis. Seria Geografia 12:65-80.

Boulton, G.S. 1979. Processes of glacier erosion on different substrata. J. Glaciology 23:15-38.

Boulton, G.S. and M. Deynoux 1981. Sedimentation in glacial environments and the identification of tills and tillites in ancient sedimentary sequences. Precambrian Res. 15:397-422.

Broster, B.E., A. Dreimanis, and J.C. White 1979. A sequence of glacial deformation, erosion, and deposition at the ice-rock interface during the last glaciation: Cranbrook, British Columbia. J. Glaciology 23:283-295.

Broster, B.E. and A. Dreimanis 1981. Deposition of multiple lodgement tills by competing glacial flows in a common ice sheet: Cranbrook, British Columbia. Arctic and Alpine Res. 13:197-204.

Broster, B.E. and S.R. Hicock 1985. Multiple flow and support mechanisms and the development of inverse grading in a subaquatic glacigenic debris flow. Sedimentology 32:645-657.

Carlisle, D. 1972. Late Paleozoic to mid Triassic sedimentary-volcanic sequence on northeastern Vancouver Island. Geol. Surv. Can., Paper 72-1, Pt. B:24-30.

De Geer, G. 1912. A geochronology of the last 12,000 years. 11th Internat. Geol. Cong. Stockholm, 1910. Compte Rendu 1:241-258.

Domack, E.W. 1983. Facies of late Pleistocene glacial marine sediments on Whidbey Island, Washington: An isostatic glacial marine sequence. In B.F. Molnia,

(ed.), Glacial marine sedimentation, p. 535-570. New York, Plenum Press.

Domack, E.W. 1984. Rhythmically bedded glaciomarine sediments on Whidbey Island, Washington. J. Sed. Pet. 54:589-602.

Dreimanis, A. 1976. Tills: Their origins and properties. In R.F. Legget (ed.), Glacial till. Roy. Soc. Can., Spec. Pub. 12:11-49.

Dreimanis, A. 1979. The problems of waterlain tills. In Ch. Schlüchter (ed.), Moraines and varves, p. 167-177. Rotterdam, Balkema.

Dreimanis, A. 1982. Work Group (1) - Genetic classification of tills and criteria for their differentiation: Progress report on activities 1977-1982, and definitions of glacigenic terms. In Ch. Schlüchter (ed.), INQUA Comm. Genesis and Lithol. Quat. Deposits. Rep. Act. 1977-1982, p. 12-31. Zurich, ETH.

Dreimanis, A. 1983. Penecontemporaneous partial disaggregation and/or resedimentation during the formation and deposition of subglacial till. Acta Geologica Hispanica 18:153-160.

Dreimanis, A. 1984. Discussion: Lithofacies types and vertical profile models; an alternative approach to the description and environmental interpretation of glacial diamict and diamictite sequences. Sedimentology 31:885-886.

Dreimanis, A. and J. Lundqvist 1984. What should be called till? Striae 20:5-10.

Evenson, E.B., A. Dreimanis, and J.W. Newsome 1977. Subaquatic flow tills: A new interpretation for the genesis of some laminated till deposits. Boreas 6:115-133.

Eyles, N., C.H. Eyles, and T.E. Day 1983. Sedimentologic and paleomagnetic characteristics of glaciolacustrine diamict assemblages at Scarborough Bluffs, Ontario, Canada. In E.B. Evenson, Ch. Schlüchter, and J. Rabassa (eds.), Tills and related deposits, p. 23-45. Rotterdam, Balkema.

Fraser, G.S. and J.C. Cobb 1982. Late Wisconsin proglacial sedimentation along the West Chicago moraine in northeastern Illinois. J. Sed. Pet. 52:473-491.

Fyles, J.G. 1963. Surficial geology of the Horne Lake and Parksville areas, Vancouver Island, British Columbia. Geol. Surv. Can. Memoir 318.

Gibbard, P. 1980. The origin of stratified Catfish Creek till by basal melting. Boreas 9:71-85.

Gravenor, C.P., V. Von Brunn, and A. Dreimanis 1984. Nature and classification of waterlain glacigenic sediments, exemplified by Pleistocene, Late

Paleozoic, and Late Precambrian deposits. Earth Sci. Rev. 20:105-166.

Gustavson, T.C., G.M. Ashley, and J.C. Boothroyd 1975. Depositional sequences in glaciolacustrine deltas. In A.V. Jobling and B.C. McDonald (eds.), Glaciofluvial and glaciolacustrine sedimentation. SEPM Spec. Pub. 23:304-320.

Haldorsen, S. 1981. Grain-size distribution of subglacial till and its relation to glacial crushing and abrasion. Boreas 10:91-105.

Harland, W.B., K.N. Herod, and D.H. Krinsley 1966. The definition and identification of tills and tillites. Earth Sci. Rev. 2:225-256.

Hicock, S.R. 1980. Pre-Fraser Pleistocene stratigraphy, geochronology, and paleoecology of the Georgia Depression, British Columbia. Ph.D. thesis, Univ. Western Ontario, London, Canada.

Hicock, S.R., A. Dreimanis, and B.E. Broster 1981. Submarine flow tills at Victoria, British Columbia. Can. J. Earth Sci. 18:71-80.

Hicock, S.R. and J.E. Armstrong 1985. Vashon Drift: Definition of the formation in the Georgia Depression, southwest British Columbia. Can. J. Earth Sci. 22:748-757.

Hicock, S.R. and A. Dreimanis 1985. Glaciotectonic structures as useful ice-movement indicators in glacial deposits: Four Canadian case studies. Can. J. Earth Sci. 22:339-346.

Karrow, P.F. 1984. Discussion: Lithofacies types and vertical profile models; an alternative approach to the description and environmental interpretation of glacial diamict and diamictite sequences. Sedimentology 31:883-884.

Kemmis, T.J. and G.R. Hallberg 1984. Discussion: Lithofacies types and vertical profile models; an alternative approach to the description and environmental interpretation of glacial diamict and diamictite sequences. Sedimentology 31:886-890.

Kruger, J. 1979. Structures and textures in till indicating subglacial deposition. Boreas 8:323-340.

Kruger, J. 1984. Clasts with stoss-lee form in lodgement tills: A discussion. J. Glaciology 30:241-243.

Lawson, D.E. 1981. Distinguishing characteristics of diamictons at the margin of the Matanuska Glacier, Alaska. Annals of Glaciology 2:78-84.

Lawson, D.E. 1982. Mobilization, movement, and deposition of active subaerial sediment flows, Matanuska Glacier, Alaska. J. Geol. 90:279-300.

Lowe, D.R. 1982. Sediment gravity flows II: Depositional models with special reference to the deposits of high density turbidity currents. J. Sed. Pet. 52:279-298.

Lowe, D.R. and R.D. LoPiccolo 1974. The characteristics and origins of dish and pillar structures. J. Sed. Pet. 44:484-501.

MacClintock, P. and A. Dreimanis 1964. Reorientation of till fabric by overriding glacier in the St. Lawrence valley. Am. J. Sci. 262:133-142.

Mackiewicz, N.E., R.D. Powell, P.R. Carlson, and B.F. Molnia 1984. Interlaminated ice-proximal glacimarine sediments in Muir Inlet, Alaska. Marine Geol. 57:113-147.

Mark, D.M. 1973. Analysis of axial orientation data including till fabrics. Geol. Soc. Amer. Bull. 84:1369-1374.

Mills, P.C. 1983. Genesis and diagnostic value of soft-sediment deformation structures - a review. Sed. Geol. 35:83-104.

Muller, E.H. 1981. Lodgement till. INQUA Comm. Genesis and Lith. Quat. Deps. Circular 21:1-6.

Muller, E.H. 1983. Dewatering during lodgement of till. In E.B. Evenson, Ch. Schlüchter, and J. Rabassa (eds.), Tills and related deposits, p. 13-18. Rotterdam, Balkema.

Muller, J.E., B.E.B. Cameron, and K.E. Northcote 1981. Geology and mineral deposits of Nootka Sound map-area, Vancouver Island, British Columbia. Geol. Surv. Can. Paper 80-16.

Orheim, O. and A. Elverhoi 1981. Model for submarine glacial deposition. Annals of Glaciology 2:123-128.

Powell, R.D. 1981. A model for sedimentation by tidewater glaciers. Annals of Glaciology 2:129-134.

Powell, R.D. 1983. Glacial marine sedimentation processes and lithofacies of temperate tidewater glaciers, Glacier Bay Alaska. In B.F. Molnia (ed.), Glacial marine sedimentation, p. 185-231. New York, Plenum.

Reineck, H.-E. and I.B. Singh 1980. Depositional sedimentary environments. New York, Springer Verlag.

Shaw, J. 1979. Genesis of the Sveg tills and Rogen moraines of central Sweden: A model of basal melt-out. Boreas 8:409-426.

Shaw, J. 1982. Melt-out till in the Edmonton area, Alberta. Can. J. Earth Sci. 19:1548-1569.

Shaw, J. 1983. Forms associated with boulders in melt-out till. In E.B. Evenson, Ch. Schlüchter, and J. Rabassa

(eds.), Tills and related deposits,
p. 3-12. Rotterdam, Balkema.

Shaw, J. 1985. Subglacial and ice marginal
environments. In G.M. Ashley, J. Shaw,
and N.D. Smith (eds.), Glacial
sedimentary environments. SEPM Short
Course 16:7-84.

Smith, N.D. and G.M. Ashley 1985.
Proglacial lacustrine environment. In
G.M. Ashley, J. Shaw, and N.D. Smith
(eds.), Glacial sedimentary
environments. SEPM Short Course
16:135-215.

Stevens, R. 1985. Glaciomarine varves in
late-Pleistocene clays near Goteborg,
southwestern Sweden. Boreas 14:127-132.

Stoker, M.S. and A. Bent 1985. Middle
Pleistocene glacial and glaciomarine
sedimentation in the west-central North
Sea. Boreas 14:325-332.

Thomas, G.S.P. and R.J. Connell 1985.
Iceberg drop, dump, and grounding structures from Pleistocene glaciolacustrine
sediments, Scotland. J. Sed. Pet.
55:243-249.

Visser, J.N.J., W.P. Colliston, and J.C.
Terblanche 1984. The origin of soft-
sediment deformation structures in
Permo-Carboniferous glacial and proglacial beds, South Africa. J. Sed. Pet.
54:1183-1196.

Walker, R.G. 1984. Turbidites and asso-
ciated coarse clastic deposits. In
R.G. Walker (ed.), Facies models.
Geosci. Can. Reprint Ser. 1:171-188.

Genetic Classification of Glacigenic Deposits, Goldthwait & Matsch (eds)
© 1988 Balkema, Rotterdam. ISBN 90 6191 694 1

# A lithofacies analysis and interpretation of depositional environments of montane glacial diamictons, Jasper, Alberta, Canada

V.M.Levson & N.W.Rutter
*Department of Geology, University of Alberta, Edmonton, Alberta, Canada*

ABSTRACT: Glacial diamictons described and sampled during regional stratigraphic studies in Jasper National Park, Alberta, Canada, are categorized using a facies approach. The facies designations are based only on objective field criteria but are designed to aid ultimately in genetic interpretations of the described deposits, which are required for meaningful stratigraphic correlations.

Twelve diamicton facies are recognized: 1) massive diamicton; 2) diamicton with steeply dipping sand lenses; 3) diamicton with plano-convex sand lenses; 4) diamicton with abundant abraded and embedded clasts; 5) stratified diamicton; 6) diamicton with circular sand lenses; 7) diamicton with normal faulted sand lenses; 8) diamicton with silt and clay beds; 9) diamicton with compressively deformed sand lenses; 10) stratified diamicton with trough-shaped sand lenses; 11) sandy diamicton with sand and gravel beds; and 12) gravelly diamicton with sand and gravel beds.

The twelve facies are grouped into two main sediment associations. Facies 1 to 6 commonly are vertically and laterally gradational with one another, suggesting a genetic relationship. This association apparently formed in or directly below the subglacial debris-rich zone of a glacier. Similarly, some or all of facies 7 to 12 commonly occur together in gradational sequences. Facies 7 is interpreted as supraglacial in origin, whereas facies 8 to 12 are believed to be ice-marginal deposits.

Genetic interpretations of the facies are validated by a general agreement between expected and observed facies sequences. Facies interpreted as ice-marginal deposits are invariably unconformably overlain by diamictons interpreted as subglacial tills. These in turn are typically overlain by sediments believed to be supraglacial and/or proglacial deposits. By analyzing vertical facies relationships and sediment associations, the stratigraphic significance of multiple diamicton sequences can be determined. Thus, the facies analysis presented here may be useful in other stratigraphic studies.

## 1 INTRODUCTION

Although many researchers have studied and described the deposits of modern glaciers in mountain environments (Sharp 1949; Boulton 1968, 1970, 1971; Shaw 1977; Boulton and Eyles 1979; Eyles 1979; Lawson 1979a, 1981a, 1981b), this paper discusses previously undescribed lithofacies of Pleistocene montane glacial deposits.

Facies designations are based on readily observable field criteria, and are consequently of use in regional stratigraphic studies where large numbers of outcrops must be described. The criteria used to distinguish facies in this study, although objective in nature, are also useful in genetic interpretation.

Descriptive approaches to the analysis

of glacial deposits have not been developed in the past, mainly because of the commonly massive and homogenous nature of tills. The facies designations described here rely heavily on the presence of stratified sediments that are often intimately associated with glacial diamictons. Such sediments often occur in small quantities as isolated bodies, and they have commonly been ignored or given little importance in the past. This reliance on associated stratified sediments in the differentiation of diamicton facies is justified, as interpretations on till genesis are often largely based on these sediments (e.g., Lawson 1979a, 1981a).

Most existing till classification systems are genetically based (Boulton 1976; Dreimanis 1976, 1982a, this volume;

Haldorsen 1982; Shaw 1982) and have three drawbacks. First, they are of limited use as field classifications mainly because genetically different tills may have similar characteristics and are commonly indistinguishable. Thus, the information required to make accurate genetic interpretations in the field is commonly not available. Second, subtle differences in till characteristics often result in major reinterpretations of till type and depositional environment. Thus, highly detailed field and laboratory observations are required before deposits can be classified. This may prohibit the use of genetic classifications in regional studies where large numbers of exposures must be examined and time may not be available for thorough descriptions. It also results in errors when interpretations must be made without adequate information. Finally, genetically based systems are continually evolving as new or more detailed information from modern glacial environments becomes available. As a result, deposits classified according to genetically based systems are continually subject to reclassification. For example, describing a deposit as a "lodgement till" in the field is not an objective approach but a genetic interpretation, and is therefore subject to revision. Even the term "till" is now seldom used in the field because of its genetic implications; it has been replaced by "diamicton." Recent studies emphasizing the descriptive facies approach (e.g., Eyles and Eyles 1983; Proudfoot 1985; Shaw 1987) have led to major reinterpretations of sediments traditionally thought of as tills deposited at the base of active glaciers.

The usage in this paper of the terms diamicton, till, and specific types of till, such as lodgement, meltout, and flow, follows the definitions given by Dreimanis (1982a, and this volume). Diamictons were identified as being of glacial origin by the presence of such features as glacially abraded (striated and faceted) clasts, clasts of erratic (distal) lithology, and bimodal or multimodal particle size distributions. These features are indicative of glacial transport, and are the main characteristics generally used to identify tills in the field (Dreimanis and Schlüchter 1985). Tills remobilized during nonglacial times may often be recognized because they usually conform to the regional slope, exhibit paleocurrents transverse to the valley, and are confined mainly to the valley side. Sediments deposited during mass movements in postglacial times are

easily recognized by characteristic geomorphic features. Mass flow deposits of nonglacial origin, deposited mainly under paraglacial conditions, comprise a significant percentage of diamictons in the Jasper region and will be the subject of a forthcoming paper.

Supraglacial, englacial, basal, subglacial, and ice-marginal are generalized genetic terms that refer to the location of till formation as discussed by Shaw (1982). The term basal is used here in a more generalized sense than subglacial, in that the former includes the debris-rich zone at or within the lower part of a glacier. Positions within the glacier that are entirely encased within ice are referred to here as englacial regardless of the distance above the glacier base. The use of the term facies follows that recommended by Walker (1984).

It is hoped that the facies analysis of glacial diamictons described here for the Jasper region will be applicable with modifications to other mountain areas, and will provide a basis for the development of nongenetic classification schemes of glacial deposits in other environments.

## 2 STUDY AREA

The study area is located in the Rocky Mountains of west-central Alberta and encompasses a large portion of Jasper National Park (Fig. 1). The Continental Divide forms the western boundary of the study area. Jasper townsite (1058 m asl) lies approximately in the center of Jasper National Park (Fig. 1). Relief in the region is high, with mountain peaks commonly rising about 1500 m above valley bottoms. Studied sections occur mainly in the Athabasca, Miette, Whirlpool, Snake Indian, and Rocky River valleys, (Fig. 1) where thick sequences of Quaternary sediments are exposed.

## 3 FACIES DESIGNATIONS

Detailed field descriptions of diamictons were made at about 75 localities in the study area (Fig. 1). Representative descriptions are available in published accounts by Bobrowsky et al. (1987), Levson and Rutter (1986), and Levson (1986) and are not duplicated here. However, to exemplify the nature of the field data, a number of measured sections from a locality with a wide range of sediment types is provided in Figure 2. From the field descriptions, the glacial diamictons were subdivided into 12 major

Fig. 1. Study area and section locations.

facies, which are described completely in Table 1. Facies numbers are for identification purposes in this paper only.

Differentiation of facies was based on characteristics that were:

1. as diagnostic of any one facies as possible;

2. easily recognizable in the field; and

3. thought to be of ultimate importance for genetic interpretations.

These criteria are similar to those outlined by Walker (1984). The main diagnostic characteristics of each facies are included in the facies title (Table 1). Although some other properties, such as pebble fabric and clast origin, may be characteristic of each facies and important for interpreting depositional environment, they are not considered diagnostic because they may be typical of more than one facies.

As in many regions, glacial diamictons in the Jasper area are massive and show little macroscopic internal structure. Consequently, facies designations were primarily based on the nature and structure of sorted and stratified sediments commonly associated with the diamictons. These sediments may occur as lenses, pods, wisps, or thin beds or laminae within the diamictons, or they may be complexly intercalated with them. Most of these forms of sorted sediment are less than one meter thick, often only centimeters or millimeters thick. Thicker sequences of stratified sediments (on the order of meters) may be interbedded with, or grade laterally into, diamicton beds. Thus, facies were identified largely on the type, amount, and geometry of these associated sorted and stratified sediments. In diamictons where characteristic internal structures such as textural or color banding and high clast contents were present, designations were also based on the internal characteristics of the diamictons themselves. These criteria are similar to those used by other authors for differentiating till types in modern environments (Boulton 1968, 1976; Shaw 1977; Eyles 1979; Lawson 1979a, 1981a).

Interpretations of the depositional environment of each facies are provided below. They are based mainly on observed sedimentary structures, but analytical data such as grain-size distribution and pebble fabric data are used to verify the interpretations. Lateral and vertical variations within diamictons and stratigraphic relationships with other depo-

119

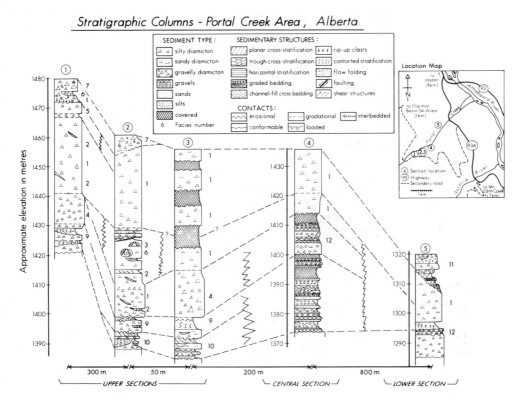

Fig. 2. Diamicton facies and associated sediments in the Portal Creek area; section locations shown on inset map (after Levson and Rutter 1986, Fig. 10).

sits were also of primary importance in making genetic interpretations. It must be emphasized that the interpretations are subject to revision and are distinct from the descriptions given on the left side of Table 1. A summary of the interpretations of each diamicton facies is provided on the right side of Table 1. Frequent reference should be made to the complete facies descriptions in Table 1 when reading the facies interpretations that follow.

### 3.1 Facies 1: Massive diamicton

Facies 1 is comprised entirely of fine grained, matrix-supported diamicton with a total absence of sorted sediments (Fig. 3, lower half). Striated and faceted clasts are abundant (Table 1). This facies characteristically has a sharp, planar, erosional lower contact, especially when overlying sands and gravels (Levson and Rutter 1986, Fig. 4).

The abundance of glacially abraded

clasts and the fine matrix textures indicate that diamictons of facies 1 are subglacial tills. Transport of debris at the base of a glacier results in comminution, producing striated clasts and relatively fine textured matrices (Haldorsen 1982). The dominance of local lithologies and the presence of strong, unimodal, pebble fabrics parallel to the valley (Fig. 4) support this interpretation.

Although the above characteristics provide good evidence for basal transport, the actual mechanism of deposition cannot be determined without further information (see discussion of facies 2 to 4). However, the strongly developed pebble fabrics parallel to the valley and the very dense, compact nature of the tills indicate that little or no secondary movement occurred after till formation. This suggests that the tills are primary or ortho-tills (Dreimanis 1982a), and that lodgement and meltout processes may have been largely responsible for their deposition (Boulton 1976; Dreimanis 1976; Shaw

120

Fig. 3. Facies 1: massive, matrix-supported diamicton (Dm); grading upwards into facies 3: massive diamicton containing plano-convex sand lenses (S); some lenses exhibit horizontal stratification (h); pick at bottom right is about 65 cm long.

1982; Haldorsen and Shaw 1982). The presence of active ice prior to deposition is indicated by the planar, erosional nature of the lower contact and by the incorporation of sediments from underlying deposits (Table 1). The above interpretation is also supported by compressive deformation in the underlying sediments and by the common presence of intraclasts believed to be rip-ups eroded by the overriding ice.

## 3.2 Facies 2: Diamicton with steeply dipping sand lenses

Diamictons of facies 1 and 2 are similar, but the latter contains rare thin lenses and beds of sand and/or gravel with strong, downvalley dips (Table 1). Diamictons of this facies are interpreted as tills deposited at the glacier base in pre-existing depressions formed by bedrock topography. The dip of sand and gravel lenses and layers and the preferred plunge shown on pebble fabric diagrams of this facies (Fig. 5) may be the result of deposition of sediment along the inclined walls of these depressions. The decrease in dip that generally occurs upsection supports this interpretation, because the slope of such depressions would decrease as they were gradually filled (Levson and Rutter 1986, Fig. 14).

The most common types of bedrock depressions in the Jasper region are narrow valleys cut by tributary streams that flow partially across major valleys. The depositional environment within these transverse tributary valleys would be similar to that in "lee-side localities" described by Haldorsen (1982). Characteristics of facies 2 diamictons in the Jasper region are similar to the lee-side tills described by Haldorsen except that matrix textures are finer and clasts less angular in the Jasper deposits. These differences are attributed to the soft shale bedrock in valley bottoms in the Jasper area, which is easily eroded, shaped, and broken down into finer particles. Till fabrics on the lee-side tills studied by Haldorsen (Fig. 6a) are also very similar to those of facies 2 (Fig. 6b). Haldorsen (1982) did not find lenses of sorted sediments in lee-side tills at Astadalen (Norway), but their presence in this facies is not considered to preclude a similar origin. Haldorsen interpreted the deposits found at lee-side localities as both subglacial meltout and lodgement tills with a gradation between the two genetic types.

Another origin for the deposition of lee-side tills was described by Boulton (1971), who observed considerable accumulations of "flow" till in cavities under Svalbard glaciers on the down glacier side of bedrock knobs. Some sorting of the till did result from water moving down the slope. Pebbles tended to lie with their long axes parallel to the slope direction. Diamictons of facies 2 have similar pebble fabric characteristics (Fig. 5) and may also have been deposited by the flowing of debris into an open subglacial cavity. If transversely oriented tributary valleys in the Jasper region were narrow enough to prevent grounding of basal ice, then subglacial cavities similar to those described by Boulton (1971) may have developed. Such cavities would become depositional sites for:

1. streams flowing at the base of the glacier, leaving lenses (channels) of sorted sediment; and

Table 1. Description of 12 major facies.

| Facies | | Main Characteristics and Sedimentary Structures | | Total Percent Sorted Sediment | Other Diamicton Properties | | | | Basal Contact |
|---|---|---|---|---|---|---|---|---|---|
| No. | Abbreviated Title | Diamicton | Sorted Sediments within diamicton | | Matrix Texture | Clast Shape and Size Distribution | Frequency of Striations | Dominant Clast Derivation | |
| 1 | Massive diamicton | Massive, unsorted, matrix supported, dense; Underlying materials often compressively deformed, eroded and incorporated in base of diamicton as matrix or intraclasts; matrix rich in silt and clay. | Sand and silt lenses absent. | Nil | Sandy-mud (50-70% silt & clay with silt generally > clay) | Mainly subangular to subrounded; Some facets; Dominantly pebbles to cobbles | Abundant (about 35 to 50% of all soft clasts, mainly shale & limestone, striated) | Local | Sharp and planar; gradational when underlain by facies 2 to 6 |
| 2 | Diamicton with steeply dipping sand lenses | Same as Facies 1 | Rare, thin, lenses and beds of sand and/or gravel with strong (10 - 30°) downvalley dips. | Less than 5 | As above | As above | As above | Local | as above |
| 3 | Diamicton with plano-convex sand lenses | Similar to Facies 1 except more variable matrix textures | Rare, horizontal, sand and gravel lenses with strongly convex upper surfaces and planar to slightly trough-shaped lower surfaces (plano-convex); often with undisturbed cross-stratification conformable to the lenses' lower boundaries) | Less than 5 | As above | As above | As above | Local | as above |
| 4 | Diamicton with abundant abraded and embedded clasts | Same as facies 1 except clasts at base of diamicton commonly partially buried (embedded) in compressively deformed under-lying sediments; all soft clasts are heavily striated parallel to a-axis; faceting on upper clast surfaces common; may be clast supported. | Absent or as in Facies 3 | Less than 5 | As above | "Barrel and Bullet" shapes abundant; Fractures and facets common; Pebble to boulder sizes. | Very abundant (about 70-90% of all soft clasts heavily striated) | Local | Sharp and planar (erosional) |
| 5 | Stratified diamicton | Horizontal diamicton strata (commonly 5 - 20 cm thick and 10's of meters wide) distinct mainly due to color and texture differences; strata unsorted, drape boulders & have diffuse boundaries. | Absent or as in Facies 3 | Less than 5 | As above but with clay generally > silt | Similar to facies 1; low total clast content | Common (about 20 -50% of all soft clasts striated) | Local | Gradational |
| 6 | Diamicton with circular sand lenses | Similar to facies 1; large clasts often protrude into lenses from encasing diamicton | Sand and gravel lenses approximately circular (in cross-section); stratification in lenses deformed often normal faulted; sand and gravel beds exhibit large textural variation, often poorly sorted and contain diamicton inclusions | Less than 10 | As above but with similar silt & clay contents | Same as facies 5 | Same as facies 5 | Mainly local | Gradational |
| 7 | Diamicton with normal faulted sand lenses | Cobbles and boulders abundant; Massive to weakly stratified due mainly to variations in clast size between beds; generally matrix supported; intermediate matrix textures; poorly compacted; clasts from high elevations common. | Irregular layers and trough-shaped lenses of normal faulted sands and gravels occur sporadically; stratification in lenses rare; poorly sorted; crude horizontal layering in diamicton accentuated by sorted layers. | Less than 20 (usually less than 10) | Sandy-mud to Muddy-sand ( 35-60% silt and clay) | Commonly angular to subangular; Cobbles and boulders unusually abundant | Striations rare | Non-local (often very far travelled and from high elevations) | Usually gradational but locally is clear and planar (erosional) |
| 8 | Diamicton with silt and clay beds | Massive, matrix supported, fine-grained, occasional folded and convoluted laminations; only 10 to 20% total clast content; diamicton complexly intercalated with silt and clay | Horizontally laminated silt and clay beds a few mm's to 10's of cm's thick; individual laminae usually laterally traceable for several meters; silts and clays contain scattered pebbles; large clasts in diamicton may protrude into underlying silts and clays and deform bedding. | 5 to 50 | Sandy-mud to Muddy-sand ( 45-65% silt and clay) | Subangular to rounded; Mainly pebbles to small cobbles | Common | Local or non-local (usually local) | Planar and sharp with small scale irregulari-ties and load structures |
| 9 | Diamicton with compressively deformed sand lenses | Matrix to clast supported; diamicton beds (1 to 2 m thick) dominate (often more than 80%); stratification accentuated by vertical variability in clast size distribution in the diamicton; angular inclusions of underlying materials common in lower parts of diamicton; | Indistinct lenses and thin diffuse beds of sorted sediment result in poorly developed stratification; trough-shaped lenses common; sorted beds dip gently or are horizontal and sometimes traceable for 10's of meters; sands and gravels commonly poorly sorted and often compressively deformed (faulted, folded and/or sheared), deformation greatest in upper part; load structures common; laminated silts and clays locally present; some large clasts deform bedding. | 10 - 20 | Muddy-sand to Silty-sand (40 - 50 % silt and clay) | Same as facies 8 | Common | Usually local | Sharp to gradational (often scoured) |
| 10 | Stratified diamicton with trough-shaped sand lenses | Diamicton less abundant and sandier than facies 9 (usually comprises 50 - 80% of the deposit); other diamicton characterisitics similar to facies 9 | Regularly shaped, well defined, plano-concave sand and gravel lenses ( 1 - 5 m wide and 0.5 - 1 m thick); little or no deformation of lenses; trough cross-beds conform to lower boundary; small scale horizontal laminae and planar cross-beds present in undisturbed form; sands and gravels moderately to well sorted. | 20 to 50 | Silty-sand to Muddy-sand (30 - 45 % silt and clay) | Same as facies 8 | Common | Same as facies 8 | Erosive with "scour-troughs" similar in dimension to lenses |
| 11 | Sandy diamicton with sand and gravel beds | Diamicton less abundant than sorted materials; matrix supported; unsorted to very poorly sorted; sandy; little or no stratification | Sands and gravels exhibit large and sudden changes in grain-size distribution, sorting, and bed orientation (dips up to 40°); frequently folded and faulted; load and fluid injection structures common; angular inclusions often present | More than 50 | Muddy-sand (25 - 35 % silt and clay) | Subangular to subrounded with some angular clasts; pebble to boulder sizes | Common to rare | Local to non-local | Highly irregular & often deformed; generally erosional |
| 12 | Gravelly diamicton with sand and gravel beds | Gravelly, generally clast supported, grading into moderately sorted gravels; higher clast contents and coarser matrix textures than facies 11 | Sands and gravels generally horizontally bedded, well to very well sorted, weakly to moderately horizontally stratified and planar & trough cross-stratified; some ripple bedding; bedding commonly dips to valley sides; clasts in gravels sometimes imbricated and rounded to well rounded. | More than 50 | Muddy-sand to Sand (15 to 30 % silt and clay) | Subangular to rounded; mainly pebbles & cobbles, few boulders | Rare | Local to non-local | Usually sharp and planar |

* Pebble fabrics are 3-D representations of the a-axis orientations of blade and prolate shaped, medium to large pebbles.
** S1 is a normalized representation of the strength of clustering around the principle eigenvector (V1).
*** See text for explanation.

122

(Table 1. continued).

| Frequency of Occurrence and Lateral Extent | Pebble Fabric * | S 1** Value | Facies Associations | Facies Number | Type of deposit and depositional processes | Position of Formation | Position of Deposition |
|---|---|---|---|---|---|---|---|
| Very common (more than 50 localities), in thicknesses of 10 m or more, laterally continuous for 10's of meters | Strong, unimodal, V1* parallel to valley, shallow upvalley plunge | 0.78 0.74 | Vertically grades into facies 2 to 6 | 1 | Basally derived till of indeterminate origin (probably mainly lodgement and meltout) | Mainly Subglacial | BASAL / SUBGLACIAL |
| Uncommon (less than 10 localities), occurs downvalley of bedrock obstructions usually a few meters thick as defined by associated lenses | Same as Facies 1 except a-axes exhibit steep downvalley plunge (5 - 15°) and fabric may be slightly bimodal | 0.67 | Vertically grades into facies 1 and 3 to 6 | 2 | Lee-side till deposited by sub-glacial flow, melt-out and/or lodgement processes | | |
| Uncommon, usually only a few meters thick as defined by associated lenses | Same as Facies 1 | 0.70 0.67 0.67 | Occurs at the base of massive diamictons (facies 1 or 2) | 3 | Meltout till (and possibly also lodgement till in some localities) deposited at the glacier base | | |
| Occurs sporadically, usually thin (less than 2 m) and discontinuous, may be represented only by one or more embedded clasts (10-15 localities) | Moderately to very strong, unimodal, transverse or parallel, horizontal or slight upvalley plunge | 0.64 | Usually underlies facies 1, 2 or 3 & unconformably overlies sorted deposits (commonly sands and gravels) or facies 8 to 12 | 4 | Lodgement till | Subglacial | |
| Uncommon, obscure, 5 to 20 meters thick, individual bands laterally traceable for 10's of meters | Unimodal, moderately strong, V1 parallel to valley, slight upvalley plunge | 0.57 0.55 0.55 | Associated with facies 6 & often conformably overlies facies 1, 2, 3, and/or 4 | 5 | Meltout till deposited at the glacier base | Englacial? | |
| Rare (<5 localities), thickness and lateral extent defined by associated lenses | Moderate to weak, unimodal to multimodal, V1 usually parallel to valley | 0.50 | Associated with facies 5 and often conformably overlies facies 1, 2, 3, and/or 4 | 6 | Meltout till formed at positions within ice probably near the glacier base | Englacial | |
| Common on surface (20-30 localities), average about 5 m's thick, laterally extensive for 100's of meters, locally thin or absent | Moderate to weak, unimodal, bimodal, or multimodal, V1 parallel or oblique to valley, high a-axis dips | 0.56 0.55 0.47 | Commonly forms the surface deposit and is closely associated with facies 10, often overlies facies 6 | 7 | Flow tills deposited by supraglacial mass movements with intermittent fluvial activity | | SUPRAGLACIAL |
| Locally common; usually < 5 m thick (diamicton beds often < 1 m thick), laterally extensive (diamicton often occurs as lenses several m wide) | No data | No data | Often associated with thick sequences of silts and clays | 8 | Sub-aquatic till flows interbedded with glacial lacustrine sediments | Mainly Frontal (Proglacial) | ICE-MARGINAL |
| Common, thickness variable, generally less than 10 m's; laterally extensive for 10's of meters | Weakly developed; V1 randomly oriented; multi-modal, dip of a-axis commonly low | 0.50 0.49 | Usually overlain by facies 1, 2, 3 or 4; and often conformably underlain by facies 10 | 9 | Proximal mass movement tills with intermittent fluvial sediments | | |
| Common (especially on the surface); thickness and extent as in facies 9 | Weak; multi-modal, V1 often parallel to valley; high, upvalley dips common on a-axes | 0.49 | Associated with facies 7 on the surface; in the sub-surface underlain by facies 8, 11, or 12 or sorted deposits & overlain by facies 9. | 10 | Subaerial till flows interbedded with stream channel deposits | | |
| Common (especially near the surface), 10's of m thick (diamicton generally 1 - 5 m thick), laterally extensive | No data | No data | Laterally gradational with facies 10 and 12 | 11 | Subaerial till flow and debris flow deposits interbedded with proximal glacial fluvial outwash | Frontal and Lateral | |
| Same as facies 11 but more common in the subsurface | No data | No data | Closely associated with thick sand and gravel deposits and facies 10 and 11. | 12 | Subaerial debris flow deposits interbedded with proximal outwash | | |

**Characteristic Trends** (Sorting / Clasts: Fabric, Glacial Markings):
- Increasing proportion of sorted materials (downward)
- Coarser mean grain size of diamicton matrix and greater sorting of associated sands and gravels (downward)
- Increasing pebble fabric strength (S1 values) (upward)
- Higher numbers of striated and faceted clasts and greater compaction of diamicton (upward)

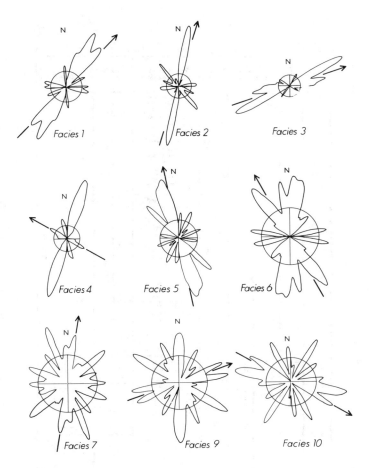

Fig. 4. Rose diagrams representing two-dimensional pebble orientations in facies 1 to 10; arrows point downvalley; no data for facies 8, 11, and 12; (25-50 a-axis orientations measured).

2. subglacial debris flows, leaving deposits similar to diamictons of facies 2.

Evidence in the Jasper region suggests that all three of the above processes (flow in a subglacial cavity, lodgement, and meltout) operated in the deposition of facies 2. A major factor controlling depositional mode would be the presence or absence of a subglacial cavity.

### 3.3 Facies 3: Diamicton with plano-convex sand lenses

Diamictons of facies 3 are similar to facies 1 but are distinguished by the presence of rare, horizontal sand and gravel lenses with strongly to slightly convex upper surfaces and planar to slightly trough-shaped lower surfaces (plano-convex

geometry) (Fig. 3).

Facies 3 diamictons are interpreted as subglacial meltout tills. Some deposition by lodgement processes may also occur, but no unequivocal evidence for this was found. The fine textures, abundance of striated clasts (Table 1), and strong fabrics parallel to the valley (Fig. 4) indicate deposition at the base of a glacier (Boulton 1976; Dreimanis 1976; Kruger 1979; Haldorsen 1982; Haldorsen and Shaw 1982). The plano-convex sand and gravel lenses that occur at the base of this facies are interpreted as subglacial stream deposits. Sand lenses have been commonly reported in meltout tills (Kruger 1979; Haldorsen 1982; Haldorsen and Shaw 1982; Shaw 1982). The geometry of the lenses is very similar to subglacial channels formed in the phreatic zone of tem-

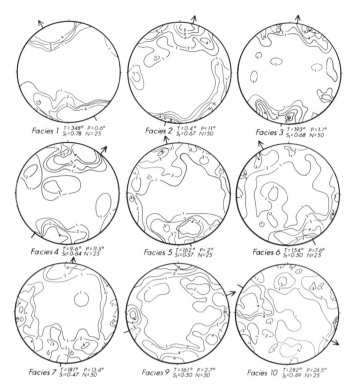

Fig. 5. Three-dimensional representation of pebble orientations in diamictons of facies 1 to 10 (no data for facies 8, 11, and 12). Equal-area projection; N is the number of clasts measured; contours represent number of points in 4% area of hemisphere for N=25 and 2% for N=50; T and P give the trend and plunge of the principal eigenvector, $V_1$; $S_1$ gives the strength of clustering around $V_1$; arrows indicate the direction of ice movement based mainly on valley orientation.

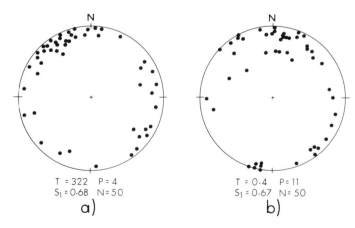

Fig. 6. Comparison of scatter diagrams from a) lee-side tills described by Haldorsen (1982); and b) from facies 2 diamicton in the Portal Creek Valley (see Fig. 5 for explanation of symbols). Note the unbalanced clustering on one side of each diagram.

125

perate glaciers (Shreve 1972).

Shaw (1982) described plano-convex lenses of sand and gravel at the base of tills in the Edmonton, Alberta area. He interpreted the lenses as the result of bed load sedimentation in subglacial tunnels. The sharp upper contact of the lenses and the draping of diamicton over them suggested that the overlying till was deposited by meltout of the debris-rich ice into which the tunnels were cut. The similarity of sand lenses of this facies to those described by Shaw (1982) suggests that their associated diamictons are also subglacial meltout tills. The dense, compact nature of the diamicton, abundance of striated clasts, and fine textures also suggest this origin (Haldorsen 1982; Haldorsen and Shaw 1982; Shaw 1982) as does the moderately well developed pebble fabric (Lawson 1979b), which shows a shallow, upvalley, preferred plunge (Fig. 5).

### 3.4 Facies 4: Diamicton with abundant abraded and embedded clasts

Although diamictons of facies 4 are similar to facies 1 diamictons, facies 4 is distinguished by the unusually high abundance of heavily striated, faceted, and embedded clasts (Table 1). Embedded clasts are those which appear to be partially buried in underlying deposits and usually have strongly faceted upper surfaces. Clasts often exhibit a preferred orientation parallel to the valley, which is sometimes so strong that it is visually recognizable in the field (Fig. 7). Underlying sediments are often compressively deformed.

This facies is interpreted as subglacial lodgement till. It forms as the result of pressure melting of debris-rich basal ice against bedrock obstructions or increased frictional drag between particles in traction in the glacier sole and subglacial sediments (Boulton 1970, 1975). The presence of abundant striated, faceted, and embedded clasts supports this interpretation (Kruger 1979). "Embedding" of clasts is interpreted as a lodgement process as indicated by the compressive deformational structures that occur in stratified sediments underneath the clasts (Boulton 1970, 1975; Kruger 1979). The common bullet and barrel shapes of embedded clasts further support this interpretation (Boulton 1978; Kruger 1984). Faceting and striations on the upper surfaces of lodged (embedded) clasts are probably the result of continued movement of debris-rich ice over the clasts after initial deposition. The

Fig. 7. Facies 4: Massive diamicton (Dm) with abundant striated, faceted, and embedded clasts; arrow indicates preferred orientation of long axis of clasts.

rarity of striations on the sides and bases of the clasts further suggests that they were lodged into the underlying sediment, leaving only the upper surfaces exposed to further abrasion.

Boulton (1975) showed that clasts in traction at the base of a glacier may become lodged as a result of collision with clasts in the underlying sediments. He noted that this process may produce clusters of clasts and nonhomogeneous tills with variations in sorting and fabric strength. Diamictons of this nature with locally high clast concentrations are typical of facies 4. Fracturing of clasts in contact with each other at the base of this facies and in immediately underlying sediment may be the result of compressive forces induced during the lodgement process and by the weight of the overlying ice.

Unimodal fabrics with strong preferred orientations parallel to the direction of ice movement are typical of facies 4 (Fig. 5) and are considered characteristic of lodgement tills (e.g., Lindsay 1970; Boulton 1971, 1976; Dreimanis 1976; Kruger and Marcussen 1976). Kruger (1979) and Rees (1983) found that strong pebble fabrics perpendicular to the ice-flow direction may also develop in lodgement tills subjected to shearing forces. Boulton (1970) noted that stone-orientation fabrics in basal debris bands developed a transverse peak as a result of compressive flow of the glacier over bedrock obstacles. The transverse fabric from a facies 4 diamicton, shown in Figure 4, may reflect the compressive flow conditions that likely existed as a result of confinement of ice in the tributary Jacques Creek valley at the sample site (Fig. 1) by the main glacier down valley.

3.5 Facies 5: Stratified diamicton

The main diagnostic characteristic of facies 5 is the presence of horizontal strata within massive diamicton otherwise similar to facies 1. The strata are all composed of diamicton and are recognizable by slight color or textural differences between layers (Fig. 8a). Matrix grain-size distributions tend to be similar between layers, with differences occurring mainly in the clast component (Fig. 8b). Strata lie in horizontal or slightly undulatory planes, and occasionally appear to drape over underlying boulders without showing any corresponding decrease in bed thickness (Fig. 8a).

Stratified tills have been recently discussed by numerous authors (e.g., Boulton 1976; Shaw 1979, 1983; Lawson 1981a; Dreimanis 1982b; Haldorsen and Shaw 1982). Stratification in many of the tills described by these authors was due to the presence of layers of sorted sediment. Glacially derived diamictons with this type of stratification can be deposited as true meltout tills with sorting resulting from flowing water derived from debris-poor ice layers (Dreimanis 1976; Shaw 1977, 1979) or by a variety of other mechanisms including subaerial sediment flow, overland sheet wash (Lawson 1979a, 1981a), and basal melting and quiet water sedimentation in water-filled cavities beneath floating ice (Gibbard 1980; Dreimanis 1982b).

In contrast, debris stratified diamictons with no sorted sediments have mainly been interpreted as subglacial meltout tills (Lawson 1979a, 1981a; Haldorsen and Shaw 1982). Although debris stratification in tills may be produced by mechanisms other than basal melting, the absence of sorted and stratified sediments largely restricts other possible origins. For example, the absence of sorting between a series of subaerial debris flows would be possible but unlikely, especially in the marginal zones of a glacier where meltwaters are abundant. Lawson (1979a) found that a layer of thinly laminated silt and sand commonly separated individual subaerial flow deposits and was critical

for their recognition in the sedimentary record. He did not observe sorted laminae forming as the result of melting of debris-poor layers of basal ice. Such laminae, therefore, should not be used as a property required for the recognition of meltout till. Some of the recent controversy regarding the recognition of meltout till (Lawson 1981a; Haldorsen and Shaw 1982) would be clarified if a distinction between sorted and unsorted layers within the stratified diamictons was made. For this reason, stratified diamictons that contained sorted sediment layers were not included in the definition of facies 5. Diamictons of this nature would have been identified as one of facies 8 to 12 depending on the thickness and texture of the sorted sediments and on the type of stratification. The strata in facies 5 are not comprised of well-sorted sediments and do not exhibit any internal sedimentary structures (bedding) as would be expected if they were fluvial in origin.

Diamictons of facies 5 are interpreted as subglacial meltout tills. The apparent stratification is believed to be the inherited product of debris stratification in the base of a glacier, preserved during the meltout process as described by Lawson (1979a, 1981a). The uniform draping of boulders by debris strata is attributed to differential subsidence caused by meltout of the ice adjacent to the boulders, as documented by Shaw (1983). A basal origin is indicated by the fine textures, abundance of striated and faceted clasts, dominance of clasts of local lithology, and the common association of this facies with facies 1 (Table 1).

This interpretation is supported by pebble fabrics from this facies, which show pebble orientations parallel to the direction of ice flow with low dips (Figs. 3 and 4). Fabrics of this nature have been described as characteristic of ancient meltout tills by numerous authors (e.g., Lindsay 1970; Boulton 1976; Haldorsen and Shaw 1982; Shaw 1982). The shallow upvalley dip of pebbles and moderate strength of the preferred a-axis orientation were observed in meltout tills at present-day glaciers by Boulton (1971) and Lawson (1979a, 1979b, and 1981a). The increased scatter of pebbles and lower dips in meltout tills compared with those in basal ice were attributed by these authors to settling and slight lateral shifting during the meltout process. The strengths of preferred pebble orientations ($S_1$ values) are clearly lower in diamictons of this facies than in other deposits interpreted as subglacial tills (facies 1 to 4), and generally higher than in other

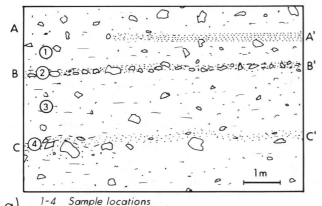

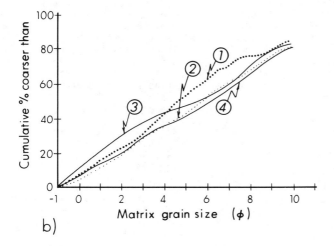

a)
1-4 Sample locations
A—A′ Discontinuous diamicton band with low clast content
B—B′ Diamicton band with high clast content
C—C′ Diamicton band draping large clasts

b)

Fig. 8. a) Schematic drawing of diamicton strata observed in facies 5; note that strata may be distinguished from the adjacent diamicton layers by low clast content, high clast content, and/or color differences. Strata may be discontinuous and commonly drape large clasts noticeably; b) Matrix grain-size distribution curves from diamicton layers at sample locations shown in Fig. 8a.

facies not believed to be subglacial tills (Fig. 9). Mark (1974) suggested that fabrics from different sedimentary environments may cluster on diagrams such as Figure 9, suggesting that facies 5 originated in a different environment (possibly higher in the ice) than facies 1 to 4.

A subglacial meltout origin for this facies is further indicated by its frequent association with facies 4, interpreted as the lodgement product of actively overriding debris-laden ice. Once

the deposition of facies 4 by lodgement ceased, debris-rich ice probably still occupied the site, and subsequent meltout would produce the stratified diamicton of facies 5 that commonly overlies facies 4.

3.6 Facies 6: Diamicton with circular sand lenses

Diamictons of facies 6 (Table 1) are characterized by associated sand and/or gravel lenses that are usually approxi-

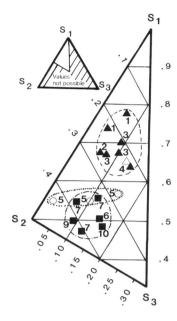

Fig. 9. Eigenvalue relationships of pebble fabrics from diamicton facies 1 to 10 (no data for facies 8, 11, and 12); Clustering of facies 1 to 4 (black triangles) and facies 6 to 10 (squares) may indicate different sedimentary environments for each cluster. Note that facies 5 (circles) occupies a slightly intermediate position. The open triangle represents an anomalously high $S_1$ value from facies 9 (see text for discussion). (Diagram follows Mark 1974.)

mately circular in cross section (Fig. 10). Oblique sections indicate that the lenses are tubular in shape. Individual sand and gravel beds within the lenses are poorly sorted, deformed, faulted, and often contain inclusions of diamicton (Fig. 10). There is a continuous gradation between sand and gravel lenses that have nearly perfect circular cross sections and those that are more irregular in shape. Beds within the latter often dip at unusually high angles.

Diamictons of this facies are interpreted as tills formed by meltout of basal debris from positions entirely encased within ice. The circular cross sections of sand and gravel lenses associated with this facies are believed to be a result of their englacial origin. Meltwaters in englacial stream tunnels near the base of temperate glaciers flow under hydrostatic pressure, forming cylindrical tunnels that are circular in cross section (Nye 1965). When bedload deposition fills the channels, tube-like deposits of sand and gravel (circular or nearly circular in cross section) develop. Such deposition may occur during waning flows prior to the abandonment of the tunnel. Channel abandonment would be common during the late stages of deglaciation, when chances for the preservation of the tunnel deposits in the sedimentary record would be higher.

Deposition within the tunnels by flowing water would be accompanied by the addition of clasts and debris from the eroding ice surfaces above and adjacent to the channel. Pieces of debris-rich ice may break off the adjacent ice surface and be buried subsequently by sediment, explaining the presence of large clasts and blocks of diamicton within the sorted sediments (Fig. 10). During meltout of the underlying ice, settling would result in deformation and normal faulting of the strata. Deformed and faulted statification and beds lying at high angles indicate extensive postdepositional movement.

Preservation of the circular cross section of the sorted lenses during meltout requires debris-rich ice. Debris concentrations up to 80% by volume have been observed at the base of some temperate glaciers (Boulton 1968, 1975). Debris concentrations of this magnitude would allow for the preservation of the circular cross sections of the sand and gravel lenses if they originally formed near the base of the glacier, allowing for minimal post-depositional disturbance. Since persistent englacial streams are concentrated in the basal portions of glacier toes, the possibility for the formation and preservation of englacial tunnel deposits is greatly increased in that zone. The more irregular and disrupted lenses probably were deposited in tunnels that overlay thicker, or more debris-poor, layers of ice. This would allow for substantial differential settling and would thus account for the irregular shape and orientation of the lenses.

The protrusion of large clasts into the upper part of the sandy lenses from the overlying diamicton (Fig. 10) suggests that the sands were deposited englacially. Clasts frozen into the glacier would occasionally protrude into englacial cavities, and deposition around them would eventually occur. Once meltout was complete, these clasts would be partially encased in both diamicton and sorted sediments.

The fine matrix textures and common presence of striated clasts of local lithology in facies 6 diamictons indicate that

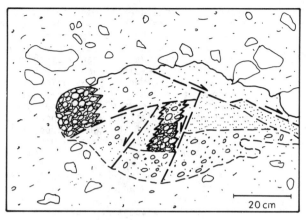

Fig. 10. Schematic drawing of faulted and deformed bedding in a near circular sand and gravel lens in massive diamicton (facies 6); Note diamicton inclusion in the lens center, large changes in grain size between beds, irregular outline of the lens, and protrusion of large clasts into the lens from the overlying diamicton.

they were basally derived. The preferred orientation of the a-axis of pebbles, parallel to the valley walls (Fig. 4), probably reflects the direction of moving ice. The high scatter of pebble orientations compared with subglacial tills (Figs. 4 and 8) may be the result of meltout of greater volumes of ice from the englacial position of formation.

Facies 6 is associated with stratified diamictons (facies 4) that are also interpreted as meltout tills. In addition, this facies commonly conformably overlies diamictons interpreted as subglacial tills. Since deposits that formed englacially would normally be expected to overlie subglacial tills, the observed and expected sequences are the same.

3.7 Facies 7: Diamicton with normal faulted sand lenses

Facies 7 diamictons are characterized by

the presence of highly disturbed (usually normally faulted) lenses and beds of sand and gravel, as well as by an abundance of boulder-sized clasts of nonlocal origin (Levson and Rutter 1986, Fig. 8). Facies 7 diamictons usually have matrix textures that vary from silty sand to sandy mud (Table 1). They may be massive or weakly stratified. Crude horizontal layering is often apparent largely as a result of variations in clast size, some layers being relatively rich in boulders and cobbles.

Diamictons of facies 7 are interpreted as supraglacial debris flow deposits. Sands and gravels associated with the diamictons are interpreted as the deposits of supraglacial streams. A supraglacial origin for the debris comprising this facies is supported by the abundance of large clasts, rarity of striated clasts, high clast angularity, dominance of clasts derived from formations out cropping only at high elevations, and the large percentages of clasts of nonlocal lithology (Table 1). Similar characteristics have been described for modern supraglacial tills by several authors (Sharp 1949; Boulton and Eyles 1979; Eyles 1979; Lawson 1979a, 1981a, 1981b).

The metamorphic erratics of the Athabasca Valley Erratics Train, commonly found in diamictons of facies 7, most likely were carried in a supraglacial position (Roed, Mountjoy, and Rutter 1967). These rocks are soft and easily disaggregated because of their well-developed schistosity, and it is therefore unlikely that they could have survived basal transport for any significant distance. The nearest outcrop to Jasper of rocks similar to those in the Athabasca Valley Erratics Train is more than 50 km to the west. Clasts of this type could not have maintained their high angularity during basal transport over such large distances, therefore supporting a supraglacial origin for facies 7.

Horizontally continuous units of clast-poor diamicton overlying relatively clast-rich layers have been observed in modern supraglacial sediment flows (Lawson 1981a), and are similar to some facies 7 diamictons. Texturally heterogeneous and poorly sorted deposits observed by Lawson (1979a) as the products of ablation processes on the Matanuska Glacier are similar to other diamictons of facies 7. Ablation deposits described by Lawson (1979a) contain lenses of sorted sediment and also have fabric characteristics similar to facies 7 diamictons. The degree of scatter of pebble a-axes about the mean in Lawson's study ($S_1$=0.550 to 0.638) is comparable to that obtained on diamictons of facies 7 ($S_1$=0.472 to 0.615), and neither shows a relationship to ice flow direction. Similarly, the variable and generally high dips of pebbles in this facies (Fig. 5) are comparable to those obtained by Lawson (1979b, 1982) in ablation deposits. Dips in the latter varied from near vertical to horizontal.

The absence of stratification within sorted sediments associated with this facies and apparent normal faulting of lenses indicate disruption and differential settling resulting from melting of the underlying ice. However, since the sorted sediments have maintained an overall horizontal orientation and some sand lenses are intact, little ice probably underlay these sediments during their last stage of formation. The amount of preservation of sedimentary structures in supraglacial tills is largely determined by the amount and rate of differential melting of the underlying ice that occurs subsequent to the last sedimentation event during till formation.

As a result of ablational slope processes, the entire supraglacial complex may have been mobilized and redeposited a number of times. Crude horizontal layering within diamictons of this facies due to variations in clast size may be the result of stacking of diamictons from separate debris flow events. This is consistent with the wide range of textures exhibited by diamictons of this facies and further suggests a supraglacial origin for them.

Deposits of this facies are often laterally and vertically gradational with deposits of facies 10, interpreted as proglacial debris flow deposits. This lateral change between supraglacial and proglacial deposits would be expected in ice marginal areas. A vertical gradation between these two facies is also likely to occur due to two factors:

1. As a glacier down-wastes in its ablation zone, progressively larger volumes of subglacially derived debris would be released on the surface and incorporated into the supraglacial deposits, resulting in a downward vertical gradation from entirely supraglacially derived debris to dominantly subglacially derived debris.

2. Once finally deposited, supraglacial tills may be covered by debris flows from adjacent ice surfaces that are topographically higher.

### 3.8 Facies 8: Diamicton with silt and clay beds

Facies 8 consists of massive diamicton beds interbedded with horizontally laminated clays and silts (Fig. 11). The diamictons are generally matrix supported and sometimes exhibit folded and convoluted silt and clay laminations (Table 1).

Diamictons of this facies are interpreted as the deposits of glacially derived debris flows that moved along the bottoms of ice-marginal lakes. The associated silts and clays are interpreted as glaciolacustrine sediments deposited between flow events. Their horizontal laminae, fine grain size, and excellent sorting indicate deposition in quiet water primarily by settling from suspension. Clasts within the clays are usually striated, indicating a glacial origin. They are too large (up to about five centimeters in diameter) in relation to the associated silts and clays to have been transported by traction currents. Consequently they are believed to be dropstones.

Fig. 11. Massive diamicton (Dm) interbedded with horizontally laminated silts
(Si) and clays (Cl); pick head is about 30
cm long.

An irregular and loaded contact at the
base of diamicton beds and deformed
bedding immediately underlying the contact
suggest rapid deposition of the diamictons
over unconsolidated deposits. Probable
flow structures (contorted and folded
laminations) within the diamictons support
this suggestion. When these characteristics are present in diamictons of
this facies, they largely preclude an origin by rainout below floating ice as
described by Gibbard (1980), Eyles and
Eyles (1983), and Miall (1983).

The sediment source for the debris flows
is believed to be glacial debris derived
from the glacier base as indicated by the
dominantly local derivation of clasts,
fine matrix textures, and abundance of
striated clasts. The debris may have accumulated either:

1. subaquatically by undermelting of the
ice margin (McCabe, Dardis, and Hanvey
1984; Powell 1983), or;

2. on the ice surface adjacent to shear
zones where basal debris was exposed,

subaerially or subaquatically, by ablation
processes (Evenson, Dreimanis, and Newsome
1977; Boulton 1968, 1971).

Some diamictons of this facies contain
abundant siltstone and shale clasts (Fig.
11), possibly derived from push moraines
in soft local bedrock (see discussion of
facies 12). Sedimentation in all of these
environments would result in a complex of
deposits similar to those described above.
A recent review of glacial, subaquatic
debris flows has been provided by Eyles,
Eyles, and Miall (1985).

More study on deposits of this facies
may allow for more detailed interpretations on the depositional environments of
the deposits. These interpretations will
be aided by information on lake parameters
such as depth and surface area drawn from
palaeogeographic reconstructions of the
individual lakes involved.

3.9 Facies 9: Diamicton with compressively
deformed sand lenses

Facies 9 consists of intercalated diamicton, sands, and gravels with compressive
deformation structures (Fig. 12).
Diamicton beds dominate this facies but
they are rarely more than one or two
meters in thickness. Sorted sediments
occur as irregularly shaped, poorly
defined layers, but they usually show a
near horizontal orientation and general
lensoid shape. Primary stratification in
the lenses is apparent but is often
faulted, folded, and/or sheared (Fig. 12).
The diamictons themselves may exhibit a
weak stratification (Table 1).

The internal structures of facies 9 are
believed to be the result of the proximity, and in some cases the overriding, of
glacial ice. The relative abundance of
diamicton, poor sorting, and the presence
of large clasts within sand and silt beds
suggest a more proximal environment than
for facies 10. The abundance of
compressive deformational features such as
faults, folds, and shear structures indicates possible pushing or overriding by
glacial ice. In addition, facies 9 is
usually overlain by deposits interpreted
as subglacial tills (mainly facies 1 or 2),
suggesting the presence of an overriding
glacier. Multimodal fabrics, such as those
exhibited by facies 9, may be due to
pebble reorientation by overriding ice
(MacClintock and Dreimanis 1964; Ramsden
and Westgate 1971). Glacial overriding may
also explain the anomalously high $S_1$ value
of one pebble fabric from facies 9 (open
triangle in Fig. 9).

Fig. 12. Facies 9: intercalated diamicton (Dm), sand (S), and gravel (G) with
compressive deformation structures (folds, thrust faults, and shear planes). Pick for
scale.

Prior to glacial overriding, diamictons of facies 9 were probably deposited by mass movements emanating from the glacier terminus region. The presence of structures interpreted as flow folds, abundant load structures, and clasts that deform underlying beds (Table 1) supports a debris flow origin for the diamictons. The multimodal, weakly developed pebble fabrics characteristic of facies 9 (Figs. 3 and 4) are typical of debris flow deposits (Marcussen 1975; Lawson 1979b, 1982).

The presence of beds and lenses of sorted sediment is critical to the genetic interpretation of diamictons of this facies. Stratigraphic relationships sometimes clearly indicate that diamictons overlying and underlying sorted sediments were deposited at different times. In adjacent areas, these diamictons are often indistinguishable, and contacts between them cannot be recognized. Thus, due to similarities in composition, a series of overlapping debris flows may only be recognized as different events by the presence of these intervening stratified sediments. Lawson (1979a, 1981a) found that most sediment flows at the margin of the Matanuska Glacier were accompanied by meltwater flowing over their surfaces. He stated that "a top layer of thinly laminated silt and sand is generally critical to identifying individual flow deposits in depositional sequences" (Lawson 1981a: 80). Thus, the thin beds of stratified sand and silt that are associated with diamictons of this facies were likely deposited by glacial meltwater flowing in sheets after deposition of underlying debris flow sediments.

Multiple superimposed flow deposits may locally be recognized by layering within the diamictons themselves. Changes in total clast content and grain-size distribution between layers could reflect different sediment sources, modes of transport, and variations in the amount of reworking of the deposits prior to final deposition. Significant changes in matrix texture over short vertical and horizontal distances (on the order of centimeters) within diamicton beds, may be the result of the erosion and incorporation of sediments in the path of the debris flows. Angular inclusions of underlying sediments within the lower portions of facies 9 diamictons are interpreted as rip-up clasts, and indicate that the flows were at least locally erosive. Lawson (1981a) found that the bulk of debris flows near Matanuska Glacier was composed of massive diamictons that locally contained inclusions of texturally or structurally distinct sediment.

3.10 Facies 10: Stratified diamicton with trough-shaped sand lenses

Facies 10 consists of poorly to moderately

133

stratified diamicton containing abundant undisturbed trough-shaped lenses of sand and gravel. Facies 10 is similar to facies 9 except that little or no faulting or deformation of stratified sediments associated with the diamictons is apparent, and the proportion of diamicton compared to facies 9 is generally low (Fig. 13). Sand and gravel lenses are abundant (usually comprising 25% to 50% of the deposits), relatively well defined, and typically plano-concave in cross section. Facies 10 diamictons are generally sandier than in facies 9 (Table 1), but other characteristics such as pebble fabric properties (Fig. 4) are similar and consequently are also interpreted as debris flows. The locally "scoured" lower contact of the diamictons indicates that the flows resulting in the deposition of facies 10 were sometimes erosive.

Deposits of facies 10 are interpreted as proglacial sediments that have been more reworked by fluvial action and probably deposited in slightly more distal locations than facies 9. Coarser diamicton matrix textures, better sorting of lenses, relative abundance of sand and gravel, and lower proportion of diamicton all support this interpretation. The high, preferential upvalley dips of clasts present in facies 10 (Fig. 5) may reflect crude imbrication in the deposits. Such imbrication may be the result of the high number of clasts in contact with each other compared to facies 9. Imbrication of the a-b plane has been observed in some debris flow deposits (Naylor 1980).

The trough-shaped lenses of sorted sediment diagnostic of facies 10 (Fig. 13) were likely deposited in proglacial stream channels that developed on the surface of previously deposited sediments. This interpretation is based on the plano-concave geometry of the lenses, the absence of faulting and deformation (indicating deposition over ice-free sediments), and the presence of bedding and laminae that conform to the trough-shaped floor of the lenses. The latter are very similar to channel fill structures observed in many subaerial streams (Reineck and Singh 1980). As surface stream channels became blocked or buried by subsequent flows, new channels formed elsewhere. This explains the random distribution of sand and gravel lenses throughout this facies.

Facies 10 occurs frequently as the uppermost unit in many exposures in the Jasper region and probably was commonly deposited during ice retreat. The abundance of meltwater expected during the retreat stage was probably responsible for the relatively large amount of sorted sediment associated with surface occurrences of facies 10.

Fig. 13. Massive diamicton (Dm) of facies 10 containing trough-shaped sand (St) and gravel (G) lenses. The near equal proportions of diamicton and sorted sediments makes this deposit transitional with facies 11. Pick for scale.

### 3.11 Facies 11: Sandy diamicton with sand and gravel beds

Facies 11 and 12 grade into each other and are defined as end members of this gradational suite. Facies 11 is characterized by unsorted to very poorly sorted sandy diamicton interbedded with sands and gravels. Sorted sediments clearly dominate over diamicton. The diamictons are generally matrix-supported, show little or no stratification, and contain some striated clasts. Vertical and horizontal changes in grain-size distribution, sorting, and bed orientation are common in the sands and gravels. Strata commonly exhibit dips on the order of 20° to 40°, and are frequently folded and faulted. The thickness of sand and gravel beds associated with facies 11 and 12 usually is on the order of meters. The beds are bounded on their upper and lower surfaces by diamicton.

Deposits of facies 11 are mainly supraglacially derived, as indicated by the clast angularity, general lack of striations, enhanced sorting, coarse textures, low density, and dominance of nonlocal, far-travelled lithologies characteristic of modern supraglacial tills (Sharp 1949; Boulton 1976; Boulton and Eyles 1979). However, some diamictons of facies 11 are dominated by clasts of local lithology, suggesting that this facies may also be locally derived.

Diamictons of facies 11 are believed to have been deposited by debris flows in ice-marginal environments. The interbedding of sorted sediments with the diamictons indicates that individual flow events were isolated by fluvial activity. The presence of deformation, load, and fluid injection structures (Table 1) suggests that the sediments were saturated, as would be expected if deposition was associated with fluvial activity. The presence of diamicton beds that conform to the geometry of adjacent stream channel deposits (Fig. 13) indicates that some debris flowed into active meltwater stream channels, temporarily or permanently interrupting fluvial sedimentation at the site. The abundance of stratified sands and gravels relative to diamicton indicates that fluvial processes dominated the depositional environment.

Poor sorting, absence of stratification, low total clast content, and presence of striated and angular clasts in some diamictons of facies 11 (Table 1) suggest that they were deposited shortly after being released from their ice source and underwent little or no resedimentation or fluvial reworking after deposition. The high variability in sorting and grain size of the sands and gravels indicates that they were deposited in streams with large fluctuations in sediment input and discharge. Such fluctuations are typical in the proximal reaches of glacial outwash streams, which are flood prone and subject to large diurnal and annual changes in discharge (Boulton and Eyles 1979).

The geomorphic association of this facies with the upper portions of terraces located along valley walls indicates that they may be lateral kame terrace deposits. The proximity of glacial ice during the deposition of facies 11 is supported by the presence of folded, faulted, and highly dipping beds (suggesting collapse caused by melting of adjacent ice or pushing by active ice).

### 3.12 Facies 12: Gravelly diamicton with sand and gravel beds

In facies 12, poorly sorted, gravelly diamictons occur as horizontal beds of uniform thickness. They are interbedded with, and grade laterally into, horizontally stratified and planar and trough cross-stratified sands and gravels (Levson and Rutter 1986, Fig. 9). Ripple bedding also occurs in the sands. Palaeocurrent directions are usually similar to those in present-day streams (i.e., parallel to the valleys), but in some areas bedding dominantly dips toward the valley side. The stratified sediments are generally well to very well sorted. Clasts are sometimes imbricated and mostly rounded to well rounded in the gravels. The diamictons are usually clast-supported. They have higher total clast contents and coarser matrix textures than diamictons of facies 11.

Diamictons of this facies are interpreted as debris flow deposits. The higher clast contents, coarser matrix textures, and better sorting suggest a greater degree of reworking than in facies 11. The moderately sorted, stratified, and imbricated gravels associated with facies 12 diamictons are interpreted as proximal outwash deposits. The proximity of a glacier is indicated by the presence of transverse palaeocurrent directions, which suggest that ice was occupying the valley center. Also, the gradational nature of these outwash deposits with the diamictons suggests that they originated in the same or adjacent depositional environment. Boulton and Eyles (1979) and Lawson (1979a, 1981a) interpret sediments that are similar in nature to facies 12 diamictons and are often associated with proximal meltwater streams as resedimented tills.

The variability in internal characteristics of diamictons of this facies is probably the result of variations in the amount of resedimentation prior to deposition. In this facies the gravelly matrix textures, high clast content, general absence of striations, and relatively high degree of sorting and clast roundness all indicate significant reworking of the diamictons by water and gravity before final deposition. Resedimentation of deposits, often resulting in the removal of fines and striations, and the concentration and rounding of clasts is very common in the marginal regions of modern glaciers (Boulton and Eyles 1979; Lawson 1979a, 1981a).

Features common to multiple superimposed flows (see discussion of facies 10) are generally not present within the diamictons, although local, weakly developed scour surfaces and some stratification within the diamictons suggest that multiple or "pulsating" events might have occurred.

For reasons similar to those discussed for facies 11, diamictons of facies 12 are believed to be mainly supraglacially derived. However, in areas where the local bedrock is soft, poorly indurated shales and siltstones, facies 12 is often dominated by clasts of these lithologies. This suggests that the origin for some of the debris flows, which resulted in the deposition of facies 12, may have been push moraines formed in the soft local bedrock. The angularity of the clasts and lack of striations suggest that it is unlikely that the debris was transported to the ice surface along shear zones before flowing as observed by Boulton (1968) on Vestspitsbergen glaciers. In general, deposits of this variety occur at lower positions in the valley than other deposits of facies 12 that have been supraglacially derived.

The significant amount of washing that may have occurred in some diamictons of facies 12 indicates that they are not "true tills" as defined by Dreimanis (1982a). They are included in this study because of their gradational nature, however, with more typical resedimented diamictons interpreted as flow tills. They may have about a 20% higher local clast content than the associated gravels, and generally have more angular clasts.

# 4 FACIES ASSOCIATIONS

A hypothetical facies sequence representing the deposits of a single glacial event is given in Figure 14. The sequence is based on actual observations on the frequency, thickness, and stratigraphic occurrence of each of the facies shown, and is therefore representative of frequently occurring vertical facies relationships. Facies 1 to 4 commonly are vertically and laterally gradational with each other, and are often conformably overlain by facies 5 and/or 6 (Fig. 14). The lower contacts of facies 1 to 4 are all sharp and planar when they overlie sorted deposits (commonly sands and gravels) or one or more of facies 8 to 12. Facies 7 usually occurs at or near the surface. It often overlies facies 5 and/or 6 and grades laterally and vertically into facies 10. Lower contacts vary from gradational to clear and undulatory, but are never sharp and planar. Facies 8 to 12 are commonly the lowest exposed diamicton facies in most sections. Facies 9 is the most common facies to lie at the erosive base of massive diamictons of facies 1 to 4, and is invariably conformably underlain by facies 10 (Fig. 14).

These observations indicate that two main facies associations occur. Facies 1 to 6 occur in a variety of gradationally associated sequences, as do facies 7 to 12 (Fig. 14). One or more of the six facies within each of these sedimentary associations may be absent at any one locality. In addition, the actual vertical sequence of facies within each association varies from place to place, although some relationships are more common (as outlined previously). The two facies associations are generally separated by erosional contacts. Facies 1 to 6 are all interpreted as subglacial tills (Table 1). The sharp, planar contact at the base of the sedimentary association of facies 1 to 4 is probably the result of erosion at the glacier base. Facies 5 and 6 are believed to form initially by meltout of debris from positions higher within the glacier than facies 1 to 4, explaining why they conformably overlie facies 1 to 4 and why pebble fabric strengths (Fig. 9) are lower than in facies 1 to 4. Facies 7 to 12 are interpreted as ice-marginal sediments (Table 1) and may have been deposited during either advance or retreat of ice in the region. Consequently, these deposits may be found both overlying and underlying diamictons interpreted as subglacial tills (facies 1 to 6). A supraglacial origin for facies 7 explains its common surface

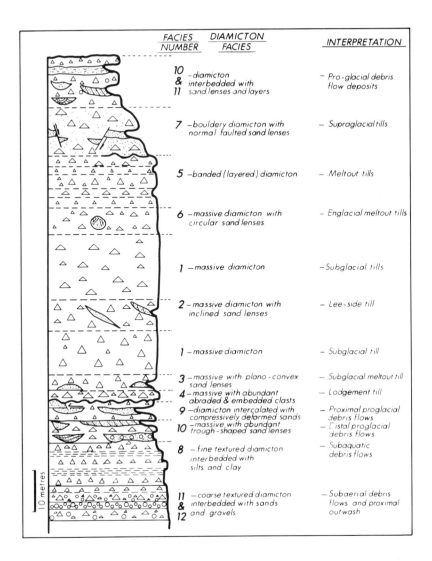

| FACIES NUMBER | DIAMICTON FACIES | INTERPRETATION |
|---|---|---|
| 10 & 11 | – diamicton interbedded with sand lenses and layers | – Pro-glacial debris flow deposits |
| 7 | – bouldery diamicton with normal faulted sand lenses | – Supraglacial tills |
| 5 | – banded (layered) diamicton | – Meltout tills |
| 6 | – massive diamicton with circular sand lenses | – Englacial meltout tills |
| 1 | – massive diamicton | – Subglacial tills |
| 2 | – massive diamicton with inclined sand lenses | – Lee-side till |
| 1 | – massive diamicton | – Subglacial till |
| 3 | – massive with plano-convex sand lenses | – Subglacial meltout till |
| 4 | – massive with abundant abraded & embedded clasts | – Lodgement till |
| 9 | – diamicton intercalated with compressively deformed sands | – Proximal proglacial debris flows |
| 10 | – massive with abundant trough-shaped sand lenses | – Distal proglacial debris flows |
| 8 | – fine textured diamicton interbedded with silts and clay | – Subaquatic debris flows |
| 11 & 12 | – coarse textured diamicton interbedded with sands and gravels | – Subaerial debris flows and proximal outwash |

10 metres

Fig. 14. Hypothetical sequence of glacial sediments deposited during a single glacial advance-retreat cycle in a mountain area. Facies 8, 11, and 12 are conformably overlain by facies 10, which grades upward into facies 9. A sharp erosional unconformity separates all the preceding deposits from facies 1 to 4, which occur as a complexly interbedded association conformably overlain by facies 5 and 6. Facies 7 commonly occurs at the top of the sequence but locally grades laterally or vertically into facies 10 and/or 11. The entire sequence is interpreted as proglacial deposits overlain successively by subglacial and supraglacial tills and proglacial debris flows. Left-hand side of column shows typical weathering profile. See Fig. 2 for legend.

occurrence as well as its lateral association with facies 10, interpreted as proglacial deposits. Similarly, the interpretation that facies 9 represents proglacial deposits subsequently overridden by glacial ice adequately explains why it is often unconformably overlain by diamictons of facies 1 to 4 and conformably underlain by facies 10.

The applicability of using facies associations to decipher the stratigraphic significance of multiple diamicton units has been demonstrated by Levson (1986) and Levson and Rutter (1986). They showed, through facies analysis, that a number of lithologically and sedimentologically distinct diamicton units in the Portal Creek area (Fig. 2) probably represent the deposits of only two distinct glacial advances. Multiple glacial events may be recognized by repetition of part or all of the expected facies sequence for a single glacial event. In particular, the successive, unconformable appearance of associations of diamictons interpreted as subglacial tills suggests that multiple glacial advances may have occurred.

## 5 SUMMARY

Twelve diamicton facies are recognized in Jasper National Park, Canada (Table 1). They are distinguished on the basis of objective field criteria, primarily the nature and structure of sorted and stratified sediments commonly associated with the diamictons. Environmental interpretations of the glacial diamictons are based on information currently available from process studies in modern glacial environments. Facies 1 to 6 are interpreted as subglacial tills mainly because of the dominance of clasts of local lithology, unimodal pebble fabrics with strong preferred orientations parallel to the valley, abundance of striated and faceted clasts, rarity of associated sand and gravel lenses, fine-grained matrices, and low clast contents (Table 1). Facies 1 is entirely massive and of indeterminate origin, but was probably deposited mainly by subglacial meltout and/or lodgement processes. Facies 2 is distinguished by the presence of sporadic, undisturbed sand and gravel lenses and beds that exhibit a preferred down-valley dip. Diamictons of this facies are believed to be the result of subglacial flow, meltout, and possibly also lodgement on sloping surfaces in lee-side localities. Plano-convex lenses within facies 3 suggest that the associated diamictons were deposited mainly by subglacial meltout. Abundant, heavily striated faceted and embedded clasts in facies 4 indicate deposition by lodgement. Diamicton strata within facies 5 are probably the inherited product of meltout of debris layers in the lower portion of a glacier. Highly disturbed, nearly circular sand and gravel lenses typical of facies 6 may have formed in positions entirely encased within ice near the glacier base. The associated diamictons were probably deposited by meltout of basal and/or englacial debris (Table 1).

Diamictons of facies 1 to 6 are invariably dominated by clasts of local lithology. In contrast, facies 7 diamictons are characterized by an abundance of boulder-sized clasts of nonlocal origin. In addition, facies 7 is distinguished by the presence of highly disturbed (usually normally faulted) lenses and beds of sand and gravel. These features indicate a supraglacial origin for facies 7 (Table 1). Massive diamicton beds of facies 8 are interbedded with horizontally laminated clays and silts and are interpreted as proglacial subaquatic debris flow deposits. Facies 9 consists of intercalated diamicton, sands, and gravels (with compressive deformation structures) interpreted as proglacial deposits overridden by a glacier. Facies 10 is characterized by diamicton with abundant, relatively well defined, trough-shaped sand and gravel lenses, interpreted as ice-marginal stream channel sediments deposited intermittently with proglacial debris flow deposits. Facies 11 is characterized by unsorted to very poorly sorted sandy diamicton interbedded with sands and gravels. These sediments are interpreted as subaerial till flow and debris flow deposits interbedded with proximal glacial fluvial outwash. In facies 12, poorly sorted, gravelly diamictons occur as horizontal beds of uniform thickness interbedded with horizontally stratified and planar and trough cross-stratified sands and gravels. They are interpreted as subaerial debris flow deposits interbedded with proximal outwash sediments (Table 1).

Two main sediment associations are recognized. Facies 1 to 6 comprise what is believed to be a subglacial sediment association (possibly derived in part from basal and even englacial debris). Facies 7 to 12 are interpreted as a supraglacial/ice-marginal sediment association (Table 1). Contacts between facies within each sediment association generally are gradational, whereas contacts between the two associations usually are erosional.

Interpretation of observed vertical facies sequences and associations are used to develop an interpretive model for a single glacial event (Fig. 14). Distal ice-marginal sediments progressively grade upward into more proximal deposits and eventually are unconformably overlain by subglacial tills. These in turn grade upward into basal, supraglacial, and/or proglacial deposits. The lower contacts of diamicton facies associations interpreted as subglacial tills invariably are sharp and planar, indicating erosion at the base

138

of an active glacier. Contacts at the base
of facies interpreted as supraglacial
and/or ice-marginal sediments vary from
conformable to erosional. In the latter
case, they were probably fluvially eroded
by meltwaters and not by glacial ice. In
complex stratigraphic sequences, multiple
glacial events may be recognized by repetition of part or all of the expected
facies sequence for a single glacial
event.

ACKNOWLEDGEMENTS

Funding for this research was provided by
the Boreal Institute and by N.S.E.R.C.
Parks Canada permitted the collection of
samples in Jasper National Park. N. Catto,
D. Liverman, and D. Proudfoot provided
very useful scientific and editorial
assistance in drafting this paper. H.A.K.
Charlesworth developed the computer
program for fabric data analysis. Field
assistance was provided by A. Gambier and
B. Levson.

REFERENCES

Bobrowsky, P.T., V. Levson, D.G.E.
Liverman, and N.W. Rutter 1987.
Quaternary geology of northwestern
Alberta and northeastern British
Columbia, Twelfth INQUA Congress field
excursion A-24. Ottawa, National
Research Council of Canada.

Boulton, G.S. 1968. Flow tills and related
deposits on some Vestspitsbergen
glaciers. Journal of Glaciology
7:391-412.

Boulton, G.S. 1970. On the deposition of
subglacial and meltout tills at the
margin of certain Svalbard glaciers.
Journal of Glaciology 9:231-245.

Boulton, G.S. 1971. Till genesis and
fabric in Svalbard, Spitsbergen. In R.P.
Goldthwait (ed.), Till: A symposium, p.
41-72. Columbus, Ohio State University
Press.

Boulton, G.S. 1975. Processes and patterns
of subglacial sedimentation, a theoretical approach. In H.E. Wright and J.T.
Mosley (eds.), Ice ages: Ancient and
modern, p. 7-43. Liverpool, Seel House
Press.

Boulton, G.S. 1976. A genetic classification of tills and criteria for
distinguishing tills of different
origin. Geographia 12:65-80.

Boulton, G.S. 1978. Boulder shapes and
grain-size distributions of debris as
indicators of transport paths through a
glacier and till genesis. Sedimentology

25:773-799.

Boulton, G.S. and N. Eyles 1979.
Sedimentation by valley glaciers, a
model and genetic classification. In Ch.
Schlüchter (ed.), Moraines and varves,
p. 11-25. Rotterdam, Balkema.

Dreimanis, A. 1976. Tills, their origin
and properties. In R.E. Legget (ed.),
Glacial till: An interdisciplinary
study, p. 11-50. R.S.C. Special
Publication No. 12. Ottawa, Royal
Society of Canada.

Dreimanis, A. 1982a. Genetic classification of tills and criteria for their
differentiation, and definitions of glacigenic terms. In Ch. Schlüchter (ed.),
p. 12-31. INQUA Commission on the genesis and lithology of Quaternary deposits, INQUA Work Group 1, Progress
report on activities 1977-1982. Zurich,
INQUA.

Dreimanis, A. 1982b. Two origins of the
stratified Catfish Creek Till at Plum
Point, Ontario, Canada. Boreas
11:173-180.

Dreimanis, A. and Ch. Schlüchter 1985.
Field criteria for the recognition of
till or tillite. Palaeogeography,
Palaeoclimatology, Palaeoecology
51:7-14.

Evenson, E.B., A. Dreimanis, and J.W.
Newsome 1977. Subaquatic flow till: A
new interpretation for the genesis of
some laminated till deposits. Boreas
6:115-133.

Eyles, N. 1979. Facies of supraglacial
sedimentation on Icelandic and Alpine
temperate glaciers. Canadian Journal of
Earth Sciences 16:1341-1362.

Eyles, C.H. and N. Eyles 1983. Sedimentation in a large lake: A reinterpretation
of the late Pleistocene stratigraphy at
Scarborough Bluffs, Ont. Canadian
Journal of Earth Sciences 11:146-152.

Eyles, C.H., N. Eyles, and A.D. Miall 1985.
Models of glaciomarine sedimentation and
their application to the interpretation
of ancient glacial sequences.
Palaeogeography, Palaeoclimatology,
Palaeoecology 5:15-84.

Gibbard, P. 1980. The origin of stratified
Catfish Creek Till by basal melting.
Boreas 9:71-85.

Haldorsen, S. 1982. The genesis of tills
from Astadalen, southeastern Norway.
Norsk Geologisk Tidsskrift 62:17-38.

Haldorsen, S. and J. Shaw 1982. Meltout
till and the problem of recognizing
genetic varieties of till. Boreas
11:261-269.

Kruger, J. 1979. Structures and textures
in till indicating subglacial deposition.
Boreas 8:323-340.

Kruger, J. 1984. Clasts with stoss-lee

form in lodgement till: A discussion. Journal of Glaciology 30:241-243.

Kruger, J. and I.B. Marcussen 1976. Lodgement till and flow till: A discussion. Boreas 5:61-64.

Lawson, D.E. 1979a. A sedimentological analysis of the western terminus of the Matanuska Glacier, Alaska. Cold Regions Research and Engineering Laboratory Report 79-9. Hanover, United States Army.

Lawson, D.E. 1979b. A comparison of the pebble orientations in ice and deposits of the Matanuska Glacier, Alaska. Journal of Glaciology 87:629-645.

Lawson, D.E. 1981a. Distinguishing characteristics of diamictons at the margin of the Matanuska Glacier, Alaska. Annals of Glaciology 2:78-84.

Lawson, D.E. 1981b. Sedimentological characteristics and classification of depositional processes and deposits in the glacial environment. Cold Regions Research Engineering Laboratory Report 81-27. Hanover, United States Army.

Lawson, D.E. 1982. Mobilization, movement and deposition of active subaerial sediment flows, Matanuska Glacier, Alaska. Journal of Geology 90:279-300.

Levson, V.M. 1986. Quaternary sedimentation and stratigraphy of montane glacial deposits in parts of Jasper National Park, Canada. M.Sc. Thesis. Edmonton, University of Alberta.

Levson, V.M. and N.W. Rutter 1986. A facies approach to the stratigraphic analysis of Late Wisconsinan sediments in the Portal Creek area, Jasper National Park, Alberta. Geographie Physique et Quaternaire 40(3):129-144.

Lindsay, J.F. 1970. Clast fabric of till and its development. Journal of Sedimentary Petrology 40:629-641.

MacClintock, P. and A. Dreimanis 1964. Reorientation of till fabric by overriding glacier in the St. Lawrence Valley. American Journal of Science 262:133-142.

Marcussen, I. 1975. Distinguishing between lodgement till and flow till in Weischselian deposits. Boreas 4:113-123.

Mark, D.M. 1974. On the interpretation of till fabrics. Geology 2:101-104.

McCabe, A.M., G.F. Dardis, and P.M. Hanvey 1984. Sedimentology of a late Pleistocene submarine-moraine complex, County Down, Northern Ireland. Journal of Sedimentary Petrology 54:716-730.

Miall, A.D. 1983. Glaciomarine sedimentation in the Gowganda Formation (Huronian), Northern Ontario. Journal of Sedimentary Petrology 53:477-491.

Naylor, M.A. 1980. The origin of inverse grading in muddy debris flow deposits, a review. Journal of Sedimentary Petrology 50:1111-1116.

Nye, J.F. 1965. Stability of a circular cylindrical hole in a glacier. Journal of Glaciology 5:567-587.

Powell, R.D. 1983. Glacial-marine sedimentation processes and lithofacies of temperate tidewater glaciers, Glacier Bay, Alaska. In B.F. Molnia (ed.), Glacial marine sedimentation, p. 185-231. New York, Plenum Press.

Proudfoot, D.N. 1985. A lithostratigraphic and genetic study of Quaternary sediments in the vicinity of Medicine Hat, Alberta. Ph.D. thesis. Edmonton, University of Alberta.

Ramsden, J. and J.A. Westgate 1971. Evidence for reorientation of a till fabric in the Edmonton area, Alberta. In R.P. Goldthwait (ed.), Till: A symposium, p. 335-344. Columbus, Ohio State University Press.

Rees, A.I. 1983. Experiments on the production of transverse grain alignment in a sheared dispersion. Sedimentology 30:437-448.

Reineck, H.E. and I.B. Singh 1980. Depositional sedimentary environments. New York, Springer-Verlag.

Roed, M.A., E.W. Mountjoy, and N.W. Rutter 1967. The Athabasca Valley erratics train. Canadian Journal of Earth Sciences 4:625-632.

Sharp, R.P. 1949. Studies of supraglacial debris on valley glaciers. American Journal of Science 247:289-315.

Shaw, J. 1977. Tills deposited in arid polar environments. Canadian Journal of Earth Sciences 14:1239-1245.

Shaw, J. 1979. Genesis of the Sveg tills and Rogen moraines of central Sweden: A model of basal melt out. Boreas 8:409-426.

Shaw, J. 1982. Meltout till in the Edmonton area, Alberta, Canada. Canadian Journal of Earth Sciences 19:1548-1569.

Shaw, J. 1983. Forms associated with boulders in melt-out till. In E.B. Evenson, Ch. Schlüchter, and J. Rabassa (eds.), Tills and related deposits, p. 3-13. Rotterdam, Balkema.

Shaw, J. 1987. Glacial sedimentary processes and environmental reconstruction based on lithofacies. Sedimentology 34:103-106.

Shreve, R.L. 1972. Movement of water in glaciers. Journal of Glaciology 11:205-214.

Walker, R.G. 1984. General introduction: Facies, facies sequences and facies models. In R.G. Walker (ed.), Facies models, 2nd ed. p. 1-11. Geoscience Canada Reprint Series No. 1. Ottawa, Geological Association of Canada.

*Genetic Classification of Glacigenic Deposits, Goldthwait & Matsch (eds)*
© *1988 Balkema, Rotterdam. ISBN 90 6191 694 1*

# Sublimation till

John Shaw
*Department of Geography, Queen's University, Kingston, Ontario, Canada*

Sublimation till requires long-term ex-
tremes of cold and aridity for its formation; therefore, its formation on Earth
is presently restricted to the Antarctic
continent. Where it does occur it has
distinctive structure, and may produce
landforms reflecting englacial debris
distribution. Sublimation till was first
predicted on theoretical grounds, and was
subsequently observed in the field. The
theoretical basis for sublimation till is
relatively simple, and given the requisite
conditions, this till will form, albeit at
a slow rate. In some respects, it is the
purest of tills, as no liquid water is
involved in its formation. It is also the
rarest of tills on Earth, although we may
speculate on its widespread distribution
on Mars.

Definition:
Sublimation till is till released by the
sublimation of debris-rich ice.

Concept:
The concept of till depositing by sublimation was introduced by Bell (1966). He was
concerned with the time taken to remove
interstitial ice from frozen debris, and
he presented the relationship

$$t = \frac{\rho_i \, \rho_a z^2}{2 \eta \, \Delta R}$$

where t=time (s), $\rho_a$ = density of air
($kg/m^3$), $\rho_i$ = density of ice ($kg/m^3$), z =
till thickness (m), $\eta$ = viscosity of air
(mPa.s), $\Delta R$ = difference in vapour density
between the top of the ice-rich debris and
the top of the ice-depleted till. It
should be noted that vapour transport
occurs only between the top of the ice-
rich debris and the atmosphere.
Consequently, sublimation till is purely a
supraglacial deposit.

The theory of sublimation till is quite
simple. Once a glacier in a polar arid
environment has a debris cover thicker
than the thickness of the active layer,
there can be no further supraglacial
melting. Removal of ice, provided that the
surface temperatures remain below
freezing, can only occur by sublimation,
the direct transition from the solid to
the vapour state. Shaw (1977) described
this process for the Dry Valleys, McMurdo
Sound region, Antarctica, and presented
photographic evidence to suggest deposition of till by sublimation near the
Antarctic Peninsula (Shaw 1985). In a
recent discussion Eyles, Eyles, and Miall
(1986) use the calculations of Robinson
(1984) as well as personal communications
with Robinson (1985) to dispute the
occurrence of sublimation as a till-
forming process. They argue that there are
probable areas of basal melting at the bed
of Taylor Glacier and, therefore, till
deposition occurs by basal melt-out. Basal
melting almost certainly occurs at Taylor
Glacier, but it does not occur around the
margin where till is being released
supraglacially. Nor is basal melting
occurring around the margins of the
Lacroix Glacier, where a ground temperature of -13°C was measured at a depth of
one metre. It should not be considered
that sublimation till is the only till
type produced in arid polar environments.
But for some debris-covered glaciers in
areas that have had mean temperatures well
below freezing for many thousands of
years, ablation by sublimation must be
significant.

Derivation:
Sublimation tills may form from debris
transported basally or englacially.

Position:
Till to the depth of the active layer may
be melt-out or mass-movement deposits.
Sublimation till may therefore occur
beneath other supraglacial till. It may
rest upon subglacial till formed by lodgement or melt-out.

Thickness:
According to Bell (1966), almost 2000

Fig. 1. Sublimation till, Vega Island, Antarctic Peninsula. The sublimation till is above layered and steeply dipping debris-rich ice. Structures in the debris-rich ice are preserved in the till.

years are required for the release of 1 m of till, and about 7000 years for 2 m. The process clearly becomes less effective as till thickness increases.

Structure:

Till release is essentially a freeze-drying process, with the result that the finest structures of debris bands are preserved in sublimation till (Fig. 1). There is some flattening of structures with loss of ice volume. In situ stratified sediment may result from piping where surface meltwater streams cut down into a region of sublimation till formation (Healy 1975).

Fabric:

Till fabrics should be like those of melt-out tills.

Grain size:

Release of till is such a passive process that the grain-size distribution of the till is not affected by deposition. The grain size will therefore reflect the entrainment and transport processes undergone by the debris.

Surface expression:

The surface expression of sublimation till should reflect the volume of debris in the original glacier. Where debris is concentrated by stacking, ridges occur. Where the volume of debris is uniformly distributed in the ice, a till plain occurs.

REFERENCES

Bell, R.A.I. 1966. A seismic reconnaissance in the McMurdo Sound region Antarctica. Journal of Glaciology 6:209-221.

Eyles, N., C.H. Eyles, and A.D. Miall 1986. Lithofacies types and vertical profile models; an alternative approach to the description and environmental interpretation of glacial diamict and diamictic sequences. Reply. Sedimentology 33:152-155.

Healy, T.R. 1975. Thermokarst -- a method of de-icing ice cored moraine. Boreas 4:19-23.

Robinson, P.H. 1984. Ice dynamics and thermal regime of Taylor Glacier, South Victoria Island, Antarctica. J. Glaciol. 30:153-160.

Shaw, J. 1977. Tills deposited in arid polar environments. Canadian Journal of Earth Sciences 14:1239-1245.

Shaw, J. 1985. Subglacial and ice marginal environments. In G.M. Ashley, J. Shaw, and H.D. Smith (eds.), Glacial sedimentary environments, p. 7-84. SEPM Short Course 16. Tulsa, Soc. Econ. Paleont. and Mineral.

*Genetic Classification of Glacigenic Deposits, Goldthwait & Matsch (eds)*
*© 1988 Balkema, Rotterdam. ISBN 90 6191 694 1*

# Watermorainic sediments: Origin and classification

Wojciech Morawski
*Geological Institute, Warsaw, Poland*

The term "watermorainic sediments" is proposed for a complex of sediments of diversified lithology, formed in contact with the ice by deposition of melt-out debris transported by mudflows and meltwater. These sediments are formed during deglaciation at the ice margin in crevasses and depressions amid dead-ice blocks, and in cavities under glacier ice. At first, glacial debris is melted out, then subjected to gravitational transport in mudflows depositing flow till as well as to washing and transporting by meltwaters. The latter processes produce stratified meltwater sediments varying in grain size from clay to gravel. Flow till and stratified meltwater sediments are usually interbedded, forming interfingering lenses or layers of various sizes. Watermorainic sediments possess specific structures and textures as well as lithologic characteristics. Complexes of this type, many meters thick, can be deposited during deglaciation in as short a time as one summer season.

In my opinion, flow till forms a component of such a complex of watermorainic sediments (Table 1). Therefore, the whole interfingering complex of sediments, developed in the same place and time and caused by the same processes, should be classified as glacial sediments instead of a single element, the flow till. Moreover, watermorainic sediments deposited by gravitation and water transport should not be classified as till in the strictest sense.

Watermorainic sediments (including flow till) should be considered as primary or original sediments. These sediments comprise a diamicton that has been melted out from ice and subjected to repeated gravitational and water transport as well as to varying sorting. Such processes continue until full stabilization, that is, deposition on a solid substrate (not on ice) and attainment of dynamic stability. Before this, the melted-out glacial debris, already forming a diamicton, is under transport and should not yet be treated as a sediment. Thus, gradual stabilization is to be considered a multiphase sedimentation process but not a result of redeposition.

On the basis of morphologic, stratigraphic, and structural location, a primary glacial watermorainic sediment (i.e., flow till) has to be strictly separated from lithologically similar but secondary deposits of slope origin, such as various redeposited types of tills. Such later redeposited sediments should be included in a genetic classification of secondary, slope deposits.

Meltwater saturates a diamicton that has melted out from ice and enables it to move as mudflows. At the same time, meltwater washes and segregates this material at its primary location and later during its flowage. The sorted material is deposited in local depressions on the ice. With continued melting of the ice, changes in the morphology of the substrate occur and the material remobilizes, flows, and is remixed. Such repeated processes wash and sort the diamicton to various degrees; fine fractions are taken away. This process sometimes results in stratification of flow tills or their gradual transition into typical aquatic sediments, formed by transport and deposition of dispersed grains.

Individual lenses or layers of stratified meltwater sediments are usually similar to typical outwash or ice-lake deposits, but a detailed analysis of their primary structures -- poor sorting, presence of fragments, rafts, lenses, and lumps of flow till -- and granulometric composition suggests a very short transport distance.

Complexes of watermorainic sediments -- alternating lenses and layers of flow till and stratified meltwater sediments -- are

Table 1. Watermorainic sediments within the genetic classification of glacial deposits.

| Facies | Process | Position |
|---|---|---|
| A. Morainic sediments | Lodgement till<br>Melt-out till<br>Sublimation till | |
| B. Watermorainic sediments | Flow till<br>Stratified meltwater sediments | Englacial<br>Subglacial<br>Supraglacial<br>Lateral<br>Proglacial |
| C. Glacioaqueous sediments | Fluvioglacial sediments<br>Limnoglacial sediments | |

commonly subjected to gravitational displacements during dewatering and before reaching a dynamic stabilization. Within such complexes, synsedimentary flow and load deformations appear, confirming their homogeneity and genetic consistence.

Watermorainic sediments formed deposits on the outside of terminal moraines and in areas of kame-and-kettle deglaciation. They also comprise, partly or entirely, various glacigenic landforms such as moraines, eskers, or kames.

Watermorainic facies are transitional between morainic facies and glacioaqueous facies. Boundaries between watermorainic sediments and morainic, fluvioglacial (outwash), and ice-lake (limnoglacial) sediments are frequently indistinct and can be difficult or even impossible to identify. A combination of geomorphologic, stratigraphic, structural, textural, and lithologic data is necessary for precise definition of these boundaries.

Genetic Classification of Glacigenic Deposits, Goldthwait & Matsch (eds)
© 1988 Balkema, Rotterdam. ISBN 90 6191 694 1

# Protalus till

W.P. Warren
*Geological Survey of Ireland, Dublin, Ireland*

Protalus ramparts (Bryan 1934), probably first described by Kinahan (1894) who termed them clocha sneachta, are now well known in the literature. They are usually specifically restricted to mounds of talus formed at the base of a snow patch. Washburn (1979) clearly excluded ridges formed at the margin of an ice sheet. That talus does collect at the margins of glaciers is obvious. It may occur as a lateral deposit that accumulated at the margin of the glacier directly from the hill slope (Warren 1979). In this case it is not a till. It may also slide, topple, roll, bounce, or fall over the glacier surface to accumulate at the glacier margin. In this case it is a till, having been deposited by or from glacier ice. Where glacially transported debris is also accumulating, the talus debris may not retain its distinctiveness, but where there is little or no other marginal deposition the talus will accumulate as a distinctive deposit identical to protalus at the margin of a snow bank.

In southwest Ireland at Cummeenmore, such deposits pass laterally into meltout deposits in a moraine ridge (Warren 1979). The process envisaged at this locality is identical to that which produces a protalus rampart at a snow patch -- frost-riven blocks fall from the backwall onto the ice surface and slide, roll, bounce, or fall to the foot of the ice canopy (Fig. 1). In this situation the ice need not transport the debris in any way. If the block rests on the ice surface for any period, however, it may be transported by the ice, and it may then resume its journey by sliding, rolling, bouncing, or falling until it is deposited. In the first case, the deposit is primarily a talus, having simply used the ice as a canopy over which it moved. In the second, the deposit moves towards Lawson's (1986) category "ice-slope scree or talus/ice-slope colluvium/dump moraine." In either case, where the block has moved over the surface

of the ice and the primary movement from source to final deposition is gravitational, I propose the term protalus till. It is a till because it is deposited from glacier ice. I use the term protalus rather than talus because of the similarity in process between the accumulation of protalus rampart deposit (protalus) and the till being defined here.

In many cases it will be impossible to distinguish between a nonglacial protalus deposit and protalus till when reconstructing depositional environments from geological evidence. The sediments will have no distinguishing features, but in many situations it is possible to conclude that the canopy/agent of deposition was glacier ice, owing to the position of the sediment relative to the source or to diagnostically glacigenic sediments. This is the process I described at the meeting of the Commission on the Genesis and Lithology of Quaternary Deposits in the Southern Pyrenees in 1983. Its definition and the term "protalus till" owe much to the discussion at that meeting.

## REFERENCES

Bryan, K. 1934. Geomorphic processes at high altitudes. Geogrl. Rev. 24:655-656.
Kinahan, G.H. 1894. The recent Irish glaciers. Ir. Nat. 3:236-240.
Lawson, D. 1986. Glacigenic mass movement deposits. Unpublished manuscript, circulated with INQUA Commission 2, Workgroup 1 (2A) Circular 28.
Warren, W.P. 1979. Moraines on the northern slopes and foreland of the MacGillycuddy's Reeks, south-west Ireland. In Ch. Schlüchter (ed.), Varves and Moraines, p. 223-236. Rotterdam, Balkema.
Washburn, A.L. 1979. Geocryology: A survey of periglacial processes and environments. London: Edward Arnold.

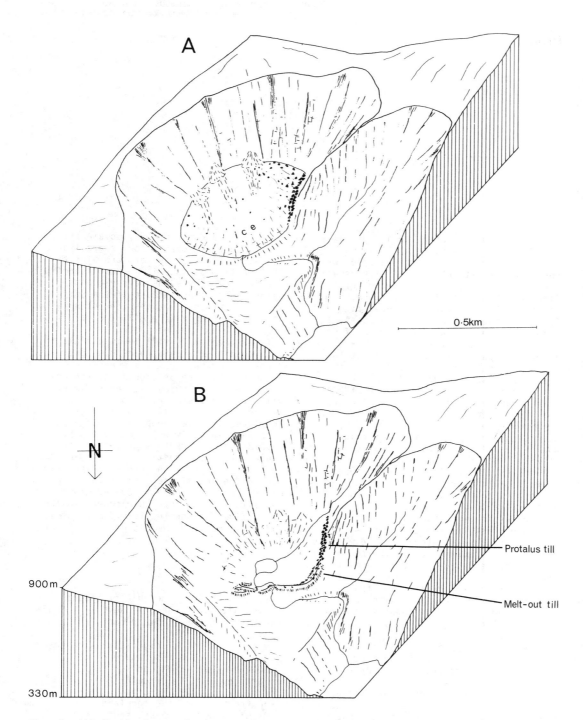

Fig. 1. (A) Small cirque glacier showing surface debris from back and side walls accumulating at the margin; (B) Deposits remaining in the cirque basin after the glacier has melted. Protalus till has accumulated in the corner of the cirque most shaded from the insolation (northeast facing). These block diagrams are drawn from a map of the Cummeenmore cirque basin and its deposits in southwest Ireland (see Warren 1979).

146

Genetic Classification of Glacigenic Deposits, Goldthwait & Matsch (eds)
© 1988 Balkema, Rotterdam. ISBN 90 6191 694 1

# Glacigenic resedimentation: Classification concepts and application to mass-movement processes and deposits

Daniel E.Lawson
*Cold Regions Research and Engineering Laboratory, Hanover, N.H., USA*

ABSTRACT: Sedimentation in the glacial sedimentary environment results from a complex interaction of processes acting on or with ice, water, and sediment. Studies of both modern glaciers and ice sheets and of the products of past glaciations indicate the importance of sedimentologic analyses using both descriptive and genetic terminology, with the latter dependent on the former. Facies and facies associations must be defined based on an assemblage of properties and related observations in order to interpret fully the glaciologic, climatologic, hydrologic, sedimentologic, and other information contained in glacigenic sedimentary sequences, and thus to reconstruct the former depositional environment.

Glacigenic depositional processes are genetically grouped as either primary or secondary. Primary processes release debris from the glacier directly and form deposits described as diamictons and classified genetically as till. Secondary processes within the glacial environment are those that cause resedimentation of glacial materials. They may mobilize, rework, transport, and resediment materials previously deposited by the primary till-forming processes or materials previously deposited by secondary processes that are being reworked. Classification of secondary deposits uses existing terminology whenever possible.

Secondary processes grouped generally as gravitational slope or mass-movement processes produce diamicts that may appear similar to those of tills or tillites. Glacigenic mass-movement processes that may be active in subaerial and subglacial locations include free fall, spall, slides, slumps, and sediment gravity flow. Diamictons commonly preserved from these processes include ice-slope colluvium, slope colluvium, and sediment flow deposits of various types. The external features, internal structure, related sedimentary properties, and stratigraphic and facies relationships allow their identification within glacigenic sedimentary sequences. Slump-related deformation and other features are also often associated with these mass-movement diamictons.

The identification and interpretation of glacigenic mass-movement deposits can provide important information on the former depositional environment. Their recognition in glacigenic sedimentary sequences is critical to correct environmental reconstructions.

## 1 CLASSIFICATION CONCEPTS

Any genetic classification for glacigenic deposits must consider the environment of glaciers or ice sheets as a sedimentary environment in the same sense as for a lake, stream, coast, or other environment. The classification must be based on the depositional processes as well as the resulting deposits. Because sedimentary processes in terrestrial or marine glacial settings are so complex and with potential for interaction with the sedimentary processes and materials of adjacent, nongla-

cial environments, it is imperative that the glacier or ice sheet not be seen only as a distinct entity. Ancient stratigraphic sequences necessarily record the interaction of environments; therefore, the classification must be consistent with those of other sedimentary environments. In addition, diamictons and diamictites are common in glacigenic sequences, but they can also be common components of certain nonglacial marine, lacustrine, and terrestrial sequences. These materials in particular require strict definitions and thorough descriptions of their properties for correct classification and genetic interpretation.

Consistency between classifications and recognition of the truly "glacial" processes of deposition are recognized in recently proposed, process-oriented genetic classifications (e.g., Lawson 1979a, 1981a; Boulton and Deynoux 1981; Dreimanis, this volume). These classifications separate or distinguish the products of primary depositional processes (resulting in the direct release of debris from glacier ice) from the products of secondary depositional processes (occurring subsequent to direct release, and reworked or resedimented). Both groups of processes may produce diamictons that have traditionally been classified as till. The primary and secondary distinction, however, separates those sediments that are truly glacial and unique to the glacial environment from those that are not unique and can originate in other sedimentary environments juxtaposed and interstratified with the glacial strata. Although the criteria used to separate or group primary and secondary processes and deposits (or similar terms) may differ, the classification concept remains the same.

In working at active Alaskan glaciers and with North American Quaternary deposits, I have found it workable to maintain a strict definition for the process and deposit groups. Only those thermal and mechanical processes that release debris from the ice and result in its immediate deposition are considered primary and form tills (Lawson 1979a, 1981a,b). All others are by definition secondary. From a sedimentologic point of view, primary processes produce sedimentary deposits that are genetically related to the glacier and thus may possess properties (e.g., sedimentary structures, grain orientations, stratification) that originate by, and thus relate directly to the mechanics, thermal regime, hydrology, or other aspects of the source glacier. Our current understanding of these aspects as related to primary deposits is not sufficient to interpret them in detail, but that is a goal for the future.

In contrast, secondary processes operate partly or fully independent of glacier ice and may occur anywhere within the glacial environment. From a sedimentologic point of view, they cause a resedimentation of glacially-derived materials or, for that matter, nonglacially-derived materials present or introduced into the glacial environment by other mechanisms, such as the inwash identified by Evenson and Clinch (1987). Secondary processes may erode, mobilize, transport, rework and/or redeposit sediments released or previously deposited by primary processes or those previously deposited by secondary processes that are again being reworked.

The sedimentary characteristics of secondary deposits are developed mainly as the result of erosion, transport, and deposition by these processes. In a strict sedimentologic sense, these deposits are nonglacial in origin. Thus, although the debris in glacier transport can be the major sediment source, secondary deposits do not provide direct information on the glacier or ice sheet. Their presence and properties can provide information on the local environment, such as stream hydrology, climate, and those factors that control the intensity, location, and importance of the depositional process. Only through analyses of their facies associations, stratigraphic relationships, and certain lithologic properties can the glacigenic environment of deposition be identified.

The resedimentation concept must therefore be recognized when interpreting glacigenic sequences in order to make correct environmental reconstructions. This concept also provides a useful means of distinguishing diamictons that can originate by multiple processes within the glacial environment. In the past, diamictons have commonly been lumped stratigraphically as till, either because of a lack of recognition of their possible diverse origins or an absence of criteria to identify these origins. The implied direct deposition from ice created by the "till" classification has led to incorrect interpretation of properties (such as fabric or mineralogy) and incorrect environmental reconstructions.

In order to apply a genetic classification, however, a purely descriptive analysis based upon multiple sedimentary properties of the materials as well as of various parameters including bed and deposit geometry, dimensions, stratigraphy and facies relationships must be made first (e.g., Lawson 1979a, 1981a). This initial analysis must not imply a genesis or origin to individual units, facies, or the depositional sequence as a whole (e.g., Lawson 1981a; Eyles, Eyles, and Miall 1983; Walker 1984). Descriptive properties, facies, and facies associations should be recorded and defined before genetic facies and interpretations are developed (e.g., Eyles, Eyles, and Miall 1983; Shaw 1987). Although this idealized separation may not be completely attainable in practice, bias must be minimized in describing facies and the sedimentary sequence, or the subsequent interpretation and environmental reconstruction may be suspect. This requirement has commonly been overlooked when dealing with Quaternary sediments where the "origin is obvious."

Field note descriptions and descriptive classifications (e.g., Pettijohn 1957; Friedman and Sanders 1978; Blatt, Middleton, and Murray 1980) are usually based upon one or more variations in physical properties, such as grain size, composition, structure, and other sedimentologic features applicable to the sequence under analysis. Such descriptive classifications must be flexible enough to account for the complexity inherent in the sediment, whether glacigenic or of other origin. Such descriptions tend to be individualized by geologists and may vary with the depositional sequence under study. Recently, Eyles, Eyles, and Miall (1983) attempted to develop a universal lithofacies code for vertical sequencing of glacigenic sediments, but such a code has limitations (e.g., Gibbard 1980; Dreimanis 1984; Eyles, Miall, and Eyles 1984; Kemmis and Hallberg 1984; Shaw 1987).

Dreimanis (this volume) discusses at length the types of genetic classifications that exist and the terminology commonly used; thus no attempt at proposing further improvement on genetic classification (Lawson 1981a) will be presented here. It is simply, but critically, important that the sedimentologic distinction between primary sedimentation and secondary resedimentation be realized when analyzing glacigenic sequences and developing facies models of the depositional environment. The sedimentologic differences require strict separation of deposits, facies, and facies associations in classifying them both descriptively and genetically.

## 1.1 Diamicton classification

Whereas secondary deposits exhibiting grain-size sorting and stratification are more easily identified, the origins of nonsorted or poorly sorted materials termed diamictons or diamictites (Flint et al. 1960a,b) or collectively referred to as diamicts (Harland, Herod, and Krinsley 1966) are less readily determined. Yet their presence and stratigraphic location may be crucial to interpreting the sedimentary environment of deposition. This may be particularly true for diamicts produced by gravitational mass-movement or slope processes that are common in other environments, and appear similar to tills or tillites. Further, secondary diamictons may be deposited within each of the subenvironments of the glacier (i.e., subglacial, ice-marginal, englacial/supraglacial, subaqueous, submarine). Thus, they may occur at stratigraphically important horizons within a glacigenic sedimentary sequence, often in direct association with diamictons deposited directly from glacier ice (till).

Diamictons deposited subaerially and, less commonly, subaqueously, by sediment gravity flow (and in some instances diamictons deposited by any mass-movement process) have historically been referred to as "flowtills/flow tills" (e.g., Hartshorn 1958; Boulton 1968, 1971; Marcussen 1973; Morawski 1976; Evenson, Dreimanis, and Newsome 1977; Boulton and Deynoux 1981; Hicock, Dreimanis, and Broster 1981; Huddart 1983; Eybergen 1987; Gustavson and Boothroyd 1987, among many others). There is, however, an ongoing controversy over use of this term (see Dreimanis, this volume). In general, the diamicts produced by mass-movement processes are secondary in origin. The question therefore arises as to whether a term that indicates uniqueness of the deposit to the glacial environment is appropriate or whether terminology consistent with that for other sedimentary environments should be used.

In the case of sediment gravity flows, recent studies have clearly demonstrated that both the mechanics of the flow process (Lawson 1982) and the characteristics of their deposits (e.g., Boulton 1968, 1971; Lawson 1979a,b, 1981b, 1982, in preparation; Kemmis, Hallberg, and Lutenegger 1981; Sharp 1982; DeJong and Rappol 1983) are the same as those of other terrestrial environments (e.g., Bagnold 1954; Hooke 1967; Johnson 1970, 1984; McRoberts and Morgenstern 1974; Varnes 1978; Pierson 1980, 1981; Keefer and Johnson 1983; Nemec and Steel 1984; Shultz 1984). Only the primary sediment and water sources differ in these environments. This similarity of process and deposit was actually recognized during much earlier investigations at active glaciers (e.g., Russell 1893; Lamplugh 1911; Tarr and Martin 1914; Gripp 1929). Recent studies of terrestrial Quaternary and ancient glacigenic sedimentary sequences (e.g., Ojakangas and Matsch 1980; Kemmis, Hallberg, and Lutenegger 1981; Haldorsen 1982; DeJong and Rappol 1983; Rappol 1983; Dardis 1985; Dowdeswell, Hambrey, and Wu 1985; Johnson et al. 1985; Muller, Franzi, and Ridge 1986; Dardis and McCabe 1987; Dreimanis,

Hamilton, and Kelly 1987; Hansel and
Johnson 1987; Sharpe 1987) have
demonstrated that sediment flow diamicts
and other secondary diamicts are separable
by sedimentary analyses from diamicts produced by primary processes.

Studies of glaciomarine, glaciola-
custrine, and tidewater settings also
demonstrate the similarity of mass-
movement processes and deposits to those
of nonglacial, subaqueous settings (e.g.,
Ferrians 1963; Hampton 1972, 1975; Lowe
1975, 1982; Walker 1975; Middleton and
Hampton 1976; Kurtz and Anderson 1979;
Hicock, Dreimanis, and Broster 1981;
Powell 1981, 1983; Hein and Walker 1982;
Cohen 1983; Domack 1983; Postma, Roep, and
Ruegg 1983; Prior, Bornhold, and Johns
1984; Postma and Roep 1985; McCabe 1986;
Eyles 1987; Mustard and Donaldson 1987).

This similarity led Gravenor, von Brunn,
and Dreimanis (1984) to propose only a
modified classification and terminology
for subaqueous glacigenic settings that is
clearly based upon nonglacial subaqueous
environments. Existing terminology -- such
as sediment flow, debris flow, and slurry
-- is used. As the following discussion
demonstrates, the processes and deposits
of mass-movement in the terrestrial glacial environment are also equivalent to
those of nonglacial terrestrial settings;
therefore, existing sedimentologic terms
should be used for consistency. The
modifier "glacigenic" can indicate the
glacial setting of deposition when
necessary, as Gravenor, von Brunn, and
Dreimanis (1984) proposed.

The nature of glacigenic mass-movement
processes under subaerial conditions is
briefly described in the next section, and
the characteristics of the resulting resedimented diamictons subsequently
summarized. Selected examples from the
subaerial and subglacial terrestrial
environments of the Matanuska and other
Alaskan glaciers will be presented, but no
attempt will be made to describe glacigenic subaqueous varieties of these secondary processes and deposits. Readers
should consult Evenson, Dreimanis, and
Newsome (1977), Rust (1977), Dreimanis
(1979, this volume), Powell (1981, 1984),
Wright and Anderson (1982), Postma, Roep,
and Ruegg (1983), Visser (1983a,b),
Gravenor, von Brunn, and Dreimanis (1984),
Broster and Hickok (1985), Eyles (1987),
Eyles, Clark, and Clague (1987), and
McCabe, Dardis, and Hanvey (1987), and
references cited therein for discussions
of the processes and character of the
resulting diamictons under different
subenvironments of subaqueous deposition.

## 2 GLACIGENIC MASS-MOVEMENT PROCESSES

Gravitationally driven mass-movement processes commonly occur within ice-marginal,
supraglacial, and certain subglacial locations where they may act as agents of erosion, transport, and deposition. Their
relative importance in developing the
sedimentary sequence may vary with a
number of interrelated factors that cannot
be discussed here (see e.g., Lawson 1979a,
1982; Sharp 1982). Complex movements
involve two or more of the processes
described below, such as slump followed by
flow, or spalling followed by disaggregation and flow, and commonly mobilize and
rework sediments of ice-cored marginal
terrain. Similarly, each type of process
may occur contemporaneously throughout the
ice-marginal and subglacial environments,
particularly during the summer melt
season, and may interact with one another
to develop complex stratigraphic sequences
with much lateral and vertical variability.
Slumps leading to flows are also common
adjacent to and within ponds and small
ice-marginal lakes in otherwise
terrestrial settings. The following list
includes process terms from the traditional classification scheme of Varnes
(1978), which is commonly used in other
geological fields.

### 2.1 Fall/Free fall

Individual particles or aggregates are
detached or released from steep slopes
composed of sediment or debris-laden ice
or overhanging ice, and descend to the
base of the slope by free fall, leaping,
bounding, or rolling (Fig. 1). Impact at
the base of the slope may partly or fully
disaggregate composite blocks of material.
Sediment cover failure may result initially from oversteepening caused by
buried ice melt, toe erosion by fluvial
processes, and selective entrainment of
fines by eolian processes. In addition,
steeply sloping debris-laden basal ice is
common along active ice margins, particularly during seasonal and longer-term
glacier advances.

### 2.2 Spall/Topple/Collapse

Following an upslope tensional fracturing
and failure, large blocks or segments of a
slope composed of unsaturated, partly consolidated sediment undergo a forward rotation about a pivot point, and subsequently
fall or slide to the base of the slope.

Figure 1. Buried ice exposed by sediment failure, Mantanuska
Glacier, Alaska. Overlying sediment gradually fractures and fails,
collapsing at base of ice-cored slope. Sediment is readily
saturated by meltwater and assimilated into sediment piled at base
of slope; these materials eventually fail and flow. Height of
slope is about 10 m.

Slope fracturing and failure commonly
result from oversteepening by melting of
buried ice (Fig. 2), or by ablation of an
ice core exposed in a steep slope.

## 2.3 Slide/Slump

Sediment moves downslope under the force
of gravity by slip along one or more
discrete surfaces. Rotational slides or
slumps rotate down and out along a roughly
concave upward surface, whereas translational slides move along roughly planar
surfaces, such as that defined by bedding
planes or the contact between ice and
overlying sediment (Fig. 3). Internally,
slide material remains relatively
coherent, whereas slump material is
deformed and disrupted. In both types of
movement, the head and toe regions are
affected by tensional and compressional
deformation, respectively. Failure may
take place because of seepage and
increased pore pressures caused by ice
melting, and by mechanisms such as toe
erosion that cause oversteepening of the
sediment mass. Both translational and
rotational types are common in subaerial
and subaqueous ice-marginal environments
and have been observed in subglacial cavities.

## 2.4 Lateral spreads/Retrogressive slumps

Movement, particularly in fine-grained,
matrix-supported sediment with little
strength, involves lateral extension and
upslope tensile fracture or shear that in
turn cause progressive failure in sediment
further upslope as toe support is lessened
or removed. After failure, the sediment
undergoes deformation and remolding as it
spreads laterally downslope. This process
is common in subaerial, oversaturated
sediments that are lying on a melting
buried ice surface with a low slope angle
(Fig. 4), and in subglacial sediments
within larger cavities near the ice
margin.

## 2.5 Sediment gravity flow

Generally, the sediment gravity flow process is the downslope movement of
sediment/water mixtures under the force of
gravity (Fig. 5). Sediment flows exhibit
gradational rheological behavior that
ranges in a general sense from plastic to
viscous to semifluid (Lawson 1982). Rates
of movement vary from almost imperceptible
to rapid, but are usually not turbulent.
In the glacial environment, sediment flows
originate either from recently disaggregated material lying on melting ice of the
ice-marginal zone or from recently

Figure 2. Slope (foreground) fractured by melting of buried ice; blocks rotating out and down to right as spall failure, Matanuska Glacier, Alaska. Slope colluvium lies as wedge at base of far slope near the photo center.

Figure 3. Slump in sediment 3 m thick, overlying stagnant glacier ice. Remolded, slowly flowing sediment in toe region, Matanuska Glacier, Alaska. Scale (photo center) of 2 m-length.

Figure 4. Tensile failure in fine-grained sediment (0.6 m thick) on stagnant ice surface of low-angle (<3°) slope; downslope movement, deformation, collapse, and remolding to left; sediment flow is then initiated. Scale: 0.5 m length.

released debris saturated by meltwater along and beneath the active ice margin. Flow mobilization occurs during and following porewater pressure elevation within recently disaggregated debris or glacigenic sediment generated by an increase in meltwater infiltration, melting of buried ice, or both (Lawson 1979a, 1982).

The mechanics of flow are complex, with several mechanisms of grain support as well as grain transport potentially operating in the same flow at the same time during movement (Lowe 1975, 1982; Middleton and Hampton 1976; Lawson 1979). These mechanisms may also change in importance as movement downslope occurs.

The factors determining which support and transport mechanisms operate in subaerial flows have not been fully identified. However, there is a strong relationship in the glacigenic environments between the amount of water and the degree of saturation of the matrix material at flow mobilization and subsequent flow behavior (Lawson 1982). A continuum of flow types has been observed. For example, in the subaerial flows initiated on buried ice of ice-marginal areas and characterized by low water content, high bulk density, and measurable shear strength, shear is localized in a thin zone at the flow's base. This flow type is essentially transitional in nature from the slump process to more viscous flows. The upper sediment does not generally deform, and

the strength of the matrix material supports larger particles (up to boulder size) during movement. As the volume of water in the matrix increases, sediment flows are characterized by a thicker basal zone undergoing shear and a well-defined, nondeforming central plug. Within the shear zone, traction and saltation of coarse bedload material, localized fluidization, transient turbulent mixing with flow over steep channel bed irregularities, and grain-to-grain interactions may occur. Flow material becomes liquefied when water content exceeds saturation throughout, continuous particle-to-particle contacts are lost, and other grain support mechanisms are inactive.

In different flow types, changes in certain physical properties and flow activity also correlate with water content variations. For example, as water content increases, the matrix density, flow thickness, and matrix mean grain size generally decrease, whereas the rate of movement, erosiveness, and degree of channelization generally increase (Lawson 1982). Flow remains laminar except where localized temporary turbulence is caused by sudden, sharp increases in channel bed slope.

3 GLACIGENIC MASS-MOVEMENT DEPOSITS

Diamictons produced by mass-movement processes may range from matrix-to clast-

153

A

B

Figure 5. Examples of subaerial sediment gravity flows: a) well-defined lobe of viscous, low-water-content flow during initial stages of deposition and dewatering. Scale 1 m long; b) lobate plug flow; coarse-textured ridges composed of source material not entirely incorporated and remolded prior to movement; "incised" in heavily oversaturated nonmoving sediment; c) channelized flow with well-developed plug zone; large cobbles in transport within plug; boulders that intercept bed are rolled and slid along it by force of flow mass; d) meltwater flowing in rills on surface of active sediment flow.

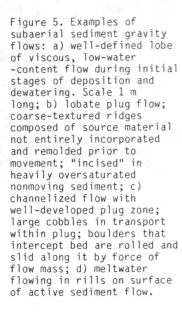

D

C

supported. Although these processes may act upon any of the materials present, the predominant grain-size distribution of these diamictons still reflects mainly the grain-size distribution of the glacier's debris. This relationship is important because size distribution may affect material properties, such as shear strength or permeability, and may thus determine which mass-movement processes are active as well as the nature of the sedimentary sequence that develops.

Local climate is important in determining rates of ablation and buried ice melt, and thus in determining the quantity and activity of meltwater that directly or indirectly interacts with these processes to develop terrestrial glacigenic sequences. The importance of this factor is clear in comparing sedimentation in the much colder and drier Antarctic (e.g., Shaw 1977; Robinson 1979) with that in warmer and wetter environments such as Alaskan coastal and inland glaciers (e.g., Lawson 1979a; Anderson, Goldthwait, and McKenzie 1986; Gustavson and Boothroyd 1987) and those of Iceland or Spitsbergen (e.g., Boulton 1972; Eyles 1979, 1983; Krüger 1982). Preservation of resedimented diamictons may likewise vary with these and other factors (e.g., Lawson 1979a:103).

The most commonly reported secondary mass-movement deposits at both active glacier margins and within Quaternary and older glacigenic sequences are those of sediment gravity flows, typically in complex association with meltwater channel and sheet flow deposits, and both primary and secondary diamictons. Lateral and vertical variability is inherent in sequences containing diamictons of mass-movement origins. Sedimentary characteristics of the various deposits primarily reflect the mechanics of mass movement and the mode of deposition; in many instances, it is the final mechanisms of movement or flow and mode of deposition that remain imprinted on the deposit.

## 3.1 Ice slope colluvium

Ice-slope colluvium (Lawson 1979a, 1981a) develops along the base of steeply sloping, debris-rich basal zone ice at the active ice margin (Fig. 6a). Less commonly, it is deposited at the base of stagnant debris-rich ice exposed as a core within ice-marginal slopes and from the roofs of subglacial cavities. Ablation of the basal ice releases debris that is then transported either by mass-movement processes of free fall, rolling, or sliding of clasts and frozen debris clots, or by thin meltwater sheet and rill flows. Thin sediment flows may also develop at the base of the slope and together, these processes interact to produce this diamicton. Most sediment accumulates as a pile at the base of the slope, but thin sheet and rill flow of meltwater may erode small (20-100 mm wide) channels and subsequently deposit thin silt and sand lenses that become interbedded within the diamicton.

### External features

The deposit is typically wedge to lenticular in cross section and extends laterally, parallel to the trend of the ice margin or degrading source slope. Dimensions usually range from tens of centimeters to several meters in thickness, up to 100-200 meters laterally, and tens of meters or more in width. Thickness and width depend largely upon the debris content of the ice and the stability of the active ice margin. The deposit surface is generally rough and irregular, mainly because of the coarse material and washing by meltwater flow (Fig. 6b).

### Internal features

Deposits range from matrix-dominated, poorly-sorted, gravelly-silty-sand to clast-dominated silty gravels. They are mainly structureless and heterogeneous, with variable density, but may contain irregular lenses and layers of well-sorted sands, silts, and clays (Fig. 7) deposited by meltwater flowing from the ablating ice face or accumulating in small pools on the deposit surface. Conversely, meltwater in the absence of significant sediment input from ablating basal zone ice (most common during the late stages of sedimentation at the ice margin) coarsens the uppermost materials, in some cases developing a gravel lag (Fig. 6b; Lawson 1981a). Similarly, localized lags may develop in low areas from meltwater flow removal of fine-grained sediment. The longest axis of pebble- through boulder-size clasts often has a weak preferred trend approximately parallel to the trend of the source slope (Lawson 1979b), although near vertical orientations may also be common. Larger clasts falling from the ice face penetrate the soft sediment, deforming strata in the same manner as ice-rafted clasts, while producing craters and sedimentary ejecta at the deposit surface. Frozen debris clots, not disaggregated by falling or toppling, may be preserved, but they are often indistinct and difficult to

A

B

Figure 6. a) Ice-slope colluvium deposited adjacent to active ice margin; basal zone ice lies left of the 1-m scale. Smooth, shiny area is oversaturated sediment released by ablation of basal ice and accumulated behind ridge of sediment. Meltwater flow away from the ice has cut small rills into this ridge. Sediment below the scale collapses as buried ice melts; b) coarse surface of ice slope colluvium actively being deposited along Matanuska Glacier, Alaska, during latter stages of sedimentation. Lowered debris content and meltwater-removal of fines lead to coarse surface materials.

Figure 7. Coarse ice-slope colluvium, Matanuska Glacier, Alaska. Internally, diamicton is mainly structureless but contains randomly dispersed lenses and discontinuous layers of sorted silt or sand, some laminated. The deposit appears coarser in texture than it is because of wind erosion of the finer matrix.

recognize later. Deformation may result because underlying ice melts, leading to oversaturation, pore-fluid expulsion, and sediment failure. Fluidization conduits and mud volcanoes may form during this dewatering process. In alpine settings, coarse, angular, supraglacially-derived sediment may become intermixed, particularly within the upper part of the deposit during the final stages of deposition.

### Stratigraphic relationships

These deposits form discontinuous linear lenses or pods within other ice-marginal and subglacial materials. They are laterally associated with other resedimented materials, such as sediment flow, channel-fill, lacustrine, and meltwater fan deposits. In vertical section, ice-slope colluvium may overlie proglacial and thin ice-marginal deposits when formed during a major ice advance. However, they are typically preserved during a period of overall ice recession coupled with annual ice marginal advances. They thus overlie and interfinger both subglacial and ice-marginal materials. The lowermost contact is nonerosional and conformable with underlying sediments.

### 3.2 Slope colluvium/Slope talus or scree/Free fall deposits

Melting of buried ice without its exposure causes spalling, toppling, fall and/or collapse of the overlying sediment cover in the ice-marginal zone (Fig. 2; Lawson 1979a). The resulting diamicton, referred to as slope colluvium, is distinct from ice slope colluvium in being internally chaotic, without the presence of lenses and discontinuous thin layers of meltwater silt and sand. Deposition of slope colluvium under dry conditions results in numerous voids up to several centimeters wide within the upper parts of the deposit; meltwater seepage at the base of the slope may saturate it, reducing its strength and ultimately increasing its compactness and density.

### External features

Deposits extend laterally, parallel to the trend of the source slope, and tend to be irregular in cross section. They range in thickness from tens to hundreds of centimeters and in length from 15 to 200 or more meters. Their overall dimensions depend mainly upon the height and length of the source slope, thickness of buried ice, and thickness of sediment covering it. The surface likewise tends to be irregular and rough, but slopewash, snowmelt runoff, and eolian processes reduce surface relief and coarsen surficial materials.

### Internal features

This diamicton is relatively loosely packed, and consists of disaggregated slope materials that may surround intact or partly fractured blocks of source material (Fig. 8). Intraformational blocks from source materials are commonly preserved in apparently random orientations, clearly shown when they possess sedimentary structures. Fractures and partings in the blocks appear fresh and angular, and may occur preferentially along bedding planes, large clast surfaces, or other material anisotropies. Uppermost slope materials are usually dry, and some wind deflation of fines may take place, but the overall grain-size distribution depends mainly on that of the source and, thus, matrix-supported diamictons will prevail when basal debris contains a significant proportion of silt. Pebble fabric appears to be random, with high angle dips possible.

### Stratigraphic associations

Slope colluvium is preserved mainly during the last phases of ice degradation and reworking in ice-marginal zones. It therefore tends to occur in the upper part of ice-marginal depositional sequences and is not typically associated with meltwater deposits. The lower contact is sharp and conformable. Sediment flow diamictons and related secondary deposits may, however, overlie slope colluvium in ice-marginal areas where the overall topographic slope is away from the active ice margin.

### 3.3 Sediment flow deposits (diamictons)

The majority of commonly preserved diamictons deposited by mass-movement processes are those of sediment gravity flow. They include deposits that have been termed flow diamicton, till flow, debris flow, mud flow or earth flow, etc. Because the flow processes involve a number of different transport mechanisms and depositional modes, flow diamictons vary considerably in their sedimentary characteristics, including the presence or absence of distinctive sedimentary structures. Usually however, such variations allow these deposits to be identified and distinguished from primary-deposited diamictons.

Subaerial and subglacial sediment flow diamictons exhibit properties that vary relatively systematically, and can be related to the water content of the sediment flow source before final deposition. When water content is lowest and sediment movement is slow and highly viscous, the materials possess strength and there is little internal deformation, allowing preservation of properties inherited from source materials. It is important to recognize that the source material may

Figure 8. Slope colluvium deposited in ice-marginal zone of Matanuska Glacier during latter stages of resedimentation. Field notebook for scale.

include just-released debris, till, previously resedimented materials, or any combination of these. Thus, properties that may be inherited by flow diamictons may or may not be primary in origin. At progressively higher water content, shearing and deformation increase in importance throughout the body of the flow, and source material properties are increasingly modified or destroyed. At the highest water content when flows actually appear fluid, strength is absent and only fine-grained sediment up to granule size may be transported. In this case, diamicton properties are entirely derived from the depositional process and the post-depositional processes of dewatering and consolidation.

In the following discussion, flow diamicton properties are described in relation to the continuum of flow processes and to properties inherited from the sediment flow source, particularly water content. Sediment flow diamicton characteristics and their origins are described in detail in Lawson (1979a:40-71; in preparation).

### External features

Dimensions of individual flow deposits are determined primarily by those of the source flow, deposits from flows with lower water content having potentially greater extent and thickness. Dimensions may also vary with the mode of deposition. Individual deposits typically range from several centimeters to 2 m or more in thickness and extend over an area as small as a few square meters to one of a thousand square meters. Coalesced flow deposits may, however, cover much larger areas; at Matanuska Glacier, coalesced flow deposits extend over areas of several hundred to several tens of thousands of square meters and develop sequences up to 10 meters or more thick.

The configuration of deposits in plan view changes with water content. Flow deposits with lower water content have a well-defined, lobate shape with one or more arcuate ridges at the surface, whereas liquefied flows develop a relatively smooth surface and a shape determined by topography and configuration (Lawson 1979a). The slope angle at which deposition takes place also decreases with increasing water content of the source flow.

Various types of features developed during movement and deposition may be preserved on the flow deposit surface. Surface features, such as flow lineations around obstructions, crenulations, and protruding blocks or aggregates, may be preserved when there is sufficient strength to the body or central plug of the flow (Fig. 9). Likewise, internally deformed lobate deposits develop from flows with low water content. In these deposits, the outermost lobe and flow-shoved marginal sediments in front of such flows can exhibit compressional deformational structures and, when saturated or oversaturated layers are overridden, diapiric-injection features develop. Pore-fluid expulsion during consolidation, particularly of liquefied flows, can occur, developing mud volcanoes on the deposit surface.

Most flow activity is accompanied by sediment-laden meltwater sheet or rill flows that deposit a thin, often discontinuous layer of laminated silt and sand on the deposit surface, which may partly or fully bury surface features. If meltwater flow is erosional, it develops anastomosing or distributary rills and channel fills. Water ponded on sediment flow surfaces may produce thin, clay-rich depressional fills. Desiccation cracks may also develop later during consolidation and dewatering, while diurnal freezing may develop needle-ice casts.

### Internal features

The internal structure results from the mechanics of grain support and transport active during movement and the modes of deposition and consolidation. Because of the continuum of flow processes, zones or horizons have been identified that relate to source flow mechanics and properties (Fig. 10), each of which may or may not be present in any given sediment flow diamicton (Lawson 1981a). Contacts between these zones, with the exception of the sharp upper contact with surface-deposited meltwater silt and sand, are generally gradual or indistinct, irregular or deformational in configuration, and nonerosional and conformable with underlying sediments.

### Basal zone

A gravel-rich layer, composed mainly of pebble- and cobble-size clasts in a structureless, silty sand matrix, is usually developed from material transported as bed load by traction and saltation in the lowermost part of the flow (Fig. 11). Some of these clasts may also have settled out during movement or deposition because of strength reduction of the flow material by localized liquefaction, temporary turbulence, or shearing. Prolate clasts sometimes show a poorly defined orientation

A

B

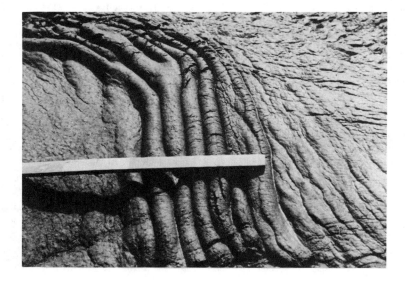

C

D

Figure 9. Examples of surficial features on sediment flow deposits: a) arcuate compressional ridges in lobe of former viscous flow (scale 2 m long); b) flow lineations around large boulder obstructing movement of a flow with well-developed plug zone. Coarse-textured material is source sediment only partly disaggregated during mobilization; c) small scale compressional folds developed in sediment obstructing movement of a sediment flow (scale 40 cm long); d) mud volcanoes deposited around pore fluid expulsion conduits during dewatering of sediment flow.

and parallel-to-slope imbrication (Lawson 1979b).

### Shear zone

The shear zone is texturally heterogeneous and develops from shearing and related mechanisms including clast interactions and dispersion. There generally appears to be less coarse material (pebble-size and larger) in this horizon than in the basal zone beneath it, if it is present. The frequent grain collisions that occur during shearing apparently account for a poorly developed fabric to prolate pebbles in this zone; blade or disc pebbles tend to lie subparallel to the flow surface and may be weakly imbricated upslope. Thick shear zones or flows completely in shear may develop inverse grading (Fig. 12). Occasionally, thin, discontinuous lenses of silty clay lie subparallel to shear and may appear "smeared" by this process. Little travelled rip-up clasts may also be present.

### Plug zone

The plug zone results from the non-deforming region within a source flow if such a zone is present. The lack of shear or other deformation produces a massive deposit that retains properties of the material (generally remolded) from which the sediment flow was derived. The strength of the matrix material maintains clasts up to boulder size in suspension during movement; these clasts can occur within the uppermost sediment of the plug zone and protrude into overlying horizons or into sediment deposited subsequent to the flow event. Aggregates and blocks of contorted, laminated silts, stratified sands, and other sediments (mainly material that was not disaggregated during remolding) are also transported or rafted here without disaggregating, and are thus preserved. Clusters of granular, non-cohesive sediment eroded from channel walls may be similarly preserved. Rafted clasts are particularly noticeable when structurally or texturally distinct from the remainder of the flow material and may protrude above the plug. Gravel-size particles of prolate to spheroidal shape, as well as aggregates, have random orientations and dispersal (Lawson 1979b).

### Dewatered horizon

This thin zone, sometimes found at the top of a flow deposit, is generally structureless and contains fewer pebble-size and larger clasts than the body of the flow (Fig. 13). Clasts apparently settle out of this zone as the result of reduced

Figure 10. Idealized
examples of sediment
flow diamictons.
Water content of
sediment flow source
prior to deposition
increases from a to
e. Numbers refer to
following zones:
basal (1), shear (2),
plug (3), dewatered
(4), meltwater (5),
and liquefied (6);
properties are
described in text.

fluid expulsion occurs during deposition
and consolidation, and various features
such as fluidization conduits may result.

Meltwater horizon

Thinly laminated silts or stratified sands
are deposited by meltwater flowing in
sheets and rills over the sediment flow
both during movement and after deposition
(Fig. 11). Meltwater pooled in
depressions, like those between
compressional arcuate ridges and blocky
plug surfaces, produces a mostly massive,
fine-grained lens on the flow surface.
Subsequent erosion by meltwater after
deposition can result in discontinuous,
laminated lenses that lie along the original flow surface. Rainwash and snowmelt
runoff can also coarsen the deposit surface and develop a thin (one- to two-
particle thick) discontinuous lag.
Commonly, the meltwater horizon is important in identifying the upper surface of
flow diamictons in stratigraphic
sequences.

Related features

Movement may cause shearing and deformation of sediments beneath and adjacent
to the sediment flow, particularly flows
of high density and low viscosity.
Overburden pressures generated by flow
masses moving onto saturated deposits can
cause soft-sediment deformation of these
readily deformable materials. Features
such as load casts, convoluted bedding,
and complexly folded strata are deformed
into the base of the flow and develop
beneath it. Dense, viscous flows moving by
shear in a thin region at their base below
a nondeforming plug develop snouts that
are preserved as flow noses within the
deposit. These noses are commonly defined
by dissimilar, compressionally-deformed
sediment shoved by the flow during the
final period of movement (Fig. 14).

In contrast to tills deposited by primary processes, sediment flow deposits do
not possess a well-defined prolate pebble
fabric (Lawson 1979b; Dowdeswell and Sharp
1986). In the nondeforming parts of the
flow, pebbles appear randomly distributed
but in the shearing or deforming horizons
beneath it, pebbles may possess a very
weak, bimodal or polymodal orientation
(Lawson 1979b). Within flows fully in
shear, a slightly stronger but still weak
orientation with a high degree of dispersion to individual clast orientations may
develop. A transverse-to-flow orientation
sometimes develops along the frontal and

strength as it dewaters (seepage and/or
fluidization), either during consolidation
or possibly during movement before deposition. Similar dewatering effects may be
observed locally in the plug and basal
zones of the flow deposit.

Liquefied zone

The liquefied zone may be a single horizon, or it can compose a complete flow
deposit. It usually consists of silty sand
or sandy silt that either results from
dewatering and consolidation after deposition, or solidifies from a liquefied flow.
Because of liquefaction, this zone is
generally structureless, but can exhibit
either distribution or coarse-tail
grading. A coarse basal layer, generally
of granule size or smaller particles
apparently develops as grains settle out
during movement and/or deposition. Pore-

Figure 11. Sediment flow diamicton illustrating gravel-rich zone
near its base, with discontinuous, laminated meltwater silts
locating the upper surface of the flow deposit. Lowermost contact
is diffuse and marked only by gravel concentration in basal horizon. Scale in cm.

Figure 12. Sediment flow diamicton showing inverse grading. Upper
flow surface marked by meltwater silt, most noticeable above
knife. Base of flow deposit lies upon thinly stratified sand.

Figure 13. Sediment flow diamicton with dewatered zone located at and above knife. Top and bottom of flow deposit are marked by stratified fine sand and discontinuous laminated silt, respectively.

Figure 14. Flow snout (left of center) with compressionally deformed sediments. Scale in cm and dm.

marginal edges of certain viscous flow lobes (e.g., Boulton 1971). However, if sediment flow diamictons have a preferred orientation, it is not related to the regional ice flow direction. The pebble fabric is generally one of the weakest in diamictons of the ice-marginal and subglacial subenvironments (Lawson 1979b; Domack and Lawson 1985; Rappol 1985; Dowdeswell and Sharp 1986), although a strong fabric has been measured by Derbyshire (1980) in thin flow lobes.

### Stratigraphic relationships

Sediment flow deposits are generally interstratified with other secondary deposits of both the ice-marginal and subglacial environments. Individual deposits are generally not laterally extensive over more than several hundred square meters, but coalesced flow deposits without well-defined boundaries between individual deposits may appear to be laterally continuous over much larger areas and preserved as such within stratigraphic sequences. Common stratigraphic associations are meltwater silt and sand strata, ice-slope colluvium, and fluvial sand and gravel deposits. Melt-out till may also be present in the same sequences, typically underlying a sequence of secondary deposits containing sediment flow diamictons. The overlying sequence, including sediment flow diamictons, may then exhibit deformation related to later melt-out of underlying ice. Thick sequences of sediment flow diamictons may develop in ice-marginal basins. Sediment flow diamictons may also overlie subglacially deposited till when deposited in subglacial cavities or along an active but receding ice margin during deglaciation.

### Slump-related features

Structural features developed by slumping may be found in association with flow deposits because this failure mechanism often precedes flow mobilization (Lawson 1982). Slumped materials may be preserved in ice-marginal deposits as resedimentation activity decreases and a relatively stable sediment cover develops.
Surface features associated with slumping may include tensional and compressional deformation structures in the former head and toe regions respectively, a rotational scarp head wall, and an irregular, chaotic blocky structure to materials partly disrupted by slumping. Surface features, however, can also be subject to slope and rain wash, snowmelt runoff, freeze-thaw processes and vegetation growth, each of which erode and modify or reduce the surface expression of the slump while coarsening surficial sediments and infilling depressions.
Internally, primary arcuate slip planes, stacked normal and thrust faults in head and toe regions, compressional (possibly overturned) folds in toe and marginal sediments, and possibly minor deformational structures below the slip plane may each be preserved. Block faulted structures may remain in head region sediment only partly mobilized by slumping, whereas the toe region may be marked by compressional structures transitional to remolded sediment flow material.

## 4 CONCLUSION

Glacigenic resedimentation by mass-movement processes produces diamictons that, from a sedimentologic viewpoint, should be distinguished from diamictons deposited directly from glacier ice. Their properties vary with the mechanics of the secondary depositional processes rather than being derived directly from the glacier source. The genetic classification of primary versus secondary glacigenic processes and deposits maintains this distinction. Subaerial mass-movement processes are common in ice-marginal and subglacial subenvironments of the terrestrial glacial environment, as well as subaqueous glacigenic environments. Their identification and interpretation provide important information about the former depositional environment and their recognition in glacigenic sedimentary sequences is critical to correct environmental reconstructions.

## REFERENCES

Anderson, P.J., R.P. Goldthwait, and G.D. McKenzie 1986. Observed processes of glacial deposition in Glacier Bay, Alaska. Inst. Polar Studies Misc. Pub. No. 236. Columbus, Ohio State Univ. Press.
Bagnold, R.A. 1954. Experiments on a gravity-free dispersion of large solid spheres in a Newtonian fluid under shear. Royal Soc. (London) Proc., Ser. A. 225:49-63.
Blatt, H., G.V. Middleton, and R.C. Murray 1980. Origin of sedimentary rocks. Englewood Cliffs, NJ, Prentice Hall.
Boulton, G.S. 1968. Flowtills and related deposits on some Vestspitsbergen glaciers. Jour. Glaciology 7:391-421.

Boulton, G.S. 1971. Till genesis and fabric in Svalbard, Spitsbergen. In R.P. Goldthwait (ed.), Till: A symposium, p. 41-72. Columbus, Ohio State Univ. Press.

Boulton, G.S. 1972. Modern arctic glaciers as depositional models for former ice sheets. Quarterly Jour. Geol. Soc. London. 128:361-393.

Boulton, G.S. and N. Eyles 1979. Sedimentation by valley glaciers; a model and genetic classification. In Ch. Schlüchter (ed.), Moraines and varves, p. 11-24. Rotterdam, Balkema.

Boulton, G.S. and M.S. Deynoux 1981. Sedimentation in glacial environments and the identification of tills and tillites in ancient sedimentary sequences. Precambrian Res. 15:397-422.

Broster, B.E. and S.R. Hicock 1985. Multiple flow and support mechanisms and the development of inverse grading in subaquatic glacigenic debris flow. Sedimentology 32:645-657.

Cohen, J.M. 1983. Subaquatic mass-flows in a high energy ice marginal environment and problems with the identification of flow tills. In E.B. Evenson, Ch. Schlüchter, and J. Rabassa (eds.), Tills and related deposits, p. 255-268. Rotterdam, Balkema.

Dardis, G.F. 1985. Till facies associations in drumlins and some implications for their mode of formation. Geog. Annaler 67A:13-22.

Dardis, G.F. and A.M. McCabe 1987. Subglacial sheetwash and debris flow deposits in late-Pleistocene drumlins, Northern Ireland. In J. Menzies and J. Rose (eds.), Drumlin symposium, p. 225-240. Rotterdam, Balkema.

DeJong, M. and M. Rappol 1983. Ice-marginal debris-flow deposits in western Allgäu southern West Germany. Boreas 12:57-70.

Derbyshire, E. 1980. The relationship between depositional mode and fabric strength in tills: Schema and test from two temperate glaciers. Univ. Adama Mickiewicza Poznaniu Seria Geografia No. 20, p. 41-48.

Domack, E.W. 1983. Facies of Late Pleistocene glacial-marine sediments on Whidbey Island, Washington: An isostatic glacial-marine sequence. In B.F. Molina (ed.), Glacial-marine sedimentation, p. 535-570. New York, Plenum Press.

Domack, E.W. and D.E. Lawson 1985. Pebble fabric in an ice-rafted diamicton. Journal of Geology 93:577-591.

Dowdeswell, J.A. and M. Sharp 1986. Characterization of pebble fabrics in modern terrestrial glacigenic sediments. Sedimentology 33:699-710.

Dowdeswell, J.A., M.J. Hambrey, and R. Wu 1985. A comparison of clast fabric and shape in Late Precambrian and modern glacigenic sediments. Jour. Sed. Petrol. 55:691-704.

Dreimanis, A. 1979. The problems of waterlain tills. In Ch. Schlüchter (ed.), Moraines and varves, p. 167-177. Rotterdam, Balkema.

Dreimanis, A. 1984. Discussion: Eyles, Eyles and Miall, 1983. Sedimentology 31:885-886.

Dreimanis, A., J.P. Hamilton, and P.E. Kelly 1987. Complex subglacial sedimentation of Catfish Creek till at Bradtville, Ontario, Canada. In J.J.M. Van Der Meer (ed.), Tills and glaciotectonics, p. 73-87. Rotterdam, Balkema.

Evenson, E.B., A. Dreimanis, and J.W. Newsome 1977. Subaquatic flow tills: A new interpretation for the genesis of some laminated till deposits. Boreas 6:115-134.

Evenson, E.B. and J.M. Clinch 1987. Debris transport mechanisms at active alpine glacier margins: Alaskan case studies. Geological Survey of Finland, Spec. Paper 3, p. 111-136.

Eybergen, F.A. 1987. Glacier snout dynamics and contemporary push moraine formation at the Turtmannglacier, Wallis, Switzerland. In J.J.M. Van Der Meer (ed.), Tills and glaciotectonics, p. 217-231. Rotterdam, Balkema.

Eyles, N. 1979. Facies of supraglacial sedimentation on Icelandic and Alpine temperate glaciers. Can. Jour. Earth Sci. 16:1341-1361.

Eyles, N. 1983. Modern icelandic glaciers as depositional models for "hummocky moraine" in the Scottish Highlands. In E.B. Evenson, Ch. Schlüchter, and J. Rabassa (eds.), Tills and related deposits, p. 47-60. Rotterdam, Balkema.

Eyles, N. 1987. Late Pleistocene debris-flow deposits in large glacial lakes in British Columbia and Alaska. Sedimentary Geology 53:33-71.

Eyles, N., B.M. Clark, and J. Clague 1987. Coarse-grained sediment gravity flow facies in a large supraglacial lake. Sedimentology 34:193-216.

Eyles, N., C.H. Eyles, and A.D. Miall 1983. Lithofacies types and vertical profile models; an alternative approach to the description and environmental interpretation of glacial diamict and diamictite sequences. Sedimentology 30:395-410.

Eyles, N., A.D. Miall, and C.H. Eyles 1984. Lithofacies types and vertical profile models: An alternative approach to the description and environmental interpretation of glacial diamict and diamictite sequences, reply. Sedimentology 31:891-898 (also 1986, 33:152-155).

Ferrians, O.N. 1963. Glaciolacustrine diamicton deposits in the Copper River Basin, Alaska. U.S. Geol. Survey Prof. Paper 475-C, p. C129-C135.

Flint, R.F. and others 1960a. Symmictite: A name for non-sorted terrigenous sedimentary rocks that contain a wide range of particle sizes. Geol. Soc. Amer. Bull. 71:507-510.

Flint, R.F. and others 1960b. Diamictite, a substitute term for symmictite. Geol. Soc. Amer. Bull. 71:1809-1818.

Friedman, G. and J. Sanders 1978. Principles of sedimentology. New York, John Wiley and Sons.

Gibbard, P. 1980. The origin of stratified Catfish Creek till by basal melting. Boreas 9:71-85.

Gravenor, C.P., V. von Brunn, and A. Dreimanis 1984. Nature and classification of waterlain glaciogenic sediments, exemplified by Pleistocene, Late Paleozoic and Late Precambrian deposits. Earth-Sci. Revs. 20:105-166.

Gripp, K. 1929. Glaciologische und geologische ergebnisse der Hamburgischen Spitzbergen-Expedition 1927. Abb. Natur. Verein Hamburg 22:147-249.

Gustavson, T.C. and J.C. Boothroyd 1987. A depositional model for outwash, sediment sources and hydrologic characteristics, Malaspina Glacier, Alaska: A modern analog of the southeastern margin of the Laurentide Ice Sheet. Geol. Soc. Amer. Bull. 99:187-200.

Haldorsen, S. 1982. The genesis of tills from Astadalen, southeastern Norway. Norsk Geologisk Tidsskrift 62:17-38.

Hampton, M.A. 1972. The role of subaqueous debris flow in generating turbidity currents. Jour. Sed. Petrology 42:775-793.

Hampton, M.A. 1975. Competence of fine-grained debris flows. Jour. Sed. Petrology 45:834-844.

Hansel, A.K. and W.H. Johnson 1987. Ice marginal sedimentation in a late Wisconsin end moraine complex. In J.J.M. Van Der Meer (ed.), Tills and glaciotectonics, p. 97-104. Rotterdam, Balkema.

Harland, W.B., K.N. Herod, and D.H. Krinsley 1966. The definition and identification of tills and tillites. Earth Science Revs. 2:225-256.

Hartshorn, J.H. 1958. Flowtill in southeastern Massachusetts. Geol. Soc. America Bull. 69:477-482.

Hein, F.J. and R.G. Walker 1982. The Cambro-Ordovician Cap Enrage Formation, Quebec, Canada: Conglomeratic deposits of a braided submarine channel with terraces. Sedimentology 29:309-330.

Hicock, S.R., A. Dreimanis, and B.C. Broster 1981. Submarine flow tills at Victoria, British Columbia. Can. Jour. Earth Sci. 18:71-80.

Hooke, R.L. 1967. Processes in arid region alluvial fans. Jour. Geology 75:438-460.

Huddart, D. 1983. Flow tills and ice-walled lacustrine sediments, the Petteril Valley, Cumbria, England. In E.B. Evenson, Ch. Schlüchter, and J. Rabassa, eds., Tills and related deposits, p. 81-94. Rotterdam, Balkema.

Johnson, A.M. 1970. Physical processes in geology. San Francisco, Freeman Cooper and Co., p. 432-572.

Johnson, A.M. 1984. Debris flow. In D. Brunsden (ed.), Slope instability, p. 257-362. New York, Wiley.

Johnson, W.H., A.K. Hansel, B.J. Socha, L.R. Follmer, and J.M. Masters 1985. Depositional environments and correlation problems of the Wedron Formation (Wisconsin) in northeastern Illinois. Illinois State Geol. Survey Guidebook 16.

Keefer, D.K. and A.M. Johnson 1983. Earthflows: Morphology, mobilization and movement. U.S. Geol. Survey Prof. Paper 1264.

Kemmis, T.J. and G.R. Hallberg 1984. Discussion: Eyles, Eyles and Miall, 1983. Sedimentology 31:886-890.

Kemmis, T.J., G.R. Hallberg, and A.J. Lutenegger 1981. Depositional environments of glacial sediments and landforms on the Des Moines Lobe, Iowa. Iowa Geol. Survey Guidebook Series No. 6.

Krüger, J. 1982. Processor og till fabric i et recent dodislandskab ved Höfdabrek-Kujökull, Island. Dansk Geol. Foren. Arsskrift 1981, 45-56.

Kurtz, D.D. and J.B. Anderson 1979. Recognition and sedimentologic description of recent debris flow deposits from the Ross and Weddell Seas, Antarctica. Jour. Sed. Petrol. 49:1159-1170.

Lamplugh, G.W. 1911. On the shelly moraine of the Sefstrom Glacier and other Spitsbergen phenomena illustrative of British glacial conditions. Yorkshire Geol. Soc. Proc. 17:216-241.

Lawson, D.E. 1979a. Sedimentological analysis of the western terminus region of the Matanuska Glacier, Alaska. Cold Regions Research and Engineering Laboratory, Hanover NH, Report 79-9.

Lawson, D.E. 1979b. A comparison of the pebble orientations in ice and deposits of the Matanuska Glacier, Alaska. Jour. Geology 87:629-645.

Lawson, D.E. 1981a. Distinguishing characteristics of diamictons at the margin of the Matanuska Glacier, Alaska. Annals of Glaciology 2:78-84.

Lawson, D.E. 1981b. Sedimentological characteristics and classification of depositional processes and deposits in the glacial environment. Cold Regions Research and Engineering Laboratory, Hanover NH, Report 81-27.

Lawson, D.E. 1982. Mobilization, movement and deposition of subaerial sediment flows, Matanuska Glacier, Alaska. Jour. Geology 90:279-300.

Lowe, D.R. 1975. Subaqueous liquefied and fluidized sediment flows and their deposits. Sedimentology 23:285-308.

Lowe, D.R. 1982. Sediment gravity flows. II. Depositional models with special reference to the deposits of high-density turbidity currents. Jour. Sed. Petrol. 52:279-297.

Marcussen, I. 1973. Studies on flow till in Denmark. Boreas 2:213-231.

McCabe, A.M. 1986. Glaciomarine facies deposited by retreating tidewater glaciers: An example from the Late Pleistocene of Northern Ireland. Jour. Sed. Petrol. 56:880-894.

McCabe, A.M., G.F. Dardis, and P.M. Hanvey 1987. Sedimentation at the margins of a late Pleistocene ice-lobe terminating in shallow marine environments, Dundalk Bay, eastern Ireland. Sedimentology 34:473-493.

McRoberts, E.C. and N.R. Morgenstern 1974. The stability of thawing slopes. Can. Geotech. Jour. 11:447-469.

Middleton, G.V. and M.A. Hampton 1976. Subaqueous sediment transport and deposition by sediment gravity flows. In D.J. Stanley and D.J.P. Swift (eds.), Marine sediment transport and environmental management, p. 197-220. New York, John Wiley.

Morawski, W. 1976. Flow tills from the area of Warsaw. In Till - Its Genesis and Diagenesis. Univ. Adam Mickiewicza Poznaniu Seria Geografia NR 12, p. 133-137.

Muller, E.H., D.A. Franzi, and J.C. Ridge 1986. Pleistocene geology of the western Mohawk Valley, New York. In D.H. Cadwell (ed.), The Wisconsinan Stage of the First Geological District, eastern New York, p. 143-157.

Mustard, P.S. and J.A. Donaldson 1987. Early Proterozoic ice-proximal glaciomarine deposition: The lower Gowganda Formation at Cobalt, Ontario, Canada. Geol. Soc. Amer. Bull. 98:373-387.

Nemec, W. and R.J. Steel 1984. Alluvial and coastal conglomerates: Their significant features and some comments on gravelly mass-flow deposits. In E.H. Koster and R.J. Steel (eds.), Sedimentology of gravels and conglomerates. Memoir Can. Soc. Petrol. Geol. 10:1-31.

Ojakangas, R.W. and C.L. Matsch 1980. Upper Precambrian (Eocambrian) Mineral Fork tillite of Utah: A continental glacier and glaciomarine sequence. Geol. Soc. Amer. Bull. 91:494-501.

Pettijohn, F.J. 1957. Sedimentary rocks. New York, Harper and Row.

Pierson, T.C. 1980. Erosion and deposition by debris flows at Mt. Thomas, North Canterbury, New Zealand. Earth Surface Processes 5:227-247.

Pierson, T.C. 1981. Dominant particle support mechanisms in debris flows at Mt. Thomas, New Zealand, and implications for flow mobility. Sedimentology 28:49-60.

Postma, G. and T.B. Roep 1985. Resedimented conglomerates in the bottomsets of Gilbert-type gravel deltas. Jour. Sed. Petrol. 55:874-885.

Postma, G., T.B. Roep, and G.H. Ruegg 1983. Sandy-gravelly mass-flow deposits in an ice-marginal lake, with emphasis on plug-flow deposits. Sedimentary Geology 34:59-82.

Powell, R.D. 1981. A model for sedimentation by tidewater glaciers. Annals Glaciology 2:129-134.

Powell, R.D. 1983. Glacial marine sedimentation processes and lithofacies of temperate tidewater glaciers. In B.F. Molina (ed.), Glacial-marine sedimentation, p. 185-232. New York, Plenum Press.

Powell, R.D. 1984. Glacimarine processes and inductive lithofacies modelling of ice shelf and tidewater glacier sediments based on Quaternary examples. Marine Geology 57:1-52.

Prior, D.B., B.D. Bornhold, and M.W. Johns 1984. Depositional characteristics of a submarine debris flow. Jour. Geology 92:707-727.

Rappol, M. 1983. Glacigenic properties of till. Ph.D. Thesis, Univ. of Amsterdam.

Rappol, M. 1985. Clast-fabric strength in tills and debris flows compared for different environments. Geologie en Mijnbouw 64:327-333.

Robinson, P.H. 1979. An investigation into the processes of entrainment, transportation and deposition of debris in polar ice, with special reference to the Taylor Glacier, Antarctica. Unpublished Ph.D. Thesis, Victoria University, Wellington.

Russell, I.C. 1893. Malaspina Glacier.

Jour. Geology 1:219-245.

Rust, B.R. 1977. Mass flow deposits in a Quaternary succession near Ottawa, Canada: Diagnostic criteria for subaqueous outwash. Can. Jour. Earth Sci. 14:175-184.

Schultz, A.W. 1984. Subaerial debris-flow deposition in the Upper Paleozoic Cutter Formation, Western Colorado. Jour. Sed. Petrol. 54:759-772.

Sharp, M.J. 1982. A comparison of the landforms and sedimentary sequences produced by surging and non-surging glaciers in Iceland. Unpublished Ph.D. Thesis, Univ. of Aberdeen, Scotland.

Sharpe, D.R. 1987. The stratified nature of drumlins from Victoria Island and southern Ontario, Canada. In J. Menzies and J. Rose (eds.), Drumlin symposium, p. 185-214. Rotterdam, Balkema.

Shaw, J. 1977. Tills deposited in arid polar environments. Can. Jour. Earth Sci. 14:1239-1245.

Shaw, J. 1987. Glacial sedimentary processes and environmental reconstruction based on lithofacies. Sedimentology 34:103-116.

Tarr, R.S. and L. Martin 1914. Alaskan glacier studies. Washington, D.C., Nat. Geog. Soc.

Varnes, D.J. 1978. Slope movement types and processes. In R.L. Schuster and R.J. Krizek (eds.), Landslides, analysis and control, p. 11-33. Washington, D.C., Nat. Acad. Sci.

Visser, J.N.J. 1983a. Submarine debris flow deposits from the Upper Carboniferous Dwyka tillite formation in the Kalahan Basin, South Africa. Sedimentology 30:411-423.

Visser, J.N.J. 1983b. The problems of recognizing ancient subaqueous debris flow deposits in glacial sequences. Trans. Geol. Soc. S. Africa 86:127-135.

Walker, R.G. 1975. Generalized facies models for resedimented conglomerates of turbidite association. Geol. Soc. Amer. Bull. 86:737-748.

Walker, R.G. 1984. General introduction. Facies, facies sequences and facies models. In R.G. Walker (ed.), Facies models (2nd ed.), p. 1-11. Geoscience Canada, Reprint Series 1.

Wright, R. and J.B. Anderson 1982. The importance of sediment gravity flow to sediment transport and sorting in a glacial marine environment: Eastern Weddell Sea, Antarctica. Geol. Soc. Amer. Bull. 93:951-963.

# 2 Indicator tracing and engineering

Genetic Classification of Glacigenic Deposits, Goldthwait & Matsch (eds)
© 1988 Balkema, Rotterdam. ISBN 90 6191 694 1

# Glacigenic deposits as indicators of glacial movements and their use for indicator tracing

Heikki Hirvas, Raimo Kujansuu & Keijo Nenonen
*Geological Survey of Finland, Espoo, Finland*

Matti Saarnisto
*Department of Geology, University of Oulu, Oulu, Finland*

ABSTRACT: The activities of Work Group 9 of the INQUA commission on Genesis and Lithology of Quaternary Deposits included distribution of a questionnaire on the research methods and procedures used in glacial indicator tracing. The results of the questionnaire are summarized, and two cases of glacigenic deposits as indicators of glacial movements in Finland are presented. The Pyhäsalmi zinc-copper mine has the largest sulphide ore subcrop in Finland. It is covered by two till beds with distinctive clay content, deposited by ice flows from clearly differing directions. The geochemical anomaly from the subcrop is of very limited extent, and the till beds differ significantly in their trace metal content. In Suomussalmi, till lithology and trace metal studies conducted within the area of a narrow Archaean greenstone belt have enabled us to resolve conflicting views on ice-transport directions.

## 1 INTRODUCTION

The title of the Work Group "Glacigenic deposits as indicators of glacial movements" was changed in 1982. Dr. Raimo Kujansuu of the Geological Survey of Finland was appointed its new president, and Dr. Matti Saarnisto, University of Oulu, Finland, its secretary (since 1983). Despite the change, the aim of the new Group, entitled "Glacigenic deposits as indicators of glacial movements and their use for indicator tracing in the search for ore deposits," was to continue the work of its predecessor, the results of which have been published as a report of the Geological Survey of Finland (Virkkala et al. 1980), but with more emphasis placed on the significance for exploration of the direction of glacial movement and the transport distances involved. Our intention was to gather information on the methods used in such studies, to develop and apply these methods, and to promote their use in practical exploration work by publishing descriptions of them and their applicability to different sites and environments.

Methods utilizing glacial geology have been successful in Finland for a long time for tracing the origins of ore boulders. Cooperation with the public has produced good results. During the period 1938-1967, 18 of the total of 23 mine-grade deposits were discovered in Finland through boulder tracing and/or samples received from members of the public (Saltikoff 1984). This approach has been less successful recently, and therefore more emphasis has been given to broadly-based scientific prospecting. This involves the application of glacial geology, including studies on glacial stratigraphy and transport of glacigenic material, together with its lithological and geochemical composition. Intensive research in these fields has been undertaken in Finland. The Finnish experience, cannot, however, be taken as a universal model, since Finland covers only a small part of the area occupied by the Fennoscandian ice sheet, and the glacial sediment cover is thin or of moderate thickness. Problems with indicator tracing are different in the marginal areas of former ice sheets, in alpine environments, and in permafrost areas.

Work Group 9 has sought to collect information on glacial transport in all types of glacial environments and to make the results known to a wider audience. The activities of the Work Group have included: 1) distribution of a questionnaire on research methods and procedures in glacial indicator tracing, 2) organization of a field workshop in Finland, August 20-29, 1985, 3) publication of a book entitled "INQUA Till Symposium" (Kujansuu and Saarnisto 1987), and 4) editing of the "Handbook of glacial indicator tracing." The handbook will serve as the final report of Work Group 9, and will review the methods and procedures which, according to present knowledge, are best applicable to practical indicator tracing.

The present report first summarizes the results of the questionnaire and presents two cases of glacial transport as examples from Finland. These cases were presented and discussed at the excursion of the INQUA Till Symposium held in Finland in 1985 (Saarnisto 1985).

## 2 QUESTIONNAIRE

Work Group 9 distributed a questionnaire on the research methods employed in indicator tracing and related problems in glaciated areas. The questionnaire included 106 questions grouped under 17 main topics, and was mailed to a list of 152 names from 27 countries. Approximately 120 replies were received, most from northern Europe and North America, as was expected. Some, however, also came from Asia and South America. The questions and summaries of replies are given below.

1. Upon what indicators do you base your decision to commence research at a given site: 1.1. bedrock, 1.2. boulders, 1.3. heavy minerals, 1.4. geochemical, 1.5. geophysical, 1.6. others.

Replies: Boulders and geochemical indicators are most commonly used. Other indicators include suitable geology as deduced from regional studies, exposure of stratigraphic succession, and vegetation contrasts.

2. Do you study morainic landforms: 2.1. in respect to their morphogenesis, 2.2. using their orientation to reconstruct directions of glacial flow, 2.3. to determine the glaciodynamics regionally or locally, 2.4. both morphologically and in terms of material composition, 2.5. in other ways.

Replies: The orientation of landforms and a combination of landform morphology and material composition are most commonly used.

3. Which glacigenic erosional forms do you study? 3.1. glacially eroded valleys and mega-grooves, 3.2. streamlined moulded bedrock forms, 3.3. roches moutonnées, 3.4. friction cracks, 3.5. striations, 3.6. others.

Replies: Various types of striation are most commonly studied together with bedrock forms. Crag and tail formations were also mentioned in some answers.

4. What aspects of till do you study?

4.1. genesis, 4.2. stratigraphy, 4.3. structure, 4.4. grain size, 4.5. stone content, 4.6. stone-size distribution, 4.7. clastic roundness, 4.8. consolidation, 4.9. colour, 4.10. other properties.

Replies: All properties are commonly investigated, but till genesis, stratigraphy, structure, grain size, and stone content are investigated more often than the rest.

5. What aspects of till fabric do you study? 5.1. traditional till fabric analyses i.e., horizontal orientation, 5.2. three-dimensional orientation, 5.3. microfabric, 5.4. anisotropy of magnetic susceptibility, 5.5. anisotropy of specific resistance, 5.6. other aspects.

Replies: Traditional till fabric is most commonly analyzed, but three-dimensional orientation is also quite commonly studied and (to a lesser extent) magnetic properties. Structural features, such as foliation, shear features, orientation of inclusions etc., are also considered.

6. What aspect of till lithology do you study and for what purposes? 6.1. stone counts, 6.2. mineral analyses, 6.3. heavy mineral analyses, 6.4. for determination of transport distance, 6.5. for determination of lithostratigraphy, 6.6. for determination of direction of transport, 6.7. for discovery of source rock of indicator boulders or rock types, 6.8. for explanation of the cause of geochemical anomalies, 6.9. for other purposes.

Replies: Stone counts are used most commonly and heavy minerals less frequently for determination of transport distance and direction and discovery of source rock of indicator boulder. Stone counts and heavy minerals are also used for establishing lithostratigraphy and for explaining the cause of the geochemical anomalies.

7. What aspects of indicator fans do you study? 7.1. form and dimensions of the fans; 7.2. their genesis, i.e. basal, englacial or supraglacial transport; 7.3. other transport conditions: 7.3.1. single-phase, i.e. monocyclic transport, 7.3.2. multiphase glacial transport, 7.3.3. complex transport, i.e. more than one transport cycle, possibly including glaciofluvial transport and/or reworking of glaciofluvial material, 7.3.4. transport by ice rafting, 7.3.5. transport by icebergs, 7.3.6. glaciofluvial transport, 7.3.7. mass wasting; 7.4. other aspects.

Replies: Apart from simple indicator fans, complex transport seems to be a common case. The importance of the three-dimensional study of fans was also pointed out.

8. What aspects of glaciofluvial formations do you study and for what purposes? 8.1. morphogenesis, 8.2. both morphology and material composition, glacial palaeohydrology, 8.4. other aspects, 8.5. reconstruction of direction of glacial flow, 8.6. other purposes.
Replies: Glaciofluvial formations are commonly used for reconstructing glacial drainage patterns and the direction of glacial flow and to a lesser extent for prospecting.

9. What aspects of glaciofluvial material do you study? 9.1. genesis, 9.2. stratigraphy, 9.3. structures, 9.4. grain size, 9.5. lithology, 9.6. mineral composition, 9.7. heavy mineral content, 9.8. transport distance and origins, 9.9. other aspects.
Replies: Genesis, stratigraphy, structures, grain size, and lithology of glaciofluvial material are the aspects most commonly studied.

10. Do you use geochemical methods, and for what purposes? 10.1. to search for new indicators, 10.2. to examine known indicators in more detail, 10.3. to determine the boundaries of boulder fans, 10.4. to trace subsurface continuations of fans, 10.5. to distinguish stratigraphical units, 10.6. other purposes.
Replies: Geochemical methods (mainly trace metal analyses) are used to examine known indicators in more detail and to distinguish between stratigraphical units. Major element composition is also analyzed in some cases in order to work out the provenance of tills.

11. Do you study geophysical properties of tills? 11.1. grain density, 11.2. total magnetic susceptibility, 11.3. susceptibility anisotropy (i.e., magnetic fabric analysis), 11.4. remanent magnetism, 11.5. electrical conductivity, 11.6. conductivity anisotropy, 11.7. resistivity sounding, 11.8. seismic sounding, 11.9. "georadar," 11.10. other aspects.
Replies: Many answers were negative although interest in the subject appears to be growing. Nevertheless, geophysical methods are not in routine use in indicator tracing today with the possible exception of measurements of magnetic parameters of till and seismic sounding.

12. Which remote sensing methods do you employ? 12.1. satellite data, 12.2. air photography, 12.3. aeroradiometry, 12.4. aeromagnetometry, 12.5. aeroelectricity, 12.6. others.
Replies: All kinds of air photographs, (i.e., black and white, colour, infra-red, and others), are commonly used, as are magnetic maps. The use of satellite data is limited to regional studies.

13. What sampling methods do you use? 13.1. surface sampling, 13.2. available sections, 13.3. systematic or random mechanical excavation, 13.4. drilling with heavy or light equipment, 13.5. other methods.
Replies: Mechanical excavation and drilling -- often with light-weight equipment -- are commonly used in addition to surface sampling, depending on the purpose of the work and the thickness of glacial deposits.

14. Do you employ the results of research on glacial geology and glacigenic formations: 14.1. for local prospecting, 14.2. for regional ore prospecting, 14.3. for planning of prospecting strategies at the national or provincial level, 14.4. for other purposes.
Replies: The results of glacial geological research are most commonly used for regional ore prospecting but also for basic information on glacial history.

15. What is your own specialized field and what other topics would you like to see taken up as the themes for workshop meetings?
See replies to 17.

16. Would you please give a list of references to your published work on these topics and to that of your co-workers.
Replies: The list of references is not complete, because many respondents did not reply to this item. We have therefore omitted the list.

17. General comments on the program and the aims of the Work Group.
Replies: Many respondents were pleased that the emphasis of the Commission had been shifted to drift prospecting and to applied fields in general. A good example is the suitability of glacial sediments as containers for waste disposal. It was also hoped that the outcome of the activities of the Work Group will help in the study of pre-Pleistocene glacial episodes. One fascinating, unresolved problem in indicator tracing comes from Brazil (Dr. A.C. Rocha-Campos): Proterozoic tillites in eastern Brazil contain diamonds, and reconstruction of the ice movement pattern may help to localize the search for ancient kimberlite intrusions there.

3 GLACIAL DEPOSITS AS INDICATORS OF GLACIAL MOVEMENTS: TWO CASE HISTORIES FROM FINLAND

During the INQUA Till Symposium, organized by Work Group 9 in Finland 1985, the Pyhäsalmi mine and the Suomussalmi site were visited in order to demonstrate the methods of glacial geological work commonly used in Finland and their applica-

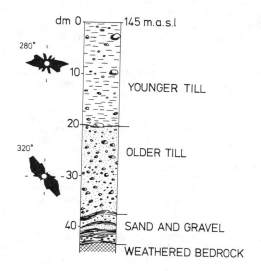

## PYHÄSALMI
## RUOTANEN

dm 0 ⎯⎯⎯⎯ 145 m.a.s.l

280°

10 — YOUNGER TILL

20 —

320°

OLDER TILL

-30-

40- SAND AND GRAVEL

WEATHERED BEDROCK

Fig. 1. Stratigraphic column from till
section of the Pyhäsalmi mine. Till fabric
is shown in rose diagrams.

tion in ore prospecting. In Pyhäsalmi the
influence of an extensive sulphide ore
subcrop on the geochemistry of the
overlying two till beds was shown, whereas
in Suomussalmi till lithology and trace
metal geochemistry was used to establish
the direction of glacial transport. This
has been a subject of debate for some time
and has influenced indicator tracing in
the local Archaean greenstone belt area,
which has a high potential for ore-grade
mineralization.

3.1 Till stratigraphy and geochemistry in
    the open pit section of the Pyhäsalmi
    mine, Pyhäjärvi, Finland

The Pyhäsalmi ore body, situated about 4
km east of the village of Pyhäsalmi in the
province of Oulu, was discovered in 1958,
when a local farmer dug a well directly
into an ore subcrop. Outokumpu Oy began to
extract ore from the site by open-pit
mining in 1962.

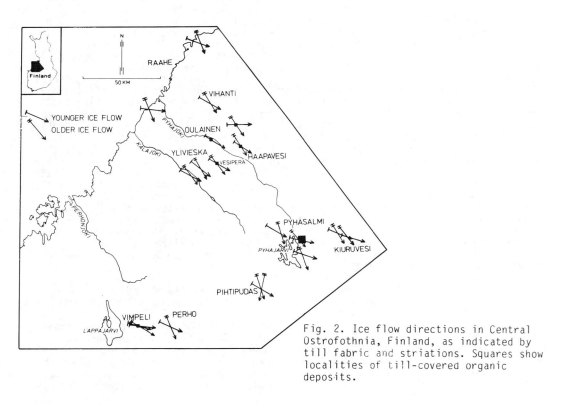

Fig. 2. Ice flow directions in Central
Ostrofothnia, Finland, as indicated by
till fabric and striations. Squares show
localities of till-covered organic
deposits.

176

The open-pit section contains evidence of two distinct till beds, a stratigraphy common in the Pyhäsalmi-Kiuruvesi area. The composition of both till beds -- including compactness, massive and fissile textures, and well-developed fabric corresponding to striations on the underlying bedrock -- indicates that both beds are primary basal tills. The upper basal till bed is an average of 2 m thick in the section, and is a brownish-grey or, in places, bluish-grey, dark silty till containing 10-20% clay of < 0.002 mm grain size with numerous massive clay fragments. Fabric analyses show that this till was deposited when glacial movement was WNW-ESE (Fig. 1), coinciding with the direction of the final phase of glacial activity in the area (Fig. 2).

The lower basal till bed is composed of a light grey compact sandy till with less than 5% clay. In places below the till there is a thin deposit of gravel and sand, but for the most part this till rests directly on weathered or unweathered bedrock. The orientation of till pebbles reveals that the direction of glacial movement was NW-SE (Fig. 1), which is the same as that of older ice flow directions in the area (Fig. 2).

According to the most recent interpretations (Nenonen 1986; Hirvas and Nenonen 1987), the youngest till and ice flow represent the Late Weichselian glaciation whereas the older till may be part of the Saalian glaciation.

Studies of the till stratigraphy in the area included a description of the glacigenic trace metal anomaly devised from the Zn-Cu-S ore at Pyhäsalmi. For this work, till samples were taken from pits dug by the excavator and were then correlated with the stratigraphy. Anomalous trace metal concentrations have also been recorded in the area east of the ore subcrop, i.e., in the direction of the last glacial movement (Fig. 3). The results suggest that the ore must have given rise to elevated Zn, Cu and S concentrations over an area of about 40 ha, the easternmost anomalous site being 550 m from the subcrop, which is in itself some 650 m long and 80 m wide, making it the largest sulphide ore subcrop in Finland. The whole subcrop was once covered by unconsolidated deposits, predominantly till. In some places it has been weathered into a gossan type of material but elsewhere polished surfaces are seen. The till anomaly is of limited extent compared with the size and orientation of the subcrop, but it does feature markedly high Zn, Cu, and S concentrations. In terms of

regional geochemical mapping, this "target area" should emerge as a one-or two-point anomaly under the current mapping scheme of 4 points/km$^2$.

Since glacial movement cuts across the ore body in slightly different directions, trace metal concentrations vary between the two till beds, the younger silty till in the northern part of the section containing up to three times the abundance of sulphur, copper, and zinc found in the older sandy till (Fig. 4).

This can be explained by the interpretation that the younger movement eroded the broader central part of the ore suboutcrop, whereas the older movement affected the narrower northern part.

3.2 Till transport in Suomussalmi, eastern Finland

Studies on glacial stratigraphy and till properties were undertaken in the province of Kainuu, eastern Finland, in order to provide a basis for ore prospecting. The area is crossed by a narrow, north-south Archaean greenstone belt, surrounded by a vast granitoid terrain. Till lithology and geochemistry have been discussed in several papers, e.g., Saarnisto and Taipale (1985) and Taipale et al. (1986). The present example of glacial transport comes from the northern end of the greenstone belt, the Suomussalmi area, where several ice flow stages have been recognized (Saarnisto and Peltoniemi 1984) (Fig. 5).

A network of 20 test pits was dug by a mechanical excavator. Thirty till fabric analyses were performed, together with 56 lithological and 136 geochemical analyses of till, The most complete stratigraphy was that of the Kaapinsalmi pit (Fig. 6; site 1/80, in Fig. 5), where four till units were identified. The lowest till (S IV), whose base remains unknown, was deposited by ice flowing from the northwest (290°-300°), across the greenstone belt. The S IV till, interpreted as basal till because of its fabric and compactness, is overlain by the S III till, which was observed in all other pits as a bed with an average thickness of about 2 m. Above the S III till in the Kaapinsalmi pit there is a heterogeneous unit of till and sorted material, designated S II and interpreted as flow till. This deposit is covered by the S I till that was found in most other pits to be less than one metre thick, except in the northernmost part of the area. The till fabric, lithology, and trace metal content of the S I, S II and S

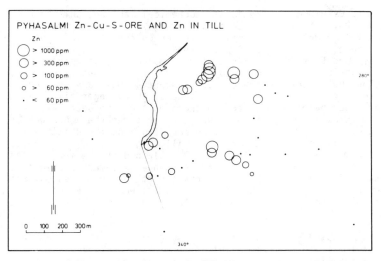

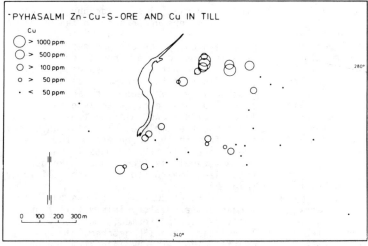

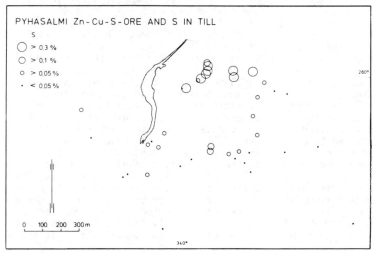

Fig. 3. Maps of the highest Zn, Cu, and S contents in the test pits on the distal side of the Pyhäsalmi Zn-Cu-S ore subcrop. Lines indicate ice flow directions.

178

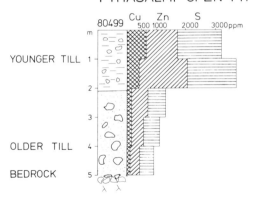

## PYHÄSALMI OPEN PIT

YOUNGER TILL

OLDER TILL

BEDROCK

Fig. 4. Zn, Cu, and S content of till beds in northern section.

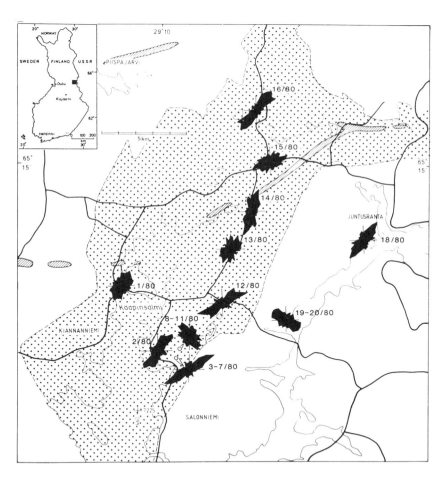

Fig. 5. Fabric of the S III till in the northern Kiantajärvi area of Suomussalmi, eastern Finland. Stippled area indicates Archaean greenstone belt (surrounded by granitoids). Eskers are shaded. Fabrics in sites 8-11/80 and 19-20/80 are interpreted as transverse.

# SITE 1/80, KAAPINSALMI

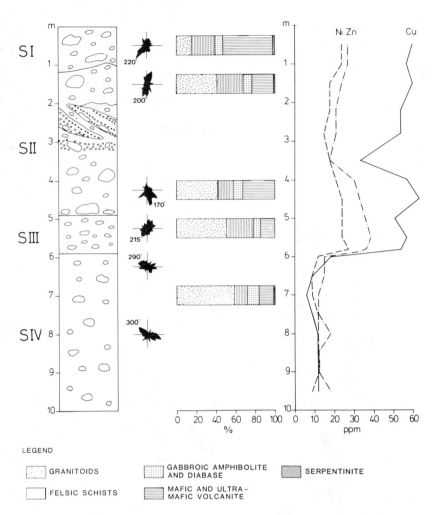

Fig. 6. Stratigraphy of the Kaapinsalmi pit, Suomussalmi, Finland, including information on till fabric and inclination, the vertical variation in trace metal content (Ni, Cu, Zn) and lithology. S I, S III, and S IV are basal tills with clear fabric and massive or fissile structures, whereas S II is composed of till and sorted material and can be interpreted as flow till. From Saarnisto and Peltoniemi, 1984.

III tills are nearly similar whereas less material originates from the greenstone belt in S IV.

Till fabric analyses on the most common till bed, S III, indicate a clear SW-NE orientation (Fig. 5). Similar fabrics emerge from the S I till; both are basal tills. The fabrics of S III till in test pits 8-11/80 and 19-20/80 are interpreted as transverse because other properties i.e., stratigraphy, compactness and

colour, are similar to those of S III till nearby. Pebble counts on the S III till (Fig. 7) indicate that the proportion of material originating from the greenstone belt (felsic schists, mafic and ultramafic volcanites, serpentinite, gabbroic amphibolite and diabase) increases towards the northeast as does the trace metal content of the S III till (Fig. 8). This indicates ice flow from the southwest because the longer the distance the ice flowed over

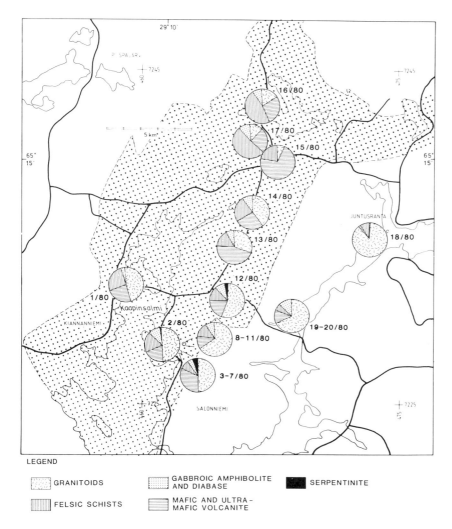

LEGEND

▨ GRANITOIDS  ▥ GABBROIC AMPHIBOLITE AND DIABASE  ■ SERPENTINITE

▥ FELSIC SCHISTS  ☰ MAFIC AND ULTRA-MAFIC VOLCANITE

Fig. 7. Lithology of the S III till (2-6-cm pebbles). Stippled area indicates Archaean greenstone belt (surrounded by granitoids). Proportion of greenstone material in the till increases towards the northeast, i.e., in the direction of glacier transport.

the greenstone belt, the greater was its influence on the tills. Variations in the underlying bedrock, however, should also be considered, especially with regard to the trace metal content of the till.

Thus, lithology and trace metal analyses on S III, (S II), and S I tills support the view that all these tills were deposited during a phase of ice flow from the southwest and not from the opposite direction as has also been suggested and applied in indicator tracing. Convincing evidence is also provided by the orientation of the esker, which is consistent

with this younger flow direction (Fig. 5), and also by striae near site 2/80.

REFERENCES

Hirvas, H. and K. Nenonen 1987. The till stratigraphy of Finland. In R. Kujansuu and M. Saarnisto (eds.), INQUA Till Symposium, Finland 1985, Geological Survey of Finland, Special Paper 3.
Kujansuu, R. and M. Saarnisto (eds.) 1987. INQUA Till Symposium. Finland 1985. Geological Survey of Finland, Special

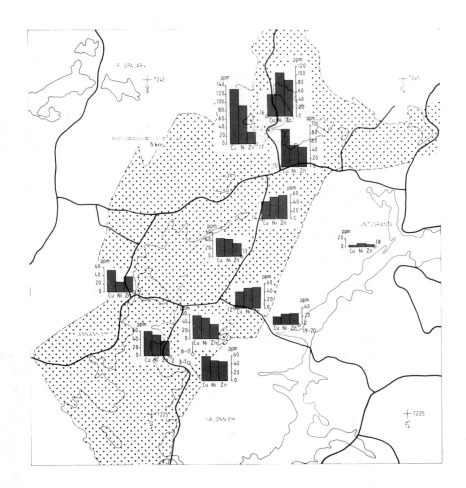

Fig. 8. Cu, Ni, and Zn geochemistry in the fine fraction (< 0.063 mm) of the S III till. Note the low metal content in sites east of the greenstone belt (stippled).

Paper 3.

Nenonen, K. 1986. Orgaanisen aineksen merkitys moreenistratigrafiassa. Summary: The significance of organic material in till stratigraphy. Geologi 38:41-44.

Saarnisto, M. (ed.) 1985. INQUA Till Symposium. Excursion guide. Geological Survey of Finland, Department of Quaternary Geology.

Saarnisto, M. and H. Peltoniemi 1984. Glacial stratigraphy and compositional properties of till in Kainuu, eastern Finland. Fennia 162:163-199.

Saarnisto, M. and K. Taipale 1985. Lithology and trace-metal content in till in the Kuhmo granite-greenstone terrain, eastern Finland. Journal of Geochemical Exploration 24:317-336.

Saltikoff, B. 1984. Boulder tracing and the mineral indication data bank in Finland. In Prospecting in areas of glaciated terrain 1984, p. 179-191. The Institution of Mining and Metallurgy.

Taipale, K., R. Nevalainen, and M. Saarnisto 1986. Silicate analyses and normative compositions of the fine fraction of till: Examples from eastern Finland. J. Sed. Pet. 56(3):370-378.

Virkkala, K., L.K. Kauranne, and H. Tanskanen 1980. Glacigenic deposits as indicators of glacial movements. Geological Survey of Finland. Report of Investigation 48.

Genetic Classification of Glacigenic Deposits, Goldthwait & Matsch (eds)
© 1988 Balkema, Rotterdam. ISBN 90 6191 694 1

# Application of glacial dynamics, genetic differentiation of glacigenic deposits and their landforms to indicator tracing in the search for ore deposits

V.-P.Salonen
*Department of Quaternary Geology, University of Turku, Finland*

ABSTRACT: The importance of glacial geological investigations in ore exploration is discussed on the basis of experiments made in Finland. The purpose of indicator tracing is to determine the length and direction of glacial transport and, finally, to find the outcrop of primary ore deposit. At individual target sites, the effectiveness of glacial processes -- erosion, entrainment, transport and deposition -- has to be analyzed in order to estimate the plausible dispersal patterns of indicator particles.

The regularities controlling the dispersal of glacially derived rock fragments have been found to be connected with large-scale glacial dynamics and with the genesis of glacial deposits and their landforms. The flow directions of the Scandinavian ice sheet in Finland have been reconstructed with the aid of striae and till fabric observations, satellite imagery of depositional landforms, and the direction of boulder trains. Areal variability of glacial transport distance for boulder fraction has been estimated by compiling data of 111 transport distance distribution determinations, 41 half-distance measurements, and length-distribution of 464 boulder trains.

## 1 INTRODUCTION

Studies on glacial dispersal trains for prospecting purposes are an important part of glacial geological research in Fenno-scandia, where they have long been used with favorable results. Until recently, the methods of indicator tracing were used mainly in Finland (Sauramo 1924; Aurola 1955; Hyvärinen et al. 1973) and Sweden (Lundqvist 1935; Grip 1953; Minell 1978); nowadays they are becoming increasingly popular in Canada as well (Shilts 1976; DiLabio 1981; Bouchard and Marcotte 1986).

Opportunities for exploiting mineable resources have diminished in the past few years. Nevertheless, indicator tracing will have value in the search for ore deposits whenever exploration is practiced in areas of glaciated terrain. The advantages of indicator tracing are its low cost and rapidity. The method is also useful for directing more expensive and detailed exploration methods such as geophysical and geochemical surveys and diamond drilling.

Indicator tracing has the additional advantage of suggesting new indications of mineralized areas in bedrock. This has been shown by the nationwide prospecting

contest organized in Finland in 1986, in which the Geological Survey of Finland and other exploration companies received over 70,000 rock samples from members of the general public. About ten percent of the samples contained ore minerals and dozens of new exploration targets were claimed. Future exploration will test the value of the new targets. Studies of ore indicators and associated glacigenic deposits and their landforms will feature prominently in the initial prospecting stage.

The aim of this paper is to draw attention to the regularities controlling the dispersal of glacially derived rock fragments. The observed regularities are connected with large-scale glacial dynamics and with the genesis of glacial deposits and their landforms. The main observations are from Finland, because summaries of boulder transport in this area have recently been published (Salonen 1986, 1987). Finland is a part of the Baltic Shield, and it is situated centrally in relation to the Scandinavian ice sheet (Fig. 1). It has been suggested that the results of the Finnish observations can be applied in other similar areas with high exploration potential, e.g., the glaciated Canadian Shield.

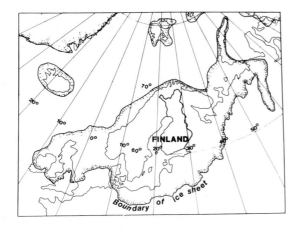

Fig. 1. Finland is at the centre of the Scandinavian ice sheet. Extent of the Weichselian glaciation is indicated according to the minimum reconstruction of Hughes et al. (1981).

## 2 FIELD WORK

The purpose of indicator tracing is to find the outcrop of a primary ore deposit. There are no problems if the amount of erratic ore boulders or smaller fragments is high. If a dispersal train can be defined, the glacial transport of the ore indicators and the provenance area in the bedrock can be estimated with high accuracy. The same applies to unambiguous glacigenic, geochemical anomaly patterns in till. The regularities of glacial dispersal are well known and can be utilized in prospecting (Kauranne 1976; Shilts 1976).

In practice, however, the situation is often more complicated. One single ore boulder situated on the surface of a glacigenic deposit may be the only hint of an orebody. In that case, closer analysis of glacigenic deposits is required to support systematic boulder tracing. In Finland, test pits dug by tractor excavator are used to study till stratigraphy and genesis, fabric and lithology (see Hirvas et al. 1977; Saarnisto and Peltoniemi 1984). Till geochemical surveys are often performed at the same time (Shilts 1984; Salminen and Hartikainen 1985).

The aim of these studies is to connect the ore indicator to glacial formations and, furthermore, to glacial processes and to the genesis of till. Modelling the processes that have led to the deposition of the indicator studied will make it easier to establish its provenance in the bedrock.

## 3 GLACIAL PROCESSES

Glacial processes -- erosion, entrainment, transport and deposition -- operate in a complex combination of interrelations between an ice sheet and the geological deposits lying beneath it. The effectiveness of individual processes has varied from area to area in the course of the glacial cycles, producing the diversity observed in indicator dispersal patterns (Fig. 2).

When the erosion rate, i.e., the production of local bedrock debris, is high, and deposition is enhanced by local pressure conditions (Boulton 1974), strong local dispersal patterns develop (Fig. 2, area A). Removal of debris is negligible, and the supply of far-transported material low.

Widespread, low-intensity anomaly patterns develop when glacial erosion associated with the effective entrainment and transport of debris has been the prevailing process in the genesis of a gla-

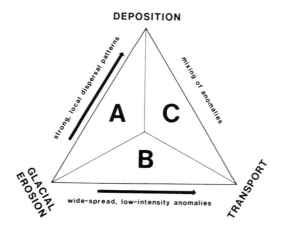

Fig. 2. Different combinations of glacial processes implying different prerequisites in indicator tracing.

cial deposit (Fig. 2, area B). If the velocity of glacier ice changes considerably from one area to another, erosion can take place because of purely thermal processes (Hughes 1981). Because basal temperature plays a key role in determining erosional activity (Drewry 1986), the patterns of basal temperatures and glacial dynamics of ice sheets control and cause regional systematicity in the dispersal patterns (see Boulton et al. 1985).

If the dominant processes are glacial transport and deposition, dispersal patterns are controlled by mixing rate (Fig. 2, area C). Mixed and diluted anomalies of more distant bedrock material may occur. On the other hand, the erosion rate is low, and hence there may be local weathered bedrock with minor transport.

## 4 A CASE HISTORY

The Main Sulphide Ore Belt is the most important target area for exploration in Finland (Fig. 3). Over 90 percent of the sulphide ore mined so far and existing in known reserves is in the deposits of the Main Sulphide Ore Belt (Kahma 1973). The bulk of the ore boulders without a known source are from this same zone (Saltikoff 1984).

The ore explorational potential has been studied in a restricted area of the Main Sulphide Ore belt (Gaál and Kuosmanen, in press). The existing ore data -- geological, petrological, geophysical, and geochemical -- from an area covering 4500 km$^2$ were collected and compared using multivariate techniques. This showed that the knowledge of glacial deposits (Fig. 3) and their genesis was also essential if the geochemical patterns were to be explained.

The study area was affected by two glacial lobes separated by an interlobate complex. The southern lobe was more active and the flow of ice caused intense drumlinization in the area (Glückert 1973). The glacial dispersal pattern is simple, and the boulder trains tend to be dense, narrow, and of intermediate (1-5 km) length (Ekdahl 1982). Glacial erosion was effective, and usually only one bed is found deposited as lodgement till.

Close to the interlobate complex, the ice flow turned towards the hydrological pressure minimum during deglaciation. Because the glacier still had great erosion capacity, hummocky boulder-rich moraines consisting of melt-out and flow tills were deposited. The boulder trains are very short (0.5-2.0 km) close to the interlobate complex.

Glacial process combinations A and B

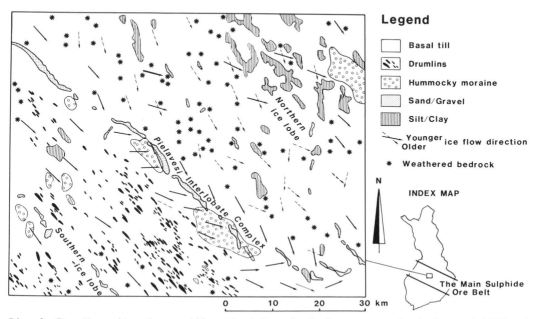

Fig. 3. The diversity of prevailing glacial geological processes leads to variability in glacial deposits and their landforms. Modified after Salonen and Tenhola (in press).

(Fig. 2) thus prevailed in the southern part of the area. Indicator tracing has been successful, and the analysis of geochemical anomalies can be based mainly on the theory of glacigenic dispersal (Nurmi 1976).

Glacial erosion was weaker in the north, and the glacier was more stagnant during deglaciation than it was south of the interlobate complex (Fig. 3). Weathered bedrock occurrences have been encountered in numerous places. The indicators of glacier ice movement show various directions, and they can possibly be connected with more than one glacial cycle (Hirvas and Nenonen 1987). The prevailing glacial processes of this northern subarea are transport and deposition (see Fig. 2, area C).

All the observations collected from this study area indicate a bipartition in the genesis and properties of till. In the area of the active ice lobe, boulder trains are narrow, overburden is thin, and glacial erosion has been effective. The till stratigraphy is simple, consisting mostly of only one till bed.

In the northern area, glacial dispersal of debris has been more difficult to establish. The glacial transport directions vary within a wide range. The surface boulder material is a mixture of fragments transported by repeated active glacial cycles and of bedrock material weathered in situ. The boulder trains are difficult to define, and the models of glacial dispersal are not sufficient to explain the glacial transport of the debris. Moreover, the geochemical dispersion is often hydromorphic in nature (Wennervirta 1968).

The whole area has marked ore exploration potential. However, conditions for successful indicator tracing are much better in the southern than in the northern part of the study area. Hence, variability caused by glacial geological processes must be taken into account when planning and conducting boulder prospecting and till geochemistry surveys. If this is done, the differentiation of glacigenic deposits and their landforms will directly serve ore exploration purposes.

## 5 LARGE-SCALE REGIONAL FEATURES OF GLACIAL ACTIVITY

The activity of the Weichselian glaciation in Finland has been well described. The Quaternary deposits of the country have recently been mapped (Kujansuu and Niemelä 1984). A summary of till stratigraphy and of glacial transport directions is soon to be published (Hirvas and Nenonen 1987). The course of deglaciation is well known (Ignatius et al. 1980), and the variations in boulder transport distance at the scale of Finland have been summarized (Salonen 1986).

The areal variability and systematics of glacial deposits in Finland have been attributed to combinations of different glacial processes. The following variables are important:
- repeated weak glacial erosion (cf. Hirvas and Nenonen 1987; Niemelä 1979)
- early stages of glaciation (Boulton 1984)
- lobe activity during deglaciation (Punkari 1984)
- long-term ice-marginal and oscillation positions
- factors associated with the local topography, underlying bedrock, or tectonics.

When the areal variability of all these factors is considered together, a set of types for the glacial geology of Finland begins to emerge (Fig. 4). It then becomes possible to establish the features most likely to have influenced glacial activity in each part of the country and to evaluate the degree of complexity in glacial transport.

In detailed field work, the crucial question in indicator tracing in the search for ore deposits is the length and direction of glacial transport. The general flow directions of the Scandinavian ice sheet have been reconstructed with the aid of striae and fabric observations (Hirvas and Nenonen 1987), satellite imagery of depositional landforms (Punkari 1984), and the directions of boulder fans and trains (Salonen 1987). The direction of glacial transport can often be determined with great accuracy during field prospecting. Determination of the transport distance tends to be one of the more difficult tasks.

Generalizations concerning the transport distances of surface boulders in Finland (Salonen 1986) may be useful in this respect. In that study, a new method was developed to describe the transport distance distribution of surface boulders using two statistics, the geometric mean (b) and the coefficient of deviation (s), a dispersion index. Systematic variations in the values of the statistics were found, and they have been compared with variable morainic landforms (Fig. 5).

For active ice hummocky moraines, the geometric mean varies between 0.4 and 3.0

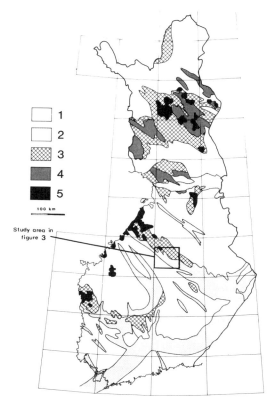

Fig. 4. Areal types for the glacial
geology of Finland:
1 - Glacial transport and origins of till
linked to ice lobe flow patterns
during deglaciation.
2 - Glacial activity associated with till
renewal and processes hampering
boulder transport in marginal and interlobate zones.
3 - Tills and glacial transport associated
with the glacial activity preceding
deglaciation.
4 - Area characterized by Peräpohjola
interstadial deposits (Korpela 1969) and
by the underlying till unit.
5 - Area characterized by repeated periods
of weak glacial erosion, preservation
of weathered bedrock, and mixing of products of several glacial cycles.

km (Fig. 5A). The geometric mean for the
transport distance in cover moraine areas
(Fig. 5B) varies between 0.8 and 10 km and
the surface boulders of drumlins have a
geometric mean of 5-17 km (Fig. 5C) as a
parameter for their transport distance
distribution.

Because the transport distance distributions are obviously associated with the
depositional morainic landforms, they can
also be connected with genetic variations
in glacial deposits. The active ice hummocky moraines and Rogen moraines are
often covered with angular boulders and
have more densely packed sandy till core.
Genetically the deposit belongs to the
last erosion/sedimentation cycle in the
transition zone of active and stagnant ice
front formed during deglaciation (see
Bouchard et al. 1984).

Cover moraine consists usually of thin
beds of basal melt-out till. It has been
deposited by relatively inactive ice.
Drumlin assemblages are subglacial deposits associated with rapid lobate flow of
glacier.

Mixed transport population (Fig. 5D) has
been encountered in ground moraine areas
especially in northern Finland. The deposit often consists of one or more lodgement till beds. It may contain rock
material transported within several glacial cycles and the transport distance (b)
of surface boulders varies within a wide
range (1.0-25 km).

To synthesize the variability of glacial
transport in Finland, the transport
distance map (Fig. 6) was compiled by com-
bining the data on glacial transport
directions and the form of ice recession
lines with measurements of glacial
transport distances. The results of 111
transport distance distributions were
extrapolated to cover the whole area by
comparing them with the length distribution of boulder trains (n=454) and half-
distance measurements (n=41) (Salonen
1986).

In the search for ore deposits, knowledge of glacial processes and their relative influence (Fig. 4) will make it
easier to show the existence of different
depositional units and their complexity in
the till stratigraphy. However, by first
estimating the transport directions and
the length of glacial transport (Fig. 6)
we can gain direct, numeric information
that will then be useful in exploration at
its tactical stage.

The above is a gross simplification of
the subject. In practice, investigations
at individual target sites require local
observations to establish the factors
actually governing glacial activity at
that scale. Furthermore, the model presented here explains only the regularities
caused by subglacial processes that are
supposed to prevail in dispersal patterns
in central areas of Quaternary ice sheets.
Nevertheless, englacial and supraglacial

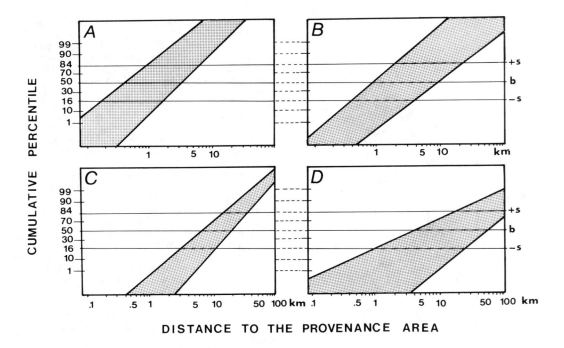

Fig. 5. The series of morainic landforms (A-D) and their generalized transport distance distributions (Salonen 1986). The transport distance increases from hummocky moraines (A) to drumlin assemblage (C) with a simultaneous decrease in their standard deviation. Mixed transport populations (D) in some ground moraine areas are characterized by variable transport distance and high value of standard deviation (gently sloping curves).

transport produces bedrock material derived from more distant provenance areas. This leads to more complicated dispersal of indicators, especially in sedimentary lowlands situated away from the shield margins (Goldthwait 1971).

## 6 SUMMARY

The crucial point in ore exploration based on indicator tracing is the establishing of the provenance areas of the indicators in bedrock. If the amount of indicator particles is high and the glacial dispersal pattern is simple, tracing itself produces information on provenance areas.

When the indicator of an orebody is a single ore boulder, a small cluster of mineralized particles, or a minor till geochemical anomaly, more detailed studies are needed. The genesis of glacial deposits, till lithology, and stratigraphy have to be interpreted before the glacial processes participating in the transport of the indicator particles can be established; the more complicated the gla-

cial history of the study site, the more detailed the genetic studies that are needed when applying the methods in ore exploration.

Understanding the reason for the variability observed in boulder transport facilitates identification of glacial processes and helps to explain their role in the transport of till material. Variability is notably high. Hence, some areas are much more promising than others for indicator tracing in the search for ore deposits.

The length and direction of glacial transport show a close relationship with glacial dynamic processes, and these data may be directly applicable in field prospecting. However, one should bear in mind that local observations are always needed to connect facts with theory.

## ACKNOWLEDGEMENTS

The manuscript was greatly improved by comments of Professor Aleksis Dreimanis and of an anonymous reviewer.

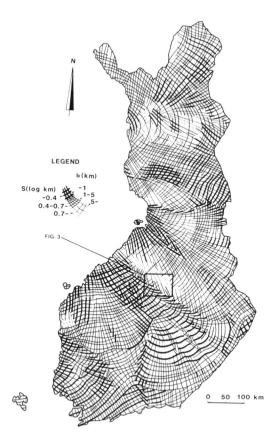

N

LEGEND

b (km)

S(log km)       -1
-0.4 -      1-5
0.4-0.7-    ,5-
0.7--

FIG. 3

0   50  100 km

Fig. 6. Areal variability of glacial transport in Finland, based mainly on surface boulder data (Salonen 1986). The thickness of the lines parallel to the main boulder transport direction is inversely proportional to the value of the geometric mean (b) of transport distance. The lines transverse to the ice flow direction indicate successive ice marginal positions during deglaciation. The boldness of the lines reflects the standard deviation (s) of the transport distance distribution. In the darkest areas, strong local dispersal patterns are expected to prevail. In the lightest areas, mixed and diluted anomaly patterns and boulder trains make indicator tracing difficult.

REFERENCES

Aurola, E. 1955. Über die Geschiebeverfrachtung in Nordkarelien. Geologinen tutkimuslaitos, Geotekn. julk. 56.

Bouchard, M.A., B. Cadieux and F. Coutier 1984. L'origine et les caracteristiques des lithofacies du till dans le secteur du Lac Albanel, Québec: Une etude de la dispersion glaciaire clastique. Chibougamou-Stratigraphy and Mineralization, ed. by J. Guha and E.H. Chown, CIM Spec. Vol. 34:244-260.

Bouchard, M.A. and C. Marcotte 1986. Regional glacial dispersal patterns in Ungava, Nouveau-Quebec. Current Research, Part B, Geol. Surv. Canada, Paper 86-1B, p. 295-304.

Boulton, G.S. 1974. Processes and patterns of subglacial sedimentation: A theoretical approach. In A.E. Wright and F. Moseley (eds.), Ice ages: Ancient and modern, p. 7-42. Liverpool, Seel House Press.

Boulton, G.S. 1984. Development of a theoretical model of sediment dispersal by ice sheets. Prospecting in areas of glaciated terrain 1984, Proceedings, 213-224. London, IMM.

Boulton, G.S., G.D. Smith, A.S. Jones, and J. Newsome 1985. Glacial geology and glaciology of the last mid-latitude ice sheets. J. Geol. Soc. London 142:447-474.

DiLabio, R.N.W. 1981. Glacial dispersal of rocks and minerals at the south end of Lac Mistassini, Quebec with special reference to the Icon dispersal train. Geol. Surv. Canada, Bull. 323.

Drewry, D. 1986. Glacial Geologic Processes. London, Edward Arnold.

Ekdahl, E. 1982. Glacial history and geochemistry of the till/bedrock interface in prospecting in the Pielavesi area of Central Finland. Prospecting in areas of glaciated terrain 1982, Proceedings, 213-227. CIM 1982.

Gaál, G. and V. Kuosmanen, in press. Exploration target selection by integration of geodata using statistical and image processing techniques: An example from Central Finland. Geol. Surv. Finland, Rep. Invest.

Glückert, G. 1973. Two large drumlin fields in central Finland. Fennia 120.

Goldthwait, R.P. 1971. Introduction to till, today. In R.P. Goldthwait (ed.), Till: A Symposium, p. 3-26. Columbus, Ohio State University Press.

Grip, E. 1953. Tracing of glacial boulders as an aid to ore prospecting in Sweden. Econ. Geol. 48:715-725.

Hirvas, H., A. Alftan, E. Pulkkinen, R.

Puranen, and R. Tynni 1977. Raportti malminetsintää palvelevasta maaperä tutkimuksesta Pohjois-Suomessa vuosina 1972-1976. Summary: A report on glacial drift investigations for ore prospecting purposes in northern Finland 1972-1976. Geol. Surv. Finland, Rep. Invest. No. 19.

Hirvas, H. and K. Nenonen 1987. The till stratigraphy of Finland. Geol. Surv. Finland, Spec. Paper 3.

Hughes, T.J. 1981. Numerical reconstruction of paleo-ice sheets. In G.H. Denton and T.J. Hughes (eds.), The last great ice sheets, p. 221-261. New York, John Wiley and Sons.

Hughes, T.J., G.H. Denton, B.G. Andersen, D.H. Schilling, J.L. Fastook, and C.S. Lingle 1981. The last great ice sheets: A global view. In G.H. Denton and T.J. Hughes (eds.), The last great ice sheets, p. 263-317. New York, John Wiley and Sons.

Hyvärinen, L., K. Kauranne, and V. Yletyinen 1973. Modern boulder tracing in prospecting. Prospecting in areas of glaciated terrain 1973, p. 87-95. London, IMM.

Ignatius, H., K. Korpela, and R. Kujansuu 1980. The deglaciation of Finland after 10,000 B.P. Boreas 9:217-228.

Kahma, A. 1973. The main metallogenic features of Finland. Geol. Surv. Finland, Bull. 265.

Kauranne, L.K. 1976. Conceptual models in exploration geochemistry. Norden 1975. J. Geochem. Explor. 5:173-420.

Korpela, K. 1969. Die Weichsel-Eiszeit und ihr Interstadial in Peräpohjola (Nördliches Nordfinnland) im Licht von submoränen Sedimenten. Ann. Acad. Sci. Fennicae Ser A. III, 99.

Kujansuu, R. and R. Niemelä 1984. Quaternary deposits of Finland 1:1,000,000. Geol. Surv. Finland.

Lundqvist, G. 1935. Blockundersökningar, historik och metodik. Sver. Geol. Unders. C 390.

Minell, H. 1978. Glaciological interpretations of boulder trains for the purpose of prospecting in till. Sver. Geol. Unders. C 743.

Niemelä, J. 1979. Suomen sora- ja hiekkavarojen arviointiprojekti 1971-78. Summary: The gravel and sand resources of Finland: An inventory project 1971-78. Geol. Surv. Finland, Rep. Invest. No. 42.

Nurmi, A. 1976. Geochemistry of the till blanket at the Talluskanava Ni-Cu-ore deposit, Tervo, Central Finland. Geol. Surv. Finland, Rep. Invest. No. 15.

Punkari, M. 1984. The relations between glacial dynamics and tills in the eastern part of the Baltic Shield. Striae 20:49-54.

Saarnisto, M. and H. Peltoniemi 1984. Glacial stratigraphy and compositional properties of till in Kainuu, eastern Finland. Fennia 162:163-199.

Salminen, R. and A. Hartikainen 1985. Glacial transport of till and its influence on interpretation of geochemical results in North Karelia, Finland. Geol. Surv. Finland, Bull. 335.

Salonen, V-P. 1986. Glacial transport distance distributions of surface boulders in Finland. Geol. Surv. Finland, Bull. 338.

Salonen, V-P. 1987. Observations on boulder transport in Finland. Geol. Surv. Finland, Spec. Paper 3.

Salonen, V-P. and M. Tenhola, in press. Quaternary geology. In G. Gaál and V. Kuosmanen (eds.), Exploration target selection by integration of geodata using statistical and image processing techniques: An example from Central Finland. Geol. Surv. Finland, Rep. Invest.

Saltikoff, B. 1984. Boulder tracing and the mineral indication data bank in Finland. Prospecting in areas of glaciated terrain 1986, Proceedings, 179-191. London, IMM.

Sauramo, M. 1924. Tracing of glacial boulders and its application in prospecting. Bull. Comm. Geol. Finlande 67.

Shilts, W.W. 1976. Glacial till and mineral exploration. In R.F. Legget (ed.), Glacial till -- an interdisciplinary study, p. 205-224. Ottawa, The Royal Society of Canada.

Shilts, W.W. 1984. Till geochemistry in Finland and Canada. J. Geochem. Explor. 21:95-117.

Wennervirta, H. 1968. Application of geochemical methods to regional prospecting in Finland. Bull. Comm. Geol. Finlande 234.

Genetic Classification of Glacigenic Deposits, Goldthwait & Matsch (eds)
© 1988 Balkema, Rotterdam. ISBN 90 6191 694 1

# On a genetic classification of tills in engineering geology

Owen L.White
*Ontario Geological Survey, Ministry of Northern Development and Mines, Toronto, Ontario, Canada, and*
*Quaternary Sciences Institute, University of Waterloo, Waterloo, Ontario, Canada*

Karen Ellis
*Ontario Geological Survey, Ministry of Northern Development and Mines, Toronto, Ontario, Canada*

ABSTRACT: Till is one of the many types of soil an engineer encounters and uses in the course of construction and it is thus embraced by the several soil classifications used by engineers. This paper looks at the possibility of applying a genetic classification of till in engineering geology.

Many investigators have attempted to provide a genetic classification of till for engineering purposes, but all have concluded the necessity of providing additional information beyond that of the basic genetic classification. Likewise, engineering classifications of soils are inadequate where tills are involved because the classifications lack any consideration of genetic factors. Thus, for a useful classification of till for engineering, description of the basic physical parameters of the material must be accompanied by an evaluation of its genetic environment.

Tills are just one of the many types of soil that an engineer uses and works with. The engineer constructs buildings on till, constructs tunnels in till, excavates below the surface of till, and uses till to construct embankments such as bridge approaches and dams.

Till is not ubiquitous but its occurrence is certainly widespread. The engineering usage of till is extensive in the northern hemisphere and is increasingly so in the southern hemisphere in Argentina, Australia, and Chile.

The engineer uses a wide range of materials and has devised systems to organize his/her knowledge and experience with those materials. Soils are no exception to this practice and, in fact, a variety of soil classification systems have been developed in engineering to meet various needs. Some of these systems have wide international application and embrace many facets of the work of the geotechnical engineer. In the Unified system, for example, the prime differentiation of the various soils is on the basis of textural analysis into coarse- and fine-grained soils with further subdivision in each category depending on the grading (coarse-grained soils) or the plasticity characteristics (fine-grained soils). This initial classification leads to additional evaluation of the various soil properties in which the engineer is directly interested.

In our consideration then, of a genetic classification of till, it is not surprising that we should look at the potential of applying a genetic classification of till to engineering practice.

John Elson (1961) was one of the first authors to link the geology of tills to their geotechnical properties -- in a paper presented to a Canadian geotechnical meeting in 1960. He showed that two major till types could be characterised in terms of their engineering properties as follows:

1) ablation till: a loose, bouldery material with a low density and high permeability and

2) lodgement or comminution till: a compact material with a high density and low permeability.

Elson also lists a deformation till referred to at times as "soft" till by some engineers and usually found as a fine-grained till with higher than usual natural water content. This deformation or soft till may occur in zones of variable size, completely enclosed by more compact till with a lower water content.

The above twofold classification of till has been used by many engineers over the years but usually with additional comments that the character of tills also varies according to:

1) the type of parent material over which the ice sheet moved before the till

was deposited and

2) the postdepositional changes that may occur (for example, Milligan 1976; Lutenegger, Kemmis, and Hallberg 1983; and Gerath 1987).

In the late 1960's and the 1970's, several investigators, particularly in the United Kingdom, spent considerable time and effort to show the relationship between glacial processes and the geotechnical properties of till in an endeavour to develop a satisfactory classification of till from an engineering or geotechnical viewpoint.

In 1971, McGown noted the various physical parameters that had been used in the past to identify till, but concluded that most were not adequate and that a useful classification of till must be based on the identification of the soil matrix together with the mode of deposition, the lithology, and the postdepositional history.

A few years later, McGown and Derbyshire (1977) suggested a classification that first identified five principal genetic till types (Table 1) before proceeding with further subdivision based on the dominant matrix material and various fabric features.

Table 1. Principal Genetic Till Types (after McGown and Derbyshire 1977)

| Till Process Characteristics | | |
|---|---|---|
| Formation | Transport | Deposition |
| Comminution | Superglacial | Flow |
| | | Ablation |
| | Englacial | |
| | | Meltout |
| | Basal | |
| | | Lodgement |
| Deformation | | |

Thus, McGown, his colleagues, and other investigators were advocating a genetic based, multi-till type classification that required input on matrix lithology and physical characteristics and fabric identification (i.e., postdepositional history). At no time has it been suggested that a genetic classification alone is adequate for a geotechnical classification of till. In fact, most authors have been quite emphatic in identifying the need for the use of criteria beyond that provided by a consideration of genetic factors. For example, if tills have their natural water content (i.e., they occur below the groundwater table), then fine-textured deformation till and flow till will always be softer than lodgement or basal meltout till.

It should also be noted that, although a genetic classification alone is not adequate to classify till for engineering purposes, the current systems of engineering soil classification are totally inadequate in dealing with soils of various genetic origins.

The Unified system mentioned above involves an initial classification based on a grain-size analysis, with further classification depending on the determination of more detail of the particle-size distribution or the plasticity characteristics. Under this system there is no differentiation between, for example, a clay of low plasticity deposited in a lacustrine environment and a clay of low plasticity found as a basal till. The current practice of the informed operator calls for the addition of the word "till" to the descriptive term where appropriate, e.g., silty clay till (CL).

The successful geotechnical engineer or engineering geologist is the one who is able to apply an appreciation of origin and geologic history to the results of the soil mechanics evaluation of a glacial deposit. The recognition of a till as having been deposited in either a basal or supraglacial environment (or even a deformation or flow environment) will certainly enhance the understanding of the density-compressibility or density-strength relationships of that material.

In conclusion, therefore, we have no option but to acknowledge that a genetic classification of till for engineering is clearly an inadequate procedure, for such a classification cannot provide the engineer with all the information needed about the material.

At the same time we must also note that conventional engineering soil classification systems are equally inadequate when applied to glacial soils. For the engineer to be fully informed about soils of glacial origin, an awareness of soil genesis is necessary for a full and realistic evaluation of the potential behaviour of the soil material.

Therefore, when glaciogenic soils are being sampled, described, and identified, the investigator should provide as a minimum for each sample the textural analysis, the Atterberg Limits (plastic limit, liquid limit, etc.), and the natural water content along with a statement, wherever possible, of the genetic origin of the soil. Where tills are involved, some attempt should be made to indicate the depositional environment of the till. As circumstances permit (and may indeed require), additional physical parameters of the soil should be determined -- for example, colour, structure, weathering, density, strength, and permeability.

## REFERENCES

Elson, J.A. 1961. The geology of tills. E. Penner and J. Butler (eds.), Proc. 14th Canadian Soil Mechanics Conference, NRCC Associate Committee on Soil and Snow Mechanics, p. 5-36. Ottawa, Technical Memo 69.

Gerath, R.F. 1987. Till; from a survey of geotechnical colleagues. Unpublished poster presentation at the Penrose Conference on Glacial Facies Models, Toronto, 1987. Geological Society of America.

Luteneger, A.J., T.J. Kemmis, and G.R. Hallberg 1983. Origin and properties of glacial till and diamictons. In R.N. Yong (ed.), Special publication on geological environment and soil properties, p. 310-331. New York, American Society of Civil Engineers.

McGown, A. 1971. The classification for engineering purposes of tills from moraines and associated landforms. Quarterly Journal of Engineering Geology 4:115-130.

McGown, A. and E. Derbyshire 1977. Genetic influences on the properties of tills. Quarterly Journal of Engineering Geology 10:389-410.

Milligan, V. 1976. Geotechnical aspects of glacial tills. In R.F. Legget (ed.), Glacial tills, p. 269-291. Royal Society of Canada, Spec. Publ. No. 12.

NOTE: This paper is published with the approval of the Director, Ontario Geological Survey.

In the course of preparing this paper, a "Bibliography on the Geotechnical Properties and Classification of Till" was compiled and used as a working document. The authors do not claim that the Bibliography is complete, but considerable efforts were made to include all relevant papers. Gaps may be extensive among the papers published in languages other than English. Most, but not all, papers listed in the Bibliography have been perused and the bibliographic details checked. A chronological listing of the items in the Bibliography is available on request to the senior author.

## BIBLIOGRAPHY ON THE GEOTECHNICAL PROPERTIES AND CLASSIFICATION OF TILL

Adam, C.H. 1985. Static electric cone penetration testing in the glacial tills of central Scotland. In M.C. Forde (ed.), Glacial tills 85, p. 75-82. Edinburgh, Engineering Technics Press Ltd.

Adams, J.I. 1961. Tests on glacial till. In E. Penner and J. Butler (eds.), Proceedings 14th Canadian Soil Mechanics Conference, p. 37-49. NRCC Assoc. Commission on Soil and Snow Mechanics. Ottawa, Tech. Memo. No. 69.

Adams, J.I. and H.S. Radhakrisna 1979. Uplift resistance of augered footings in fissured clay. Canadian Geotechnical Journal 8:452-462.

Al-Shaikl-Ali, M.M.H. 1978a. The behaviour of Cheshire Basin lodgement till in motorway construction. Proceedings Clay Fills, p. 15-23. London, Institute of Civil Engineering.

Al-Shaikl-Ali, M.M.H. 1978b. Full scale pile testing to failure to determine the effect of fissuring in stiff boulder clay in Cheshire. The engineering behaviour of glacial materials. Proc. Symp. at University of Birmingham 21-23 April 1975, 2nd edition, p. 229-234. Norwich, Geo Abstracts.

Alderman, J.K. 1959. The geotechnical properties of glacial deposits of North-West England. Ph.D. thesis, University of Manchester.

Anderson, J.G.C. and C.F. Trigg 1970. Geotechnical factors in the redevelopment of south Wales valleys. Proceedings of the Conference on Civil Engineering Problems of the South Wales Valleys, Cardiff, April, 1969.

Anderson, W.F. 1972. The geotechnical properties of the till of the Glasgow region. Ph.D. thesis, University of Strathclyde.

Anderson, W.F. 1974. Factors influencing the measured properties of Glasgow region till. Ground Engineering 7:20-27.

Anderson, W.F. and D.G. McKinlay 1978. Tests to find the modulus of deformation of till. The engineering behaviour of glacial materials. Proc. Symp. at University of Birmingham 21-23 April 1975, 2nd edition, p. 165-178. Norwich, Geo Abstracts.

Andrews, J.T. and D.I. Smith 1970. Statistical analysis of till fabric: Methodology, local and regional variability. Quarterly Journal of Geological Society of London 125:503-542.

Arber, N.R. 1985. Excavation redesign due to groundwater problems. In M.C. Forde (ed.), Glacial tills 85, p. 83-86. Edinburgh, Engineering Technics Press Ltd.

Armstrong, J.E. 1961. Soils of the coastal area of southwest British Columbia. In R.F. Legget (ed.), Soils in Canada, p. 22-32. Royal Society of Canada, Sp. Publ. No. 3.

Atkinson, J.H., P.I. Lewin, and C.L. Ng 1985. Undrained strength and overconsolidation of a clay till. In M.C. Forde (ed.), Glacial tills 85, p. 49-54. Edinburgh, Engineering Technics Press Ltd.

Balstrup, T. 1969. Danske Erfaringer Av Moraneler: Morandag 1969, Symposium anordnat av Svenska Geotekniska Foreningen den 3 december, Reprints and Preliminary Reports, p. 107-114. Swedish Geotechnical Institute, No. 39.

Batchelor, A.J. and M.C. Forde 1985. Relationship between undisturbed and remoulded effective stress shear strength parameters. In M.C. Forde (ed.), Glacial tills 85, p. 187-191. Edinburgh, Engineering Technics Press Ltd.

Batchelor, A.J., M.C. Forde, and B.H.V. Topping 1985. Slope stability design in glacial tills. In M.C. Forde (ed.), Glacial tills 85, p. 219-226. Edinburgh, Engineering Technics Press Ltd.

Bedell, P.R. 1967. The effect of overconsolidation on the modulus of elasticity of clay till. M. Eng. Sc. thesis, University of Western Ontario, London, Ontario.

Bell, A.L., B.J. Gregory, and A. McCann 1984. Short-term stability of cuts in a glacial till. Abstracts: IV International Symposium on Landslides, p. 103. Toronto, Canadian Geotechnical Society.

Bernander, S. 1969. Synpunkter Pa Moranens Barighet Samt Hur Deformationsegenskaperna Paverkar overbyggnadens Konstruktion: Morandag 1969, Symposium anordnat av Svenska Geotekniska Foreningen den 3 december, Reprints and Preliminary Reports, p. 95-106. Swedish

Geotechnical Institute, No. 39.

Bernell, L. 1957. The properties of moraines. Proceedings 4th International Conference on Soil Mech. and Fdn. Eng. 2:286-290. London.

Bernell, L. 1969. Moranens Egenskaper Som Byggnadsmaterial: Morandag 1969, Symposium anordnat av Svenska Geotekniska Foreningen den 3 december, Reprints and Preliminary Reports, p. 19-32. Swedish Geotechnical Institute, No. 39.

Bevan, O.M. and D.B. Parkes 1978. Tunnelling in glacial materials in the British Isles. The engineering behaviour of glacial materials. Proc. Symp. at University of Birmingham 21-23 April 1975, 2nd edition, p. 212-220. Norwich, Geo Abstracts.

Bou, Wei-Tseng 1975. Instability of the bluff slopes at the Elgin area pumping station, North Shore of Lake Erie, M. Sc. thesis, University of Western Ontario, London, Ontario.

Boulton, G.S. 1968. Flow tills and related deposits on some Vestspitsbergen Glaciers. Journal of Glaciology 7:391-412.

Boulton, G.S. 1970. On the deposition of subglacial and melt-out tills at the margins of certain Svalbard glaciers. Journal of Glaciology 9:231-245.

Boulton, G.S. 1971. Till genesis and fabric in Svalbard Spitsbergen. In R.P. Goldthwait (ed.), Till: A symposium, p. 41-72. Columbus, Ohio State University Press.

Boulton, G.S. 1975. Process and patterns of subglacial sedimentation: A theoretical approach. In A.E. Wright and F. Mosley (eds.), Ice ages ancient and modern. Liverpool, Seal House Press.

Boulton, G.S. 1976a. Criteria for distinguishing between tills of different types. In W. Stankowski (ed.), Till -- its genesis and diagenesis, p. 176-185. INQUA Symp. Poland 1975. UAM Geografia, No. 12.

Boulton, G.S. 1976b. The development of geotechnical properties in glacial tills. In R.F. Legget (ed.), Glacial till, p. 292-303. Royal Society of Canada, Sp. Publ. No. 12.

Boulton, G.S. 1976c. A genetic classification of tills and criteria for distinguishing tills of different origin. In W. Stankowski (ed.), Till -- its genesis and diagenesis, p. 65-80. INQUA Symp. Poland 1975. UAM Geografia, No. 12.

Boulton, G.S. 1978a. Boulder shapes and grain-size distribution of debris as indicators of transport paths through a glacier and till genesis. Sedimentology 25:773-799.

Boulton, G.S. 1978b. The genesis of glacial tills -- a framework for geotechnical interpretation. The engineering behaviour of glacial materials. Proc. Symp. at University of Birmingham 21-23 April 1975, 2nd edition, p. 52-59. Norwich, Geo Abstracts.

Boulton, G.S. 1980. Genesis and classification of glacial sediments. In W. Stankowski (ed.), Tills and glacigene deposits, p. 15-17. INQUA Symp., U.K. 1977. UAM Geografia, No. 20.

Boulton, G.S. and D.L. Dent 1974. The nature and rates of post-depositional changes in recently deposited till from south-east Iceland. Geografiska Annaler 56A:121-134.

Boulton, G.S. and M. Deynoux 1981. Sedimentation in glacial environments and identification of tills and tillites in ancient sedimentary sequences. Precambrian Research 15:397-422.

Boulton, G.S. and N. Eyles 1979. Sedimentation by valley glaciers: A model and genetic classification. In Ch. Schlüchter (ed.), Moraines and varves, p. 11-23. Proceedings INQUA Symp. on Genesis and Lithology of Quaternary Deposits, 1978. Rotterdam, Balkema.

Boulton, G.S. and M.A. Paul 1976. The influence of genetic processes on some geotechnical properties of glacial tills. Quarterly Journal of Engineering Geology 9:159-194.

Boulton, G.S., D.L. Dent, and E.M. Morris 1974. Subglacial shearing and crushing and the role of water pressure in tills from south-east Iceland. Geografiska Annaler 56A:135-145.

Brennan, M.J. and K. Ryan 1985. A method for estimating the compressibility factor of a glacial till and its use in pavement design. In M.C. Forde (ed.), Glacial tills 85, p. 71-73. Edinburgh, Engineering Technics Press Ltd.

Breth, H. 1967. Calculation of the shearing strength of a moraine subjected to landsliding due to reservoir filling in Kauner Valley, Austria. Proceedings Geotechnical Conference, Oslo 1:171-174.

Broms, B.B. 1973. The geotechnical aspects of moraines. Bulletin of the Geological Institute, University of Uppsala, New Series 5:51-60.

Broster, B.E. 1986. Till variability and compositional stratification: Examples from the Port Huron lobe. Canadian Journal of Earth Sciences 23:1823-1841.

Browne, M.A. 1985. The tills of central Scotland in their stratigraphical context. In M.C. Forde (ed.), Glacial

tills 85, p. 11-24. Edinburgh, Engineering Technics Press Ltd.

Bukin, P.A. 1968. Characteristics of dams constructed with moraine soils at the hydroelectric developments of the Kovda Cascade. Hydrotechnical Construction 2:112-118.

Busbridge, J.R. 1968. In-situ geotechnical characteristics of glacial till in the Glasgow area. M. Sc. thesis, University of Strathclyde.

Carter, P., G. Wallace, and G. Cochrane 1985. Till geology and earthworks -- a Scots quair. In M.C. Forde (ed.), Glacial tills 85, p. 67-70. Edinburgh, Engineering Technics Press Ltd.

Carter, T.G. 1983. The site investigation and engineering characterization of glacial and glacilacustrine materials. Unpublished Ph.D. thesis, University of Surrey.

Carter, T.G. 1985. Some engineering implications of glaciofluvial process classification to the construction of the Brenig Dam. In M.C. Forde (ed.), Glacial tills 85, p. 1-10. Edinburgh, Engineering Technics Press Ltd.

Casagrande, A. 1948. Classification and identification of soils. Transactions, American Society of Civil Engineering 113:901-930.

Chamberlain, T.C. 1894. Proposed genetic classification of Pleistocene glacial formations. Journal of Geology 2:517-538.

Chamberlain, T.C. 1895. The classification of American glacial deposits. Journal of Geology 3:270-277.

Chaplow, R. 1982. Engineering geomorphological investigations of a possible landslide, Killiecrankie Pass, Scotland. Engineering Implications of Earth Surface Processes, Eng. Geomorphology. Reg. Conf. of Eng. Group of Geol. Soc. at Birmingham Univ. -- Abstracts, p. 24.

Cherry, J., D. Desaulniers, E. Frind, P. Fritz, D. Gevaert, R. Gillham, and B. LeLievre 1979. Hydrogeologic properties and pore water origin and age: Clayey till and clay in South Central Canada. Workshop in Low-flow Permeability Measurements in Largely Impermeable Rocks. Paris, OECD Nuclear Energy Agency and Int. Energy Agency.

Christiansen, E.A. 1977. Engineering geology of glacial deposits in southern Saskatchewan. 30th Canadian Geotechnical Conference, Saskatoon, Sask., Canadian Geotechnical Society, p. 1-30.

Chryssafopoulos, H.W.S. 1964. Identification of young tills and study of some of their engineering properties in the greater Chicago area. Ph.D. thesis, University of Illinois.

Chryssafopoulos, H.W.S. 1963. An example of the use of engineering properties for differentiation of young glacial till sheets. Proceedings 2nd Panamerican Conference on Soil Mech. and Fdn. Eng. 2:35-43.

Cleaves, A.B. 1963. Engineering geology characteristics of basal tills, St. Lawrence Seaway project. In D. Trask and A. Kiersch (eds.), Engineering geology case histories, Geological Society of America, No. 4:52-57.

Cochrane, S.R. 1985. The clay contamination of road stone layers on glacial till subgrades. In M.C. Forde (ed.), Glacial tills 85, p. 87-91. Edinburgh, Engineering Technics Press Ltd.

Cocksedge, J.E. and D.W. Hight 1975. Some geotechnical aspects of road design and construction in tills. The engineering behaviour of glacial materials. Proc. Symp. at University of Birmingham 21-23 April 1975, 2nd edition, p. 197-206. Norwich, Geo Abstracts.

Colback, P.S.B., T.G. Carter, and D.J. Eastaff 1975. Investigation of the glacial deposits forming an integral part of the Brenig Dam and its foundation. Proceedings 11th Regional Meeting Geological Society, p. 27-31.

Conlon, R.J., R.G. Tanner, and K.L. Coldwell 1971. The geotechnical design of the townline road-rail tunnel. Canadian Geotechnical Journal 8:299-314.

Conway, B.W., A. Forster, and K.J. Northmore 1982. The glacial/periglacial legacy of mass movement in the South Wales coalfield. Engineering implications of earth surface processes. Eng. Geomorphology. Reg. Conf. of Eng. Group of Geol. Soc. at Birmingham Univ. -- Abstracts, p. 26.

Cotton, R.D. and E. Derbyshire 1982. Application of glacial process studies in large scale site investigation: The case of the Kielder Dam, Northumberland, England. Engineering implications of earth surface processes. Eng. Geomorphology. Reg. Conf. of Eng. Group of Geol. Soc. at Birmingham Univ. -- Abstracts, p. 22.

Crawford, C.B. and K.N. Burn 1962. Settlement studies on the Mt. Sinai Hospital, Toronto. Engineering Journal, Dec. 1962. Canada, p. 31-37.

Cregger, D.M., P.J. Majeski, and T.Y. Chang 1983. Geotechnical characterization of Ohio Valley soils. In R.N. Yong (ed.), Special publication on geological environment and soil properties, p. 387-404. American Society of Civil Engineering.

Cubitt, J.M., D.E. Andrews, and B. Denness 1978. Automatic identification and evaluation of geotechnical zones for till. Bulletin of the Association of Engineering Geology 15:355-374.

Culley, R.W. 1971. Effect of freeze-thaw cycling on stress-strain characteristics and volume change of a till subjected to repetitive loading. Canadian Geotechnical Journal 8:359-371.

Dahl, A.R., D.T. Davidson, and C.J. Roy 1960. Petrography and engineering properties of loess and Kansan till in southern Iowa. In D.T. Davidson (ed.), Geologic and engineering properties of Pleistocene materials in Iowa. Iowa State University Bulletin 59:93-132.

Dahl, R., K. Berg, and R. Nalsund 1981. Stabilitetsforholdene i skraninger med morene og lignende jordarter; Geol. Inst. NTH, Report 17.

Dascal, O. 1982. Compaction practice for dam cores at Hydro-Quebec. Transportation Research Record, T.R.B., Washington D.C., No. 897.

Davis, A.G. 1985. Non-destructive testing of piles founded in glacial tills -- analysis of over 30 case histories. In M.C. Forde (ed.), Glacial tills 85, p. 193-212. Edinburgh, Engineering Technics Press Ltd.

Davitt, S. and G.A. Bonner 1977. The construction of road embankments with boulder clay at Stillorgan Road, Dublin: An Foras Forbartha, Teo R.C. 130.

DeJong, J. and M.C. Harris 1971. Settlement of two multistory buildings in Edmonton. Canadian Geotechnical Journal 8:217-235.

DeJong, J. and N.R. Morgenstern 1973. Heave and settlement of two tall building foundations in Edmonton, Alberta. Canadian Geotechnical Journal 10:261-281.

DeLory, F.A., A.M. Crawford, and M.E.M. Gibson 1979. Measurements on a tunnel lining in very dense till. Canadian Geotechnical Journal 16:190-199.

Denby, G.M., C.A. Costa, G.W. Clough, and R.R. Davidson 1981. Laboratory and pressuremeter tests on a stiff clay. Proc. 10th International Conference on Soil Mech. and Fdn. Eng. Stockholm, 15-19 June, 1981, 1:577-580.

Denness, B. 1974. Engineering aspects of the chalky boulder clay at the new town of Milton Keyes in Buckinghamshire. Quarterly Journal of Engineering Geology 7:297-309.

Deoglas, D.J. 1968. Grain-size indices, classification and environment. Sedimentology 10:83-100.

Depiante, E. 1983. Geotechnical studies at

Anfiteatro Moraine, Rio Limay Valley, Argentina. In E.B. Evenson, Ch. Schlüchter, and J. Rabassa (eds.), Tills and related deposits, p. 205-209. Rotterdam, Balkema.

Derbyshire, E. 1980. The relationship between depositional mode and fabric strength in tills: Schema and test from two temperate glaciers. In W. Stankowski (ed.), Tills and glacigene deposits, p. 41-48. INQUA Symp., U.K. 1977. UAM Geografia, No. 20.

Derbyshire, E. In press. Till properties and glacier regime in parts of high Asia: Karakoram and Tien Shan. The evolution of the East Asian environment. University of Hong Kong.

Derbyshire, E. and P.F. Jones 1980. Systematic fissuring of a matrix-dominated lodgement till at Church Wilne, Derbyshire, England. Geological Magazine 117:243-254.

Derbyshire, E. and A. McGown 1973. On the properties of some modern glacial tills. Congress of the International Union for Quaternary Research (III), Christchurch, N.Z., No. 9, p. 79.

Derbyshire, E., M.J. Edge, and M. Love 1985. Soil fabric variability in some glacial diamicts. In M.C. Forde (ed.), Glacial tills 85, p. 169-175. Edinburgh, Engineering Technics Press Ltd.

Derbyshire, E., A. McGown, and A.M. Radwan 1976. Total fabric of some till landforms. Earth Surface Processes 1:17-26.

Draney, D. 1982. Geotechnical properties of strip mined areas in Northcentral Missouri. In Abstracts: Annual Meeting of the Missouri Academy of Science, Pt. Lookout, Mo. 16:166.

Dreimanis, A. 1961. Tills of Southern Ontario. In R.F. Legget (ed.), Soils in Canada, p. 80-96. Royal Society of Canada, Sp. Publ. No. 3.

Dreimanis, A. 1969. Selection of genetically significant parameters for investigation of tills. Zesz. Nauk. Univ. Im. A. Mickiewicza Posnaniu. Geografia, No. 8, p. 15-29.

Dreimanis, A. 1970. Criteria for distinction of till from other diamictons. Discussion. AMQUA Abstracts, First Meeting, Yellowstone Park and Bozeman, p. 36.

Dreimanis, A. 1971. Procedure of till investigation in North America: A general review. In R.P. Goldthwait (ed.), Till: A symposium, p. 27-37. Columbus, Ohio State Univ. Press.

Dreimanis, A. 1976. Tills: Their origin and properties. In R.F. Legget (ed.), Glacial till, p. 11-48. Royal Society of Canada, Sp. Publ. No. 12.

Dreimanis, A. 1980. Terminology and development of genetic classification of material transported by glaciers. In W. Stankowski (ed.), Tills and glacigene deposits, p. 5-10. INQUA Symp., U.K. 1977, UAM Geografia, No. 20.

Dreimanis, A. 1982. Genetic classification of tills and criteria for their differentiation. In Ch. Schlüchter (ed.), INQUA Progress Report on Activities 1977-1982, and Definitions of Glacigenic Terms. Zurich, Honggerberg.

Dreimanis, A. and J. Lundqvist 1984. What should be called till? Striae 20:5-10.

Dreimanis, A. and Ch. Schlüchter 1985. Field criteria for the recognition of till or tillite. Paleogeography, Paleoclimatology, Paleoecology 51:7-14.

Dreimanis, A. and U.J. Vagners 1971. The effect of lithology upon texture of till. In Yatsu and Falconer (eds.), Research methods in Pleistocene geomorphology, p. 66-82. 2nd Guelph Symposium.

Drozdowski, E. 1977. Ablation till and related indicatory forms at the margins of Vestspitsbergen Glaciers. Boreas 6:107-114.

Dusseault, M.B. and A.G. Vorauer 1986. Geomechanical investigation of near-surface fractures in clay tills. Part A: Fracture mapping at three sites in Southwestern Ontario. Proceedings, Technology Transfer Conference, Ministry of Environment, Toronto, Ontario, p. 266-291.

Easterbrook, D.J. 1964. Void ratios and bulk densities as a means of identifying Pleistocene till. Geological Society of America Bulletin 75:745-750.

Eden, W.J. 1976. Construction difficulties with loose glacial till on Labrador Plateau. In R.F. Legget (ed.), Glacial till, p. 391-400. Royal Society of Canada, Sp. Publ. No. 12.

Edil, T.B., D.M. Mickelson, and L.J. Acomb 1977. Relationship of geotechnical properties to glacial stratigraphic units along Wisconsin's Lake Michigan shoreline. 30th Canadian Geotechnical Conference, Saskatoon, Sask., Canadian Geotechnical Society, part II, p. 36-54.

Eerola, M. 1984. Moreenin ja muiden heikompilaatuisten materiaalien kayttomahdollisuudet tierakenteissa; Maarakennus ja kuljetus 5:238-243.

Eisenstein, Z. and L.V. Medeiros 1983. A deep retaining structure in till and sand. Part II: Performance and analysis. Canadian Geotechnical Journal 20:131-140.

Eisenstein, Z. and K.L. Sorensen 1986. Tunnelling for the south LRT extention in Edmonton, Alberta. In K.Y. Lo, J.H.

Palmer, and C.M. Yeun (eds.), Recent advances in Canadian underground technology. Proceedings 6th Canadian Tunnelling Conference, p. 291-316.

Eisenstein, Z. and S. Thomson 1978. Geotechnical performance of a tunnel in till. Canadian Geotechnical Journal 15:332-345.

Eisenstein, Z., F. El-Nahhas, and S. Thomson 1981. Strain field around a tunnel in stiff soil. Proceedings 10th International Conference on Soil Mech. and Fdn. Eng. Stockholm, 15-19 June, 1981, 1:283-288.

Elson, J.A. 1961. The geology of tills. In E. Penner and J. Butler (eds.), Proceedings 14th Canadian Soil Mechanics Conference, NRCC Assoc. Comm. on Soil and Snow Mechanics, p. 5-36. Ottawa, Tech. Memo. 69.

Engqvist, P. and T. Olsson 1974. Forsok att ur aterhamtningskurvor berakna moranbunnars kapacitet. Preliminar Rapport, Vannet i Norden 4:21-34.

Engqvist, P., T. Olsson, and T. Svensson 1978. Pumping and recovery tests in wells sunk in till. Nord. Hydraulic Conference, Hanasaari Cultural Centre, Finland, 1:42-51.

Evans, S.G. 1982. Landslides and surficial deposits in urban areas of British Columbia: A review. Canadian Geotechnical Journal 19:269-288.

Evenson, E.B. 1971. The relationship of macro- and microfabric of till and the genesis of glacial landforms in Jefferson County, Wisconsin, p. 345-366.

Eyles, N. and J.A. Sladen 1981. Stratigraphy and geotechnical properties of weathered lodgement till in Northumberland, England. Quarterly Journal of Engineering Geology 14:129-141.

Eyles, N., J.A. Sladen, and S. Gilroy 1982. A depositional model for stratigraphic complexes and facies superposition in lodgement tills. Boreas 11:317-333.

Faillace, G.A. and M.L. Silver 1975. Effect of sampling on the dynamic stress strain properties of till. The engineering behaviour of glacial materials. Proc. Symp. at University of Birmingham 21-23 April 1975, 2nd edition, p. 141-148. Norwich, Geo Abstracts.

Farrel, E.R. and R.W. Kirwan 1981. Deformation of glacial till under repeated loading. Proc. 10th International Conference on Soil Mech. and Fdn. Eng. Stockholm, 15-19 June, 1981, 3:195-200.

Fisher, R. 1983. Modulus of elasticity of a very dense glacial till determined by plasticity tests. Canadian Geotechnical Journal 20:186-191.

Fletcher, M.S. and R.A. Nicholls 1984. A buried valley in the Orwell Estuary. Quarterly Journal of Engineering Geology 17:283-288.

Fookes, P.G., D.L. Gordon, and I.E. Higginbottom 1978. Glacial landforms, their deposits and engineering characteristics. The engineering behaviour of glacial materials. Proc. Symp. at University of Birmingham 21-23 April 1975, 2nd edition, p. 18-51. Norwich, Geo Abstracts.

Fookes, P.G., L.W. Hinch, and J.C. Dixon 1972. Geotechnical considerations of the site investigation for Stage IV of Taff Vale Truck Road to South Wales. 2nd Regional Congress, Cardiff 1-25: British National Commission. Permanent International Assoc. Road Congress.

Fookes, P.G., L.W. Hinch, M.A. Huxley, and N.E. Simons 1978. Some soil properties in glacial terrain -- the Taff Valley, South Wales. The engineering behaviour of glacial materials. Proc. Symp. at University of Birmingham 21-23 April 1975, 2nd edition, p. 93-116. Norwich, Geo Abstracts.

Francis, E.A. 1975. Glacial sediments: A selective review. In A.E. Wright and F. Mosely (eds.), Ice ages: Ancient and modern, p. 43-68. Liverpool, Seal House Press.

Fraser, S. 1984. Stability analysis of till embankments. M. Sc. Research Project, University of Dublin.

Funnel, B.M. and P.F. Wilkes 1976. Engineering characteristics of East Anglian Quaternary deposits. Quarterly Journal of Engineering Geology, p. 145-157.

Gardemeister, R. 1967. On the soil and its constructional properties at the Saimaa Canal construction in Finland. Engineering Geology 2:107-115.

Gens, A. and D.W. Hight 1979. The laboratory measurements of design parameters for a glacial till. Proceedings 7th European Conference on Soil Mech. and Fdn. Eng. Brighton, U.K. 2:57-65.

George, H. 1983. Late Quaternary history and engineering geology of the Elk River Valley, Southeastern British Columbia. Ph.D. thesis, Queen's University, Kingston, Ontario.

Goldsworthy, M.H. 1986. Performance of a deep cofferdam around a collapsed tunnel in glacial clays. Ground Engineering, October, 1986, p. 14-18.

Gooding, A.M. 1973. Characteristics of Late Wisconsin tills in Eastern Indiana. Indiana Geological Survey Bulletin,

Bloomington, No. 49.

Gordon, D.L. 1971. A review of geological materials of glacial regions, their engineering properties and site investigation. M.Sc. thesis, University of London.

Gravenor, C.P. 1953. The origin of drumlins. Journal of Science 251:674-681.

Gravenor, C.P. 1974. The Yarmouth drumlin field, Nova Scotia, Canada. Journal of Glaciology 13:45-54.

Gravenor, C.P. and L.A. Bayrock 1961. Glacial deposits of Alberta. In R.F. Legget (ed.), Soils in Canada, p. 33-50. Royal Society of Canada, Sp. Publ. No. 3.

Grisak, G.E. 1975. The fracture porosity of glacial till. Canadian Journal of Earth Sciences 12:513-515.

Grisak, G.E. and J.A. Cherry 1975. Hydrogeologic characteristics and response of fractured till and clay confining a shallow aquifer. Canadian Geotechnical Journal 12:23-43.

Grisak, G.E., J.A. Cherry, J.A. Vonhof, and J.P. Blumele 1976. Hydrogeologic and hydrochemical properties of fractured till in the interior plains region. In R.F. Legget (ed.), Glacial till, p. 304-335. Royal Society of Canada, Sp. Publ. No. 12.

Grumiching, J. and L. Dubreuil 1982. A new method of geotechnical reconnaissance of non-plastic tills by rotary coring. Conference Papers 35th Canadian Geotechnical Conference, Montreal, Sept. 1982, Canadian Geotechnical Society, p. 504-518.

Gyger, M., M. Muller-VonMoos, and C. Schindler 1976. Untersuchungen und Klassifikation spat- und nacheiszeitlicher Sedimente aus dem Zurichsee. Schweiz. Mineral. Petrogr. Mitt. 56:387-406.

Haldorsen, S. 1981. Grain-size distribution of subglacial till and its relation to glacial crushing and abrasion. Boreas 10:91-105.

Haldorsen, S. 1983. The characteristics and genesis of Norwegian tills. In J. Ehlers (ed.), Glacial deposits in North-West Europe, p. 11-17. Rotterdam, Balkema.

Haldorsen, S. and J. Shaw 1982. The problem of recognizing melt-out till. Boreas 11:261-277.

Haldorsen, S., P.D. Jenssen, J. Koler, and E. Myhr 1983. Some hydraulic properties of sandy-silty Norwegian tills. Acta Geologica Hispanica 18:191-198.

Hanrahan, E.T. 1977. Irish glacial till: Origin and characteristics. Dublin,

National Institute for Physical Planning and Construction Research.

Hanrahan, E.T. and M. Phillips 1976. Problems of tunnelling in Dublin boulder clay. Proceedings of 5th I.C.S.M.F.E., Vienna.

Hansbo, S. 1969a. Faltstudier Av Sattningar I Moran Och Moranlera: Morandag 1969. Symposium anordnat av Svenska Geotekniska Foreningen den 3 december. Reprints and Preliminary Reports, Swedish Geotechnical Institute, No. 39, p. 67-78.

Hansbo, S. 1969b. Oversikt Over Pagaende Moranforskning Vid Cth: Morandag 1969. Symposium anordnat av Svenska Geotekniska Foreningen den 3 december. Reprints and Preliminary Reports, Swedish Geotechnical Institute, No. 39, p. 47-50.

Hansen, J.A., D.T. Davidson, and C.J. Roy 1960. Geologic and engineering properties of till and loess, Southeast Iowa. In D.T. Davidson (ed.), Geologic and engineering properties of Pleistocene materials in Iowa. Iowa State University Bulletin 59(28):133-166.

Hanson, W.E., J.M. Healy, and L.J. Dondanville 1968. The use and performance of Quaternary materials in the Loud Thunder Dam. In R.E. Bergstrom (ed.), Proceedings Symp. on the Quaternary of Illinois, p. 145-149. Illinois State Geological Survey.

Harland, W.B., K.N. Herod, and D.H. Krinsley 1966. The definition and identification of tills and tillites. Earth Science Review 2:225-250.

Harrington, E.J. 1978. Determination of time independent heave characteristics of mixed glacial materials. The engineering behaviour of glacial materials. Proc. Symp. at University of Birmingham 21-23 April 1975, 2nd edition, p. 235-238. Norwich, Geo Abstracts.

Harris, C. 1977. Engineering properties, groundwater conditions and the nature of soil movement on a solifluction slope in North Norway. Quarterly Journal of Engineering Geology 10:27-43.

Harrison, P.W. 1957. A clay till fabric: Its character and origin. Journal of Geology 65:275-308.

Harrison, W. 1958. Marginal zones of vanished glaciers reconstructed from the pre-consolidation pressure values of overridden silts. Journal of Geology 66:72-95.

Hartford, D. 1985. Properties and behaviour of Irish glacial till. In M.C. Forde (ed.), Glacial tills 85, p. 93-98. Edinburgh, Engineering Technics Press Ltd.

Hartlen, J. 1969. Hallfasthetsegenskaper Hos Nagra Skanska Moranleror: Morandag 1969, Symposium anordnat av Svenska Geotekniska Foreningen den 3 december. Reprints and Preliminary Reports, Swedish Geotechnical Institute, No. 39, p. 57-66.

Hartlen, J. 1973. The geotechnical characteristics of moraine clays related to their structure. Bulletin of the Geological Institute. University of Uppsala, new series 5:61-68.

Hartshorn, J.H. 1958. Flowtill in Southeastern Massachusetts. Bulletin of the Geological Society of America 69:477-482.

Haug, M.D., E.K. Sauer, and D.G. Fredlund 1977. Retrogressive slope failures at Beaver Creek, south of Saskatoon, Saskatchewan. Canadian Geotechnical Journal 14:288-301.

Helenelund, K.V. 1964. Moreenimaalajien kantavuusominaisuuksista. (Summary: On the bearing capacity of glacial till). Valtion Tekn. Tutkimusl. Tied., Sarja III, No. 79, p. 1-113.

Helenelund, K.V. 1969. Geotekniska Moranundersokningar I Finland: Morandag 1969, Symposium anordnat av Svenska Geotekniska Foreningen den 3 december. Reprints and Preliminary Reports, Swedish Geotechnical Institute, No. 39, p. 33-46.

Hendry, J.J. 1982. Hydraulic conductivity of a glacial till in Alberta. Groundwater 20:162-169.

Higginbottom, I.E. and P.G. Fookes 1970. Engineering aspects of periglacial features in Britain. Quarterly Journal of Engineering Geology 3:85-117.

Hinch, L.W. and P.L. Martin 1976. The foundation for an oil storage tank on Quaternary soils near Bridgewater, Somerset. Quarterly Journal of Engineering Geology 9:237-254.

Hogberg, E. 1969. Kompressionsforsok Pa Moran I Jatteodometer: Morandag 1969, Symposium anordnat av Svenska Geotekniska Foreningen den 3 december. Reprints and Preliminary Reports, Swedish Geotechnical Institute, No. 39, p. 51-56.

Hutchinson, J.N., S.H. Sommerville, and D.J. Petley 1973. A landslide in periglacially disturbed Etruria marl at Bury Hill, Staffordshire. Quarterly Journal of Engineering Geology 6:377-404.

Insley, A.E. and S.F. Hillis 1965. Triaxial shear characteristics of a compacted glacial till under unusually high confining pressures. Proceedings 6th International Conference on Soil Mech.

and Fdn. Eng. Montreal 1:244-248.

Jackli, H. 1962. Moranen als Baugrund und Baustoff; Strasse u. Verkehr -- Solothurn (Vogt-Schild AG), Nr. 9.

Jacobsen, M. 1967. The undrained shear strength of preconsolidated boulder clay. Proc. Geotechnical Conference, Oslo. (Also in: Danish Geotech. Inst. Bull. 26, 1968), p. 119-122.

Jacobsen, M. 1969. Laboratoriemalinger Pa Moraneler I Danmark: Morandag 1969. Symposium anordnat av Svenska Geotekniska Foreningen den 3 december. Reprints and Preliminary Reports, Swedish Geotechnical Institute, No. 39, p. 115-120.

Jacobsen, M. 1970. Strength and deformation properties of preconsolidated moraine clay. Geotekn. Inst. Bull. Copenhagen, No. 27, p. 21-45.

Jardine, W.G. and D.G. McKinley 1969. The origin and engineering properties of boulder clay in the Glasgow area. Regional Engineering Geology Conference. Strathclyde, Geological Society of London.

Johansson, H.G. 1983. Till and road construction. In J. Ehlers (ed.), Glacial deposits in North-West Europe, p. 115-122. Rotterdam, Balkema.

Jones, P.F. and E. Derbyshire 1983. Late Pleistocene periglacial degradation of Lowland Britain: Implications for civil engineering. Quarterly Journal of Engineering Geology 16:197-210.

Jorgensen, P. 1977. Some properties of Norwegian tills. Boreas 6:149-157.

Jorgensen, P. 1978. Compaction and permeability at Proctor Optimum for some Norwegian tills. Nor. Geot. Inst. 121(3):1-3.

Judd, A.G. 1980. The use of cluster analysis in the derivation of geotechnical classifications. Bulletin of the Association of Engineering Geology 17:198-211.

Jurgaitis, A. 1980. Genetic classification of glaciofluvial deposits and methods of their investigations. In W. Stankowski (ed.), Tills and glacigene deposits, p. 85-93. INQUA Symp., U.K. 1977. UAM Geografia, No. 20.

Karlsson, R. and L. Viberg 1967. Ratio C/P' in relation to Liquid limit and plasticity index, with specific reference to Swedish clays. Proceedings Geotechnical Conference, Oslo, p. 43-47.

Karrow, P.F. 1976. The texture, mineralogy and petrography of North American tills. In R.F. Legget (ed.), Glacial till, p. 83-98. Royal Society of Canada, Sp. Publ. No. 12.

Karrow, P.F. 1981. Till texture in drumlins. Journal of Glaciology

27:497-502.

Kauranne, L.K. 1960. Moreeni-murske. (Summary: Glacial till - a product of crushing). Geologi (Helsinki) 12:72-74.

Kazi, A. and J.L. Knill 1969. The sedimentation and geotechnical properties of the Cromer till between Happisburg and Cromer, Norfolk. Quarterly Journal Geology 2:67-87.

Kazi, A. and J.L. Knill 1973. Fissuring in glacial lake clays and tills on the Norfolk Coast, United Kingdom. Engineering Geology Journal 7:35-48.

Kehew, A.E. 1981. Near surface geology and geotechnical conditions of Minot area, North-central North Dakota. Geological Society of America, 15th Annual Meeting, with the North-Central Section of the Paleontology Society and Pander Society 13(6):283.

Keller, K.C., G. van der Kamp, and J.A. Cherry 1986. Fracture permeability and groundwater flow in clayey till near Saskatoon, Saskatchewan. Canadian Geotechnical Journal 23:229-240.

Kellomaki, A. and P. Nieminen 1986. Adsorption of water on the fine fractions of Finnish tills. Bulletin of the Geological Society of Finland 58, part 2:13-19.

Kelly, M. 1983. A geotechnical investigation of the glacial and post-glacial deposits in northern Ireland. Honours Project, Department of Civil Engineering, Queen's University, Belfast.

Kemmis, T.J. 1981. Importance of the regelation process to certain properties of basal tills deposited by the Laurentide ice sheet in Iowa and Illinois. Annals of Glaciology 2:147-152.

Kemmis, T.J., G.R. Hallberg, and A.J. Lutenegger 1979. Geotechnical implications of till sedimentation and stratigraphy in the Midwest. Abstract: 1979 Assoc. Engineering Geologists National Meeting, Chicago.

Kim, Y.D. 1969. Deformation characteristics of the St. Clair clay till. Ph.D. thesis, University of Western Ontario, London, Ont.

Kirwan, R.W. and M. Daniels 1961. The shear strength of compacted boulder clay. Proceedings 5th International Conference on Soil Mech. and Foundation Eng. 1, div. 1-3A:197-199. Paris.

Kirwan, R.W., E.R. Farrell, and M.L.J. Maher 1979. Repeated load parameters of a glacial till related to moisture content and density. Proceedings 7th European Conference on Soil Mech. and Fdn. Eng. 2(7):69-74. Brighton, U.K.

Kivekas, E.K. 1946. Zur kenntnis der mechanischen, chemischen und mineralogischen Zusammensetzung der finnischen Moranen. Acta Agralia Fennica 60(2):1-122.

Kjaernsli, B. and I. Torblaa 1961. Compaction of moraines in three foot layers. Transactions, Proceedings 7th International Congress on Large Dams, Rome 4:365-377.

Klohn, E.J. 1965. The elastic properties of a dense glacial till deposit. Canadian Geotechnical Journal 2:116-140.

Knill, J.L. 1974. The application of engineering geology to the construction of dams in the United Kingdom. La Geologie de L'Ingenieur. Societe Geologique de Belgique, p. 113-147.

Knutsson, G. 1966. Grundvatten i moranmark. Svensknaturvetenskap arshok, p. 236-249.

Knutsson, G. 1971. Studies of ground water flow in till soils. Geol. For. Stockh. Forh. 93:1-22.

Konovalar, P.A. and N. Rudnitskii 1964. On the coefficient of change of the modulus of soil deformation. Soil Mech. and Fdn. Eng. 3:167-170.

Korpela, K. 1977a. Moreenin geneettiset tyypit rakennusgeologisen luokittelun pohjana. (Summary: The genetic types of moraines in engineering-geologic classification). Rakennusgeologinen yhdistys ry: n Julkaisuja 11/80:1-5.

Korpela, K. 1977b. On the engineering-geologic characteristics of sorted till. Turun yliopiston maaperageologian osaston julkaisuja 23.

Kruger, J. 1979. Structures and textures in till indicating subglacial deposition. Boreas 8:323-340.

Krygowski, B., J. Rzechowski, and W. Stankowski 1969. Project of classifying glacial tills. Bull. Soc. Amis Sciences et Lettres Poznan, Ser. B. 23:141-154.

Kwan, D. 1971. Observations of the failure of a vertical cut in clay at Welland, Ontario. Canadian Geotechnical Journal 8:283-298.

Lafleur, J., A. Cummins, and S. Chiche 1982. Self-filtration of tills submitted to hydraulic gradients. Conference Papers 35th Canadian Geotechnical Conference, Montreal, Sept. 1982, Canadian Geotechnical Society, p. 50-62.

Lake, J.R. and G.C. Woodford 1958. The use of morainic material in road construction. Journal of the Institute of Highway Engineers 5:42-56.

Lamar, J.E. and J.C. Bradbury 1968. Past and present uses of Illinois Quaternary mineral materials. In R.E. Bergstrom (ed.), Proc. Symp. on the Quaternary of Illinois, p. 150-156. Illinois State

Geological Survey.

Landim, P.M.B. and L.A. Frakes 1968. Distinction between tills and other diamictons based on textural characteristics. Journal of Sedimentary Petrology 38:1213-1223.

Lane, W.N. 1968. Anisotrophy and soil fabric of Welland Ontario silty clay. M.E. Sc. thesis, University of Western Ontario.

Lavrushin, J.A. 1980. Vital problems of till sedimentogenesis. In W. Stankowski (ed.), Tills and Glacigene deposits, p. 19-40. INQUA Symp., U.K. 1977. UAM Geografia, No. 20.

Law, K.T. and C.F. Lee 1981. Initial gradient in a dense glacial till. Proceedings 10th International Conference on Soil Mech. and Fdn. Eng. Stockholm, 15-19 June, 1981, 1:441-446.

Lawson, D.E. 1981. Sedimentological characteristics and classification of depositional processes and deposits in the glacial environment. Report 81-27 U.S. Army Corps of Engineering Laboratory, Hanover, N.H.

Lefebvre, G., J. Locat, J.J. Pare, and O. Dascal 1979. Detailed study of the compressibility of a varved clay deposit as related to foundation settlement. Proceedings 32nd Canadian Geotechnical Conference: The Behaviour of Soft Soils, p. 3.128-3.148.

Legget, R.F. 1942. An engineering study of glacial drift for an earth dam, near Fergus, Ontario. Journal of Economic Geology 37:531-556.

Legget, R.F. 1966. Soils in Canada -- a brief review. Proceedings 6th International Conference on Soil Mech. and Fdn. Eng. Montreal, 8-15 Sept. 1965, 3:198-203. Univ. of Toronto Press.

Legget, R.F. and M.W. Bartley 1953. An engineering study of glacial deposits at Steep Rock Lake, Ontario. Can. Journal of Economic Geology 48:513-540.

Legget, R.F. and R.M. Hardy 1961. Engineering significance of soils in Canada. In R.F. Legget (ed.), Soils in Canada, p. 218-229. Royal Society of Canada, Sp. Publ. No. 3.

Lewin, P.I. 1985. Reversible behaviour in cyclic undrained triaxial tests on a clay till. In M.C. Forde (ed.), Glacial tills 85, p. 99-101. Edinburgh, Engineering Technics Press Ltd.

Lewin, P.I. and J.J.M. Powell 1985. Patterns of stress strain behavior for a clay till. Proceedings 11th International Conference Soil Mech. and Fdn. Eng., San Francisco, p. 553-556.

Lewis, J.D. 1983. The engineering properties of glacial tills in the Taff Valley, South Wales. Ph.D. thesis (CNAA), The Polytechnic of Wales.

Lewis, J.D. and G.O. Rowlands 1985. A review of the methods for the determination of the permeability of glacigenic and associated deposits. In M.C. Forde (ed.), Glacial tills 85, p. 39-47. Edinburgh, Engineering Technics Press Ltd.

Lind, B. and M. Nyborg 1986. The influence of sediment structures on hydraulic conductivity in till. Progress Report 1. Chalmer's Technical University, Gothenbourg, Report 80.

Lindroos, P. 1976. Moreenien luokittelusta ominaispintaalan perusteella. (Summary: Classification of till by specific surface area). Geologi 28:17-21.

Lindroos, P. and P. Nieminen 1982. Maaperakartoituksen uusi moreeniluokitus. Geologi 34:65-67.

Linell, K.A. and H.F. Shea 1960. Strength and deformation characteristics of various glacial tills in New England. Proceedings ASCE Research Conference on Shear Strength of Cohesive Soils, p. 275-314.

Little, J.A. 1984. Engineering properties of glacial tills in the vale of St. Albans. Ph.D. thesis, The City University.

Little, J.A. and J.H. Atkinson 1985. Some engineering properties of Anglian tills in the vale of St. Albans. In M.C. Forde (ed.), Glacial tills 85, p. 213-218. Edinburgh, Engineering Technics Press Ltd.

Lo, K.Y. 1977. Investigation of the possible relationship between rate of softening and recurrence of instability in bluff slopes - phase A. Report to Canadian Centre for Inland Waters, Burlington, Ont.

Loiselle, A.A. and J.E. Hurtubise 1976. Properties and behaviour of till as construction material. In R.F. Legget (ed.), Glacial till, p. 346-363. Royal Society of Canada, Sp. Publ. No. 12.

Love, M.A. and D. Derbyshire 1985. Microfabric of glacial soils and its quantitative measurement. In M.C. Forde (ed.), Glacial tills 85, p. 129-133. Edinburgh, Engineering Technics Press Ltd.

Low, W.I. and A.P. Lyell 1967. Portage Mountain Dam III. Development of construction control. Canadian Geotechnical Journal 4:184-228.

Lundahl, B. 1969. Ekonomiska Och Tekniska Aspekter Pa Djupgrundlaggning Pa Moran: Morandag 1969. Symposium anordnat av Svenska Geotekniska Foreningen den 3 december. Reprints and Preliminary

Reports, Swedish Geotechnical Institute, No. 39, p. 79-94.

Lundin, L. 1982. Soil moisture and ground water in till soil and the significance of soil type for runoff. UNGI Report, Uppsala, Sweden 66:1-216.

Lutteneger, A.J., T.J. Kemmis, and G.R. Hallberg 1983. Origin and properties of glacial till and diamictons. In R. Yong (ed.), Special publication on geological environment and soil properties, p. 310-331. American Society of Civil Engineers.

MacArthur, A. 1969. An investigation into the use of ablation tills in highway construction. M.Sc. thesis, University of Strathclyde.

MacDonald, A.B. and E.K. Sauer 1970a. Engineering properties of till. Physical environment of Saskatoon. Christiansen (ed.), Publ. Sask. Research Council/NRCC, p. 53-55.

MacDonald, A.B. and E.K. Sauer 1970b. The engineering significance of Pleistocene stratigraphy in the Saskatoon area, Saskatchewan. Canadian Geotechnical Journal 7:116-126.

MacNeil, R.H. 1965. Variation in content of some drumlins and tills in Southwestern Nova Scotia. Maritime Sediments 1:16-19.

Mahaney, W.C. 1985. Morphology and composition of two Late Wisconsin soils forming in till and lacustrine deposits Scarborough Bluffs area, south-central Ontario. Geographie Physique et Quaternaire 23:307-314.

Malinowski, J. 1961. The engineering geological investigation of the Quaternary in Poland. Quaternary of Central and Eastern Europe, Warsaw (Inst. of Geol.), p. 713-733.

Marcussen, I.B. 1975. Distinguishing between lodgement till and flow till in Weichselian deposits. Boreas 4:113-123.

Markland, A. and J.J.M. Powell 1985. Field and laboratory investigations of the clay tills at the building research establishment test site at Cowden Holderness. In M.C. Forde (ed.), Glacial tills 85, p. 147-168. Edinburgh, Engineering Technics Press Ltd.

Marsland, A. 1977. The evaluation of the engineering design parameters for glacial clays. Quarterly Journal of Engineering Geology 10:1-26.

Marsland, A. and J.J.M. Powell 1980. Cyclic load tests on 865 mm diameter plates in a stiff clay till. In G. Pande et al. (eds.), Soils under cyclic and transient loading, p. 837-847. Rotterdam, Balkema.

Matheson, D.S. 1970. A tunnel roof failure in till. Canadian Geotechnical Journal 9:313-317.

Matheson, D.S., J.G. MacPherson, and Z. Eisenstein 1986. Design and construction aspects of the Nipawin drainage tunnel. In K.Y. Lo, J.H. Palmer, and C.M. Yeun (eds.), Recent advances in Canadian underground technology, p. 221-235. Proceedings 6th Canadian Tunnelling Conference.

Mathews, W.H. 1960. Deformations of soils by glacier ice and the influence of pore pressures and permafrost. Transactions, Royal Society of Canada, vol. 54, series 3, p. 27-36.

Matich, M.A.J. and M.C. Douglas 1967. Stability of cuts in overconsolidated clay on the Don Valley Parkway. Proceedings 5th Australia-New Zealand Conference on Soil Mech. and Fdn. Eng., Auckland, p. 288-293.

May, R.W. and A. Dreimanis 1976. Compositional variability in tills. In R.F. Legget (ed.), Glacial till, p. 99-119. Royal Society of Canada, Sp. Publ. No. 12.

May, R.W. and S. Thompson 1978. The geology and geotechnical properties of till and related deposits in the Edmonton, Alberta area. Canadian Geotechnical Journal 15:362-370.

McAnoy, R.P.L., A.C. Cashman, and D. Purvis 1982. Cyclic tensile testing of a pile in glacial tills. Proceedings 2nd International Conference on Numerical Methods in Offshore Piling, Austin, Texas.

McGown, A. 1971. The classification for engineering purposes of tills from moraines and associated landforms. Quarterly Journal of Engineering Geology 4:115-130.

McGown, A. 1975. Genetic influences on the nature and properties of basal meltout tills. Ph.D. thesis, University of Strathclyde.

McGown, A. 1985. Construction problems associated with the tills of Strathclyde. In M.C. Forde (ed.), Glacial tills 85, p. 177-186. Edinburgh, Engineering Technics Press Ltd.

McGown, A. and E. Derbyshire 1977. Genetic influences on the properties of tills. Quarterly Journal of Engineering Geology 10:389-410.

McGown, A. and A.M. Radwan 1974. Cutting slopes in fissured Scottish boulder clay. Proceedings 2nd International Congress of International Assoc. Engineering Geology, Sao Paulo, Brazil.

McGown, A. and A.M. Radwan 1975. The presence and influence of fissures in boulder clays of west central Scotland. Canadian Geotechnical Journal 12:84-97.

McGown, A., W.F. Anderson, and A.M. Radwan 1978. Geotechnical properties of the tills in west central Scotland. The engineering behaviour of glacial materials. Proc. Symp. at University of Birmingham 21-23 April 1975, 2nd edition, p. 81-91. Norwich, Geo Abstracts.

McGown, A., L. Barden, S.H. Lee, and P. Wilby 1974. Sample disturbance in soft alluvial Clyde Estuary clay. Canadian Geotechnical Journal 11:651-660.

McGown, A., Saldivar-Sali, and A. Radwan 1974. Fissure patterns and slope failures in boulder clay at Hurlford, Ayrshire. Quarterly Journal of Engineering Geology 7:1-26.

McGuffey, V., J. Iori, Z. Kyfor, and D. Athanasiou-Grivas 1981. Statistical geotechnical properties of Lockport clays. New York State Department of Transportation. Soil Mechanics Bureau, Albany, N.Y., 809:54-60.

McKeown, J.D. and D.S. Matheson 1979. Use of glacial till and outwash silty sand as construction material for dam cores at Long Spruce generating station. Canadian Geotechnical Journal 16:363-378.

McKinlay, D.G. and W.F. Anderson 1974. Glacial till testing on an improved pressuremeter. Civil Engineering Journal 47-53.

McKinlay, D.G. and W.F. Anderson 1975. Determination of the modulus of deformation of a till using a pressuremeter. Ground Engineering 8:51-54.

McKinlay, D.G., A. McGown, A.M. Radwan, and D. Hossain 1978. Representative sampling and testing in fissured lodgement tills. The engineering behaviour of glacial materials. Proc. Symp. at University of Birmingham 21-23 April 1975, 2nd edition, p. 129-140. Norwich, Geo Abstracts.

McKinlay, D.G., M.J. Tomlinson, and W.F. Anderson 1974. Observations on the undrained strength of a glacial till. Geotechnique 24:503-516.

Medeiros, L.V. and Z. Eisenstein 1983. A deep retaining structure in till and sand. Part 1: Stress path effect. Canadian Geotechnical Journal 20:120-130.

Menzies, J. 1979. The mechanics of drumlin formations with particular reference to the change in porewater content of the till. Journal of Glaciology 22:373-384.

Menzies, J. 1984. Drumlins: A bibliography. Norwich, Geo Books.

Menzies, J. 1986. Inverse-graded units within till in drumlins near Caledonia, southern Ontario. Canadian Journal of

Earth Sciences 23:774-786.

Meyerhof, G.G., J.D. Brown, and G.D. Mouland 1981. Prediction of friction pile capacity in a till. Proceedings 10th International Conference on Soil Mech. and Fdn Eng. Stockholm, 15-19 June, 1981, 2:777-780.

Mickelson, D.M., L.J. Acomb, and T.B. Edil 1979. The origin of preconsolidated and normally consolidated tills in eastern Wisconsin, USA. In Ch. Schlüchter (ed.), Moraines and varves, p. 179-187. Proceedings INQUA Symposium on Genesis and Lithology of Quaternary Deposits 1978. Rotterdam, Balkema.

Milligan, V. 1976. Geotechnical aspects of glacial tills. In R.F. Legget (ed.), Glacial till, p. 269-291. Royal Society of Canada, Sp. Publ. No. 12.

Morgan, C.C. and M.C. Harris 1967. Portage Mountain Dam II: Materials. Canadian Geotechnical Journal 4:142-183.

Muller, E.H. 1982. Dewatering during lodgement of till. INQUA Symposia on the Genesis and Lithology of Quaternary Deposits, U.S.A. 1981/Argentina 1982.

Murphy, D.J., D. Koutsoftas, J.N. Covey, and J.A. Fischer 1973. Dynamic properties of hard glacial till. ASCE Proceedings, June 1978, p. 636-659.

Myhr, E. 1982. Virkningen av Otta pa grunnvatn, markvatn og avling. Dept. of agricult. hydrotechnics, Agricult. Univ. of Norway, Report No. 2, p. 1-98.

Nieminen, P. 1982. The specific surface area of Finnish till. XI INQUA Congress, Moscow, August 1982: Abstracts, p. 234. Moscow.

Nieminen, P. and A. Kellomaki 1982a. Moreenin hienoaineksen houkoisuudesta. (Porosity of Fine Fractions of Tills); Tampereen teknillinen korkeakoulu. Rakennustekniikan osasto. Rakennusgeologia, Raportti 10.

Nieminen, P. and A. Kellomaki 1982b. Veden adsorptio moreenien hienoainekseen. (Adsorption of Water on the Fine Fraction of Finnish Tills); Tampereen teknillinen Korkeakoulu, Rakennustekniikan osasto, Rakennusgeologia, Rapportti 9.

Nieminen, P. and A. Kellomaki 1984. Porosity of the fine fractions of Finnish tills. Bulletin of the Geological Society of Finland 56:221-226.

Nordberg, L. and S. Modig 1974. Investigations of effective porosity of till by means of a combined soil moisture/density gauge. Isotope Techniques in Groundwater Hydrology 2:313-340. Vienna, International Atomic Energy Agency.

Norman, J.W. 1969. Photo interpretation of

boulder clay areas as an aid to engineering geological studies. Quarterly Journal of Engineering Geology 2:149-158.

Odier, M., P. Egger, and F. DesCoeurdres 1981. Monitoring of a shallow tunnel (in French). Proceedings 10th International Conference on Soil Mech. and Fdn. Eng. Stockholm, 15-19 June, 1981, 1:329-334.

Ogunbadejo, T.A. 1973. Physico-chemistry of weathered clay crust formation. Ph.D. thesis, University of Western Ontario.

Ogunbadejo, T.A. and R.M. Quigley 1974. Compaction of weathered clays near Sarnia, Ont. Canadian Geotechnical Journal 11:642-646.

Olsson, T. 1977. Ground water in till soils. Striae 4:13-16.

Orr, T.L.L., D. Hartford, and R.W. Kirwan 1981. Effects of particle size distribution on the behaviour of till. Proceedings 10th International Conference on Soil Mech. and Fdn. Eng. Stockholm, 15-19 June, 1981, 1:329-334.

Pare, J.J., J.G. Lavallee, and P. Rosenberg 1978. Frost penetration studies in glacial till on the James Bay hydroelectric complex. Canadian Geotechnical Journal 15:473-493.

Pare, J.J., N.S. Verma, A.A. Loiselle, and S. Pinzariu 1984. Seepage through till foundations of dams of Eastmain-Opinaca-La Grande Diversion. Canadian Geotechnical Journal 21:75-91.

Paul, M.A. 1980. The compaction of soil: A geological and geotechnical analysis. The Nature and Origin of Soil Compaction, D. Carrol (Chpn), Soils Discussion Group, Northumberland, U.K., No. 16, p. 63-82.

Paul, M.A. 1981. A geotechnical model for the process of supraglacial deposition. In M.A. Paul (ed.), Soil mechanics in Quaternary science, p. 73-86. Quaternary Research Association.

Pawluk, S. and L.A. Bayrock 1969. Some characteristics and physical properties of Alberta tills. Research Council of Alberta, Bull. 26.

Peck, R.B. 1968. Problems and opportunities -- technology's legacy from the Quaternary. In R.E. Bergstrom (ed.), Proceedings Symp. on the Quaternary of Illinois, p. 138-144. Illinois State Geological Survey.

Peck, R.B. and W.C. Reed 1965. Engineering properties of Chicago subsoils. University of Illinois, Engineering Experimental Station, Bull. 423.

Peer, G.A. 1983. Tough glacial till calls for redesign. Heavy Construction News, August 15, 1983, p. 14-15.

Pepler, S.W.E. and I.P. MacKenzie 1976.

Glacial till in Winter Dam construction. In R.F. Legget (ed.), Glacial till, p. 381-390. Royal Society of Canada, Sp. Publ. No. 12.

Peters, J. and J. McKeown 1976. Glacial till and the development of the Nelson River. In R.F. Legget (ed.), Glacial till, p. 364-380. Royal Society of Canada, Sp. Publ. No. 12.

Pitts, J. 1982. Geomorphological observations as aids to the design of coast protection works on a part of the Dee Estuary. Engineering Implications of Earth Surface Processes. Eng. Geomorphology. Reg. Conf. of Eng. Group of Geol. Soc. at Birmingham Univ., Abstracts, p. 17.

Ponniah, D.A. and R. McAnoy 1985. Pile jacking in glacial tills. In M.C. Forde (ed.), Glacial tills 85, p. 137-140. Edinburgh, Engineering Technics Press Ltd.

Pour-Naghshband, G.R. 1978. Clay mineralogy and geotechnical properties of Tarras clay, basin clays and tills from some parts of Schleswig-Holstein. Meyniana Conference, Geologisches Institut der Universitat Kiel 30:55-60.

Powell, J.J.M. and I.M. Uglow. In press. Comparisons of results from Menard, self boring and push-in pressuremeters in a stiff clay till. Proceedings 2nd International Conference on Offshore Site Investigations, Society for Underwater Technology, London, March 1985.

Powell, J.J.M., A. Marsland, and A.N. Al-Khafagi 1983. Pressuremeter testing of glacial clay tills. Proceedings International Symposium on In-Situ Testing, Paris. Bull. International Assoc. Engineering Geologists 27:373-378.

Prest, V.K. and J. Hode-Keyser 1977. Geology and engineering characteristics of surficial deposits, Montreal Island and vicinity, Quebec. Geological Survey of Canada Paper No. 75-27.

Prudic, D.E. 1982. Hydraulic conductivity of a fine-grained till, Cattagaraugus County, New York. Groundwater 20:194-204.

Quigley, R.M. 1975. Weathering and changes in strength of glacial till. In E. Yatsu, A. Ward, and F. Adams (eds.), Mass Wasting: Guelph Symposium on Geomorphology, p. 117-131. Norwich, Geo Abstracts.

Quigley, R.M. and T.A. Ogunbadejo 1973. Soil weathering, soil structure and engineering properties, Sarnia clay crust. University of Western Ontario, Faculty of Eng. Science, Soil Mechanics

Research Report SM-2-73.

Quigley, R.M. and T.A. Ogunbadejo 1976. Till geology, mineralogy and geotechnical behaviour, Sarnia, Ont. In R.F. Leggett (ed.), Glacial till, p. 336-345. Royal Society of Canada, Sp. Publ. No. 12.

Quigley, R.M. and D.B. Tutt 1968. Stability -- Lake Erie North Shore bluffs. Proc. 11th Conference on Great Lakes Research, International Assoc. Great Lakes Research, p. 230-238.

Quigley, R.M., P.J. Gelinas, W.T. Bou, and R.W. Packer 1977. Cyclic erosion-instability relationships. Lake Erie North Shore bluffs. Canadian Geotechnical Journal 14:310-323.

Radhakrishna, H.S. and T.W. Klym 1974. Geotechnical properties of a very dense glacial till. Canadian Geotechnical Journal 11:396-408.

Radhakrishna, H.S., C.F. Lee, and T.W. Klym 1977. Laterally loaded piers in glacial till. 30th Canadian Geotechnical Conference, 1977, Saskatoon, Sask. Canadian Geotechnical Society, p. IV-14 to IV-32.

Radwan, A.M. 1974. The presence and influence of fissures in boulder clays of west central Scotland. Ph.D. thesis, University of Strathclyde.

Raukas, A. 1977. Procedures of till investigation in European countries and possibilities of their unification. Till -- its genesis and diagenesis. Uniwersytet Im. Adama Michiewicza Poznaniu. UAM Geografia, No. 12, p. 186-212.

Robinson, K.E. 1977. Design and construction of small dams using glacial till. Proceedings 30th Canadian Geotechnical Conference, 1977, Saskatoon, Sask., p. VII-25 to VII-31.

Roderick, G.L. 1978. Properties of some glacial soils in Wisconsin. The engineering behaviour of glacial materials. Proc. Symp. at University of Birmingham 21-23 April 1975, 2nd edition, p. 67-74. Norwich, Geo Abstracts.

Rosenberg, P. and N.L. Journeaux 1978. Load bearing slurry trench wall supported by glacial till. Canadian Geotechnical Journal 15:430-434.

Rosenberg, P. and C. Lupien 1977. An attempt to predict foundation settlement for foundations supported in glacial till. 30th Canadian Geotechnical Conference, 1977, Saskatoon, Sask. Canadian Geotechnical Society, p. VI-1 to VI-12.

Rukhina, E.V. 1980. Genesis and subdivision of glacial deposits. In W. Stankowski (ed.), Tills and glacigene deposits, p. 11-14. INQUA Symp., U.K. 1977. UAM Geografia, No. 20.

Rutka, A. 1961. Correlation of engineering and pedological soil classification in Ontario. In R.F. Legget (ed.), Soils in Canada, p. 183-192. Royal Society of Canada, Sp. Publ. No. 3.

Rutter, N.W. and S. Thomson 1982. Effects of geology on the development of Edmonton, Alberta, Canada. In R.F. Legget (ed.), Reviews in engineering geology -- geology under cities. Geological Society of America 5:55-61.

Rzechowski, J. 1969. Genetic classification of morainic deposits. Kwartaln. Geolog. 13:459-478.

Sauer, E.K. 1974. Geotechnical implications of Pleistocene deposits in southern Saskatchewan. Canadian Geotechnical Journal 11:359-373.

Sauer, E.K. 1978. The engineering significance of glacier ice-thrusting. Canadian Geotechnical Journal 15:457-472.

Sauer, E.K. and E.A. Christiansen 1985. A landslide in till near Warman, Saskatchewan, Canada. Canadian Geotechnical Journal 22:195-204.

Sauer, E.K. and C.L. Monismith 1968. Influence of soil suction on behaviour of a glacial till subjected to repeated loading. Highway Research Record, H.R.B., Washington, D.C., No. 215, p. 8-23.

Sauer, E.K. and N.F. Weimer 1978. Deformation of lime modified clay after freeze-thaw. ASCE Transportation Engineering Journal 104:201-212.

Schindler, C. 1974. Zur Geologie des Zurichsees. Eclogae Geol. Helv., Basel (Birkhauser) 67/1:163-196.

Schlüchter, Ch. 1980. Bemerkungen zu einigen Grundmoranen-vorkommen in den Schweizer Alpen; Zeitschrift fur Gletscherkunde u. Glazialgeologie - Innsbruck (Wagner) 16/2:203-212.

Schlüchter, Ch. 1981. Bemerkungen zur Korngrossenanalyse der Lockergesteine, insbesondere der Moranen, in der Schweiz. Verh. naturwiss. Ver. Hamburg (NF) 24(2):155-160.

Schlüchter, Ch. 1983. Die Bedeutung der angewandten Quartargeologie fur die eiszeitgeologische Forschung in der Schweiz. Physische Geographie - Zurich (Univ. Zurich) 11:59-72.

Schlüchter, Ch. 1984. Geotechnical properties of Zubo sediments. Contr. Sedimentology Stuttgart (Schweizerbart) 13:135-140.

Scott, J.S. 1976. Geology of Canadian tills. In R.F. Legget (ed.), Glacial till, p. 50-66. Royal Society of Canada, Sp. Publ. No. 12.

Scott, J.S. and D.A. St. Onge 1969. Guide to the description of till. Geological Survey of Canada, Paper 68-6.

Serrett, R.J. and T.B. Edil 1982. Groundwater flow systems and stability of a slope. Groundwater 20:5-11.

Shaw, J. 1977. Tills deposited in arid polar environments. Canadian Journal of Earth Sciences 14:1239-1245.

Shaw, J. 1982a. Forms associated with boulders in melt-out till. INQUA Symposium on the Genesis and Lithology of Quaternary Deposits, USA 1981/Argentina 1982, p. 3-12.

Shaw, J. 1982b. Melt-out till in the Edmonton area, Alberta, Canada. Canadian Journal of Earth Sciences 19:1548-1569.

Shaw, J. 1985. Subglacial and ice marginal environments. In G.M. Ashley, J. Shaw, and N.D. Smith (eds.), Glacial sedimentary environments, p. 7-84. Society of Paleontologists and Mineralogists, Tulsa, Oklahoma.

Shepps, V.C. 1958. Size factors: A means of analysis of data from textural studies of till. Journal of Sedimentary Petrology 28:482.

Shilts, W.W. 1975. Common glacial sediments of the shield, their properties, distribution, and possible uses as a geochemical sampling media. Journal of Geochemical Exploration 4:189-199.

Singh, P.N., S.V. Tatioussian, and C.G. Flagg 1983. A study of the geotechnical properties of Milwaukee area soils. In R. Yong (ed.), Special publication on geological environments and soil properties, p. 269-309. American Society of Civil Engineering.

Skempton, A.W. and J.D. Brown 1962. A landslide in boulder clay at Selset, Yorkshire. Geotechnique 11:280-293.

Sladen, J.A. 1979. Weathering and its effects on the geotechnical properties of tills in southeast Northumberland. M.Sc. thesis, University of Newcastle Upon Tyne.

Sladen, J.A. and W. Wrigley 1983. Geotechnical properties of lodgement till. In N. Eyles (ed.), Glacial geology, p. 184-212. Oxford, Pergamon Press.

Soderman, L.G. and Y.D. Kim 1970. Effect of groundwater levels on stress history of the St. Clair clay till deposit. Canadian Geotechnical Journal 7:173-187.

Soderman, L.G. and R.M. Quigley 1965. Geotechnical properties of three Ontario clays. Canadian Geotechnical Journal 2:167-189.

Soderman, L.G., T.C. Kenny, and A.K. Loh 1960. Geotechnical properties of glacial clays in Lake St. Clair region of Ontario. Proceedings 14th Canadian Soil Mechanics Conference, NRCC, Assoc. Comm. on Soil and Snow Mechanics, Ottawa, Technical Memorandum No. 69, p. 55-90.

Soderman, L.G., Y.D. Kim, and V. Milligan 1968. Field and laboratory studies of modulus of elasticity of a clay till. Highway Research Record, H.R.B., Washington, D.C., No. 243, p. 1-11.

Soliman, N. 1983. Effect of geologic history on the design parameters of heavily overconsolidated till. In R. Yong (ed.), Special Publication on Geological environments and soil properties, p. 246-268. American Society of Civil Engineering.

Stankowski, A. 1980. Stratigraphic and regional variation of glacial tills in Poland in the light of clay minerals investigations. In W. Stankowski (ed.), Tills and glacigene deposits, p. 57-65. INQUA Symp., U.K. 1977. UAM Geografia, No. 20.

Stroud, M.A. and F.G. Butler 1978. The standard penetration test and the engineering properties of glacial materials. The engineering behaviour of glacial materials. Proc. Symp. at University of Birmingham 21-23 April 1975, 2nd edition, p. 117-128. Norwich, Geo Abstracts.

Sun, K.H.D. 1975. The soil mechanics and clay mineralogy of glacial till in a portion of Southwestern Ohio. M.Sc. thesis, Miami University, Oxford, Ohio.

Sutcliffe, F.H. 1965. The till cofferdam in the St. Lawrence River. Canadian Geotechnical Journal 11:261-273.

Svensson, T. and G. Andersson 1974. Forsok till faltbestamning av effektiv porositet i moran. Vannet i Norden, 1975, 4:37-42.

Swedish Geotechnical Institute 1973. Moranlere dagar 1972. Linkoping, Swedish Geotechnical Institute.

Tarbet, M.A. 1973. Geotechnical properties and sedimentation characteristics of tills in southeast Northumberland. Ph.D. thesis, University of Newcastle Upon Tyne, Dept. of Geology.

Terzaghi, K. 1955. Influence of geological factors on the engineering properties of sediments. Harvard University, Soil Mechanics Series, No. 50.

Thomson, S. and F. El-Nahhas 1980. Field measurements in two tunnels in Edmonton, Alberta. Canadian Geotechnical Journal 17:20-33.

Thomson, S., R.L. Martin, and Z. Eisenstein 1982. Soft zones in the glacial till in downtown Edmonton. Canadian Geotechnical Journal 19:175-180.

Thorburn, S. and W.N. Reid 1973. Stability

of slopes in lodgement till within the Glasgow district. Civil Engineering and Public Works Review, U.K., April 1973, 68(801):321-325.

Thurner, H. 1969. Kompressions-Och Skjuvapparat For Moranmaterial: Morandag 1969. Symposium anordnat av Svenska Geotekniska Foreningen den 3 december. Reprints and Preliminary Reports, Swedish Geotechnical Institute, No. 39.

Trow, W. and J. Bradstock 1972. Instrumented foundations for two 43-storey buildings on till, metropolitan Toronto. Canadian Geotechnical Journal 9:290-303.

Van Husen, D. 1986. Bau-und Hydrogeologische Bedeutung eiszeitlicher Vorgange. Mitt. Ges. Geol. Bergbaustud. Oesterr. - Wein 33:23-45.

Vandine, D.F. 1980. Engineering geology and geotechnical study of Drynoch landslide, British Columbia. Geological Survey of Canada, Paper No. 79-31.

Vaughan, P.R. and H.J. Walbancke 1973. Pore pressure changes and the delayed failure of cutting slopes in overconsolidated clay. Geotechnique 23:531-539.

Vaughan, P.R. and H.J. Walbancke 1975. The stability of cut and fill slopes in boulder clay. The engineering behaviour of glacial materials. Proc. Symp. at University of Birmingham 21-23 April 1975, 2nd edition, p. 209-219. Norwich, Geo Abstracts.

Vaughan, P.R., D.W. Hight, V.G. Sodha, and H.J. Walbancke 1978. Factors controlling the stability of clay tills in Britain. Proc. Clay Fills, Institute of Civil Engineering, London, p. 205-218.

Virkkala, K. 1969. Classification of Finnish tills according to grain size (Finnish with English summary). Terra Cognita 81:273-278.

Von Moos, A. 1953. Der Baugrund der Schweiz.; Schweiz. Bauzeitung -- Zurich (Jean Frey AG) 50:3-12.

Vorren, T.O. 1977. Grain-size distribution and grain-size parameters of different till types on Hardangervidda, south Norway. Boreas 6:219-227.

Waters, D.B. and G.C. Woodford 1956. Morainic deposits in road construction: An assessment of the present position. Department of Science and Industrial Research, Note 2755.

Weinert, H.H. 1967. Tillite in road construction. Proceedings 4th Regional Conference for Africa on Soil Mech. and Fdn. Eng. Capetown, South Africa, p. 169-173.

West, T.R. and D.L. Warder 1983. Geology of Indianapolis, Indiana, USA. Bulletin of the Association of Engineering Geology 20:105-124.

Whalley, W.B. 1978. Abnormally steep slopes on moraines constructed by valley glaciers. The engineering behaviour of glacial materials. Proc. Symp. at University of Birmingham 21-23 April 1975, 2nd edition, p. 60-66. Norwich, Geo Abstracts.

White, G.W. 1972. Engineering implications of stratigraphy of glacial deposits. 24th International Geological Congress Section 13, Engineering Geology, p. 76-82.

White, O.L. 1961. The application of soil consolidation tests to the determination of Wisconsin ice thickness in the Toronto region. M.A.Sc thesis, University of Toronto.

White, O.L. 1975. Quaternary geology of the Bolton Area, Southern Ontario. Ministry of Natural Resources, Ontario Division of Mines, Geological Report 117.

White, O.L. 1982. Toronto's subsurface geology. In R.F. Legget (ed.), Reviews in engineering geology -- geology under cities. Geological Society of America 5:119-124.

Whyte, I.L. 1985. The angle of shearing resistance for granular tills and its relevance to slope stability. In M.C. Forde (ed.), Glacial tills 85, p. 25-30. Edinburgh, Engineering Technics Press Ltd.

Widdis, T.F. and H.G. Claphams 1981. End of construction failure of a slope in a stiff boulder clay. Proceedings 10th International Conference on Soil Mech. and Fdn. Eng. Stockholm, 15-19 June, 1981, 3:445-450.

Williams, R.E. and R.N. Farvolden 1967. The influence of joints on the movement of groundwater through glacial till. Journal of Hydrology 5:163-170.

Wright, M.D. 1982. The distribution and engineering significance of superficial deposits in the Upper Clydach Valley, South Wales coalfield. Engineering Implications of Earth Surface Processes: Eng. Geomorphology. Reg. Conf. of Eng. Group of Geol. Soc. at Birmingham Univ. - Abstracts, p. 28.

Wroth, C.P. and D.M. Wood 1978. The correlation of index properties with some basic engineering properties of soils. Canadian Geotechnical Journal 15:137-145.

Zeman, A.J. 1983. Relations between erosion resistance and geotechnical properties of cohesive sediments. Proceedings 26th Conference on Great Lakes Research, Oswega, N.Y., 26:47.

# 3  Glacial deposits and landforms in general

Genetic Classification of Glacigenic Deposits, Goldthwait & Matsch (eds)
© 1988 Balkema, Rotterdam. ISBN 90 6191 694 1

# On the comparison and standardization of investigation methods for the identification of genetic varieties of glacigenic deposits

A.Raukas
*Geological Institute, Academy of Sciences, Tallinn, Estonian SSR, USSR*

S.Haldorsen
*Department of Geology, Agricultural University of Norway, Aas, Norway*

D.M.Mickelson
*Department of Geology and Geophysics, University of Wisconsin, Madison, Wis., USA*

ABSTRACT: The unification of field and laboratory methods and standards of data processing makes it easier to exchange data and to interpret the genesis of glacigenic sediments. In some cases there are international standards that are applied in most countries. Because of different traditions, different aims of the investigations, and different geology, a complete unification is hardly possible. Computers are now applied by a large group of scientists in the field of glacial geology. Great differences in the types of computers make it difficult to work out standards of computer processible files. However, exchange of programs and data sets among smaller groups of scientists is of great interest.

## 1 INTRODUCTION

The improvement and unification of field and laboratory methods used for the investigation of glacigenic deposits serves as one of the most important tasks set before the INQUA Commission on Genesis and Lithology of Quaternary Deposits. In this paper we review only one aspect -- methods for the identification of genesis of glacigenic sediment. Between 1972 and 1978 the Work Group on the standardization of field and laboratory methods of investigation of glacigenic deposits (2B) distributed questionnaires to obtain detailed data on laboratory and field equipment used in the investigation of glacigenic deposits in different countries. The results were published both in English (Raukas 1976; Raukas, Mickelson, Dreimanis 1978; Raukas 1982) and in Russian (Raukas, Mickelson, Dreimanis 1979). For some regions local syntheses were published (e.g., Dreimanis 1971; Goldthwait 1973; Haldorsen 1975; Johansson 1976), as well as a summary of methods used for fluvioglacial deposits (Jurgaitis 1982, 1984). Summaries of till description methods in Canada are given in Scott and St.-Onge (1969) and Scott (1976).

Since the original questionnaire several international and regional symposia have been held to discuss recommendations for the unification of sampling methods, field and laboratory investigation procedures, and the standardization of primary granu-lometric boundaries (e.g., Oslo 1975; Warsaw 1975 and 1980; Tallinn 1985). Results of the symposia are partly published (Boulton 1976; and others in Stankowski 1976; Raukas 1980), but more informal conclusions are not. For the XI INQUA Congress in Moscow in 1982 a detailed report was compiled of work group activities in 1977-1982 (Raukas 1982). The same topics were also discussed at the Xth Congress of INQUA in 1977 in Birmingham, United Kingdom. In the discussion below we attempt to summarize our perception of the opinions of many researchers in this field.

There is agreement that depending on the aims of an investigation (e.g., stratigraphy, deposit genesis, engineering geology) and local geological differences between areas, different techniques and methods are appropriate. Because some materials lend themselves to different types of analysis, and because of prolonged traditions, a complete unification of investigation methods, particle-size boundaries (e.g., Shea 1973), other boundaries, and equipment is hardly possible, and sometimes even unnecessary or harmful. We should, however, be able to compare results of various investigations despite different aims, and every publication should report the methods and particle-size boundaries used. If these are not reported, the data are nearly valueless for comparison to other areas or studies.

In elaborating optimal methods to be applied to the study of glacigenic deposits, the following principles were accepted to serve as the basis:

1. The set of methods used depends on the aim of the investigation.

2. A variety of methods should be attempted in order to present as many possible answers to stratigraphical, sedimentological or applied problems.

3. The results should be achieved by optimal use of time and money. In other words, methods producing the most direct answers are preferable.

4. The methods applied should be correlatable with the techniques used for the study of other types of sediment as well as with the methods used in other areas.

5. As far as possible the method should consider the established traditions of the region, because a large internally consistent data base can sometimes be used to answer questions not approachable in other ways. On the other hand, attempts should always be made to try new approaches to the study of materials.

6. Different investigation methods give variable amounts of useful information. For this reason they should be divided into primary, secondary, and unessential ones. Generally, the determination of which methods are most important is done by literature review or test analyses of a small number of samples.

The identification of genetic varieties of glacigenic deposits is one of the most challenging tasks in an investigation, and there is a wide choice of geomorphological and lithological methods available. A majority of investigators in INQUA Work Group 2B believe that the following set of procedures should be carried out when trying to determine the genesis of glacigenic sediments:

1. Determination of the position of the outcrop, its geomorphological setting, and identification of the landform present.

2. Detailed layer-by-layer description of the outcrop, including description of primary and secondary structures, nature of contacts, etc.

3. A detailed investigation of coarse fractions, including size frequency, clast lithology, and clast fabric.

4. Sampling of matrix for laboratory investigations.

5. Laboratory investigations.

6. Mathematical and graphic processing of data.

# 2 DISCUSSION

## 2.1 Geomorphic setting

A complete study of the genesis of glaci-

genic deposits starts with geomorphological reconnaissance of the area to establish the depositional setting, recognizing that the modern landscape may not reflect the depositional setting of all materials in a given outcrop. Aerial methods including satellite photography are effective, especially while studying remote or inaccessible areas. Generally, however, photo scales of 1:20,000 or larger are best taken at times of the year when foliage is at a minimum. An important item in geomorphological investigations is the study of the character of the bedrock relief (Tavast and Raukas 1982), because together with lithologic composition of bedrock, it determines many significant properties of deposits (composition, structure and texture, etc.). The detailed morphologic characteristics (slope, orientation) of the surface on which the study outcrop or borehole is situated should be examined and a description of the whole landform complex should be provided. The size, orientation, and shape of the individual landform and the occurrence of syn- and postdepositional glaciodynamic, soil creep, slopewash, and water freeze-thaw phenomena should be described as well.

## 2.2 Detailed outcrop description

Then follows a detailed textural and structural analysis of the section. This includes the description of each bedding element and the character of boundaries with underlying and overlying layers. The texture of each unit should be estimated in the field. In addition, any sediment or structures present should be measured, including primary bedding features as well as deformation due to dewatering, loading, or overriding ice (e.g., Hicock and Dreimanis 1985). Codes such as those used by Eyles and Miall (1984) may be useful in getting complete descriptions of each unit. If present, the bedding and burial conditions of paleontological remains should be described in detail, a visual estimate can be made of the degree of weathering and oxidation, and HCl can be used to measure depth of leaching and to recognize buried leached zones.

The determination of color should always be performed on moist or wet material, because it facilitates comparison of the results obtained. For the sake of objective estimation use a Munsell color chart and, if possible, field express photometres, which present the color information in the form of figures or graphs corresponding to the reflective power of the visible part of the light spectrum

(Dobrovolskyi and Tchupachina 1980).

Clasts in glacigenic deposits can be studied in great detail in situ with frame and volume methods. In the first case, all the pebbles and cobbles occurring in an area of 0.25, 0.5, or 1.0 $m^2$ on the outcrop surface are drawn, and it is then determined which part of the framed area they cover. The ratio of different rock types and their orientation is also determined (Orviku 1958).

Although time-consuming, a more precise volume method (e.g., Raukas 1962) can be used. In this method, not only the number but also the volume of clasts of various lithologies are calculated on a unit volume (1 $m^3$). For these purposes excavations with a volume 0.15-1.0 $m^3$ (depending on the abundance of clasts) are performed in the unweathered part of till. A detailed discussion of clast lithologic analysis is given by Bridgeland (1986). The orientation of clasts is determined during excavation; then they are removed. If long axes of clasts are more-or-less horizontal, azimuth is measured on 25, 50 or 100 pebbles, depending on purpose. Long-axis plunge should be measured as well. Next, clasts are separated according to their petrographic composition, size, and class of roundness. Then the volume of each fraction, including large clasts, is determined by displacement of water.

For studies of genesis, valuable information can be obtained from the shape and roundness of clasts. In western Europe roundness is most often measured by the rounding scale of Cailleux (1954); in North America the Powers (1953) scale is most popular (Raukas, Mickelson, and Dreimanis 1978). In the Soviet Union use is made of the following subjective five-point scale: perfectly rounded, well rounded, rounded, poorly rounded and angular. Mean indices of roundness are then calculated. Often measures of shape that compare the lengths of three major axes are also established, but they provide less information than the roundness of the clasts because of lithologic influence.

Special attention should be paid to the recognition of indicator boulders, which are easily identifiable under field conditions and have limited aerial distribution, often traceable up-ice to their source. They provide valuable information about the flow direction and origin of ice and help to differentiate terrestrial and subaquatic tills. Recently F. Kaerlein (1985) has published a bibliography of indicator boulder investigations in North Europe.

Some investigators are still of the opinion that the best way to differentiate subglacial meltout or lodgement and supraglacial meltout tills is by comparing geotechnical properties, mainly the density of the material. Unfortunately, the density of till depends, as do many other geotechnical properties, on the granulometric and clay mineral composition. Thus there is not always direct correlation with the genesis of glacigenic deposits. Other factors that affect stress history -- drainage history, postglacial movement, or freeze-thaw -- are also important.

Attention should be paid to the potential occurrence of blocks of previously deposited till or other sediment carried in the till and to interbeds of fluvial sediment. Mistaking an erratic block of previously deposited till for the actual till in that location can create correlation problems. Likewise, included sediments can often argue for or against lodgement or meltout, depending on their characteristics. If included blocks are present, one assumes that lodgement could only have taken place on layers thinner than the block.

## 2.3 Sampling

Considering the possibility of progressive change in composition of glacigenic deposits, spot-samples provide an objective picture of the deposit over a large area. The method of channel sampling is used for solving special problems only. Generally, vertical channel samples with a length of 0.5 to 1.0 m are preferred; however, it depends on the purpose of the samples.

The sample size for laboratory analysis is determined by the aim of the investigation. To obtain reliable results on standard investigations the following sample sizes are recommended by Raukas and Reintam (1965): 0.5 kg for the determination of sand fraction, 3 kg for gravel, and 90 to 250 kg for pebble fraction. A sample of 50-200 g is sufficient for the elucidation of silt-clay components. For maintaining natural moisture content, if it is necessary, the samples are paraffined or placed into special plastic boxes or bags where moisture cannot be lost.

## 3 LABORATORY METHODS

In view of the great variation in the composition of tills and associated sediments, the study of till matrix is of less importance for genetic purposes, but it plays an important role in solving stratigraphic and other problems. Serebrianniy (1980) reviews much of the literature in this area, and a summary of methods is given in Raukas, Mickelson, and Dreimanis (1979). Of great importance for distinguishing terrestrial and subaquatic tills is the investigation of macro- and micro-fossils such as shells of molluscs, ostracods, Foraminifera and diatoms, as well as pollen and spores. They should indicate whether the marine flora and fauna occur in tills in primary conditions or if they have been redeposited. Sometimes methods such as fluorescent microscopy, which enables one to separate shells and pollen grains of different age on the basis of their physical characteristics, are useful. Good results in this case may be obtained by electron spin resonance (ESR) dating of subfossil shells (Ikeya and Ohmura 1981; Henning and Grun 1983; Hutt et al. 1985).

## 4 ANALYSIS OF RESULTS

Work Group 8, "Standards of computer processible files of glacigenic data," has studied methods of statistical treatment and the graphic presentation of results. The present generation of scientists in glacial geology represents a transition between the generation of geologists for whom computers were a new type of equipment and the next generation for whom the computer will be a necessary research tool, especially for data manipulation and statistical analysis (Fig. 1).

Until now, the most common data type for computerization has been grain-size data, the most widely used parameter measured in laboratory studies of glacigenic sediments. The input data in the computerization are obtained from sieving, pipette, and hydrometer analyses. Grain-size curves, presentation of sand-silt-clay percentages in triangular diagrams, and two-dimensional graphic presentations of all kinds of statistical parameters (median, mean, sorting, skewness, kurtosis, etc.) are the common outputs and have given good results in several cases. Unfortunately, the use of grain-size analysis equipment connected with an automatic data collection unit and three-dimensional graphic presentations is rather uncommon at present.

Computer programs have been used at least from the early part of the 1970's (e.g., Mark 1973) for the analysis of particle orientation in glacial sediments. Work Group activity has concerned both pebble orientation measured in the field and microfabric analysis carried out in the laboratory. In the field, both two-

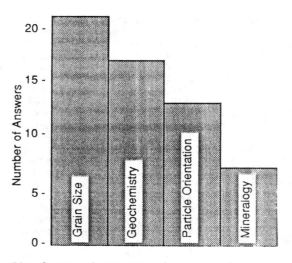

Fig. 1. Use of computers for study of glacigenic sediments. Number of answers gives the number of work group members (total 40) who use computers for analyses of the respective parameters.

and three-dimensional analysis tends to be most common. The use of computer programs to handle these data has accelerated this trend, because manual plotting of three-dimensional fabric data is very time-consuming. Analysis by means of the eigenvector method seems to be of great interest at the moment, and it requires the use of computers. Microfabric measurements are performed with ordinary petrographic microscope, with measurements of the anisotropy of magnetic susceptibility, and with electron microscope studies. Stereosets of SEM-photos can be treated in the same way as air-photos, and data programs make it possible to carry out three-dimensional statistical tests.

Glacial geologists often use computers for analyzing geochemical data and for interpreting mineralogical analysis, including X-ray data, but they have not been used much for genetic investigations of tills. Most of the commercial programs are written in FORTRAN and BASIC, which are the languages suitable for the vast majority of computers. Geology textbooks with computer programs and diskettes are available (e.g., Campbell 1985; Kinzelbach 1986). It would be of great interest to have textbooks with computer programs for methods used in the study of glacial sediments as well. The informal exchange of programs has been of practical importance, and increasingly compatible computers will make this exchange easier in the future.

## 5 SUMMARY

Depending on the mode of origin of glacigenic deposits, several types of till or its facies and subfacies have been recognized: their number ranges from two to about twenty (Dreimanis 1982). Though theoretically more than 20 varieties are possible, in the field it is often difficult to distinguish the two types most commonly cited in literature: a) ablation till and b) basal (or lodgement) till (Dreimanis 1976).

In most cases our presumptions about the geological development of ancient glaciers are based on the geological activity of modern glaciers. However, the study of tills in areas covered by modern continental glaciers is difficult, and the areas available for study are rather limited. The study of present mountain glaciers can sometimes provide valuable information on the genesis of glacial deposits as well. For example, the analyses performed in the Central Caucasus and in the Altai Mountains (Serebryanny and Orlov 1982) show that basal debris is finer than superglacial debris because of enrichment with fine particles from abrasion in basal till. Great differences also occur in the roundness of clasts. In superglacial debris the fragments show no signs of glacial transport; in basal debris the number of such angular fragments does not exceed 20%. In basal debris the clasts are more isometric and less flattened than in superglacial ones. Mineralogic and petrographic analyses show that in superglacial debris the erratic rock particles prevail, whereas in basal debris fragments of local bedrock predominate.

The same conditions seem to be characteristic of lowland regions. We cannot automatically extend the data obtained from one mountain area to other locations, however. For instance, supraglacial tills are coarse-grained in Precambrian Shield, but very fine in the prairie region of North America (Dreimanis 1976).

We should realize that in every region glacigenic deposits have their own lithological and other characteristics. In different regions different properties are important for stratigraphic and genetic studies. Every scientific conclusion that we have adopted as individuals is a product of our individual experience and is only partly true. In each region we might find specific conditions of deposition different than other areas and, therefore, we must not automatically impose our conclusions from one area on other environments. In every region we should look for local trends, thoroughly compare them with the results of neighboring regions, and then interpret them in the light of published hypotheses. In order to get better unifying theories, we need comparable field and laboratory methods and standards for processing of the data. The Commission on Genesis and Lithology of Quaternary Deposits, although it has not unified investigation methods, has suggested how to make the results of till investigations more comparable.

## REFERENCES

Boulton, G.S. 1976. Some proposals for standardization of particle size analysis and textural distinctions for till. In W. Stankowski (ed.), Till, Its Genesis and Diagenesis. Univ. A. Mickiewicza, Poznan. Ser. Geogr. 12:208-212.

Bridgeland, D.R. 1986. Clast lithological analysis. Quatern. Res. Assoc., Technical Guide No. 3:1-207.

Cailleux, André 1954. Limites dimensionnelles et noms des fractions granulométriques. Bull. Soc. Géol. France Ser. 6:4, 643-646.

Campbell, G.S. 1985. Soil physics with basic. Developments in Soil Science 14. Amsterdam, Elsevier.

Dreimanis, A. 1971. Procedures of till investigations in North America: A general review. In R.P. Goldthwait (ed.), Till, A Symposium, p. 27-37. Columbus, Ohio State Univ. Press.

Dreimanis, A. 1976. Towards standardization of particle size analysis. In W. Stankowski (ed.), Till, Its Genesis and Diagenesis. Univ. A. Mickiewicza, Poznan. Ser. Geogr., 12:202-203.

Dreimanis, A. 1982. Commission's activities during the intercongress period 1977-1982. In Ch. Schlüchter (ed.), Commission on Genesis and Lithology of Quaternary Deposits, Report on Activities 1977-1982, p. 5-11. Zurich, ETH.

Dobrovolskyi, V.V. and R.P. Tchupachina 1980. Quantitative characteristics of the coloring of glacial deposits (Russian). In A. Raukas (ed.), Methods of the Field and Laboratory Investigations of Glacial Deposits, p. 50. Abstracts of the Symposium Tallinn.

Eyles, N. and A.D. Miall 1984. Glacial facies. In R.G. Walker (ed.), Facies models, p. 15-38. Toronto, Geological Assoc. of Canada.

Goldthwait, R.P. 1973. Till investigations in North America. In Till, Abstracts of Symposium 10, IX INQUA Congress, Christchurch, New Zealand, p. 1.

Haldorsen, S. 1975. Nordisk brennmorene-forskning. En oversikt. Kvartaernytt 1:5-13.

Henning, G.J. and R. Grun 1983. ESR-dating in Quaternary geology. Quat. Sci. Rev. 2:157-238.

Hicock, S.R. and A. Dreimanis 1985. Glacio-tectonic structures as useful ice-movement indicators in glacial deposits: four Canadian case studies. Can. Jour. Earth Sci. 22:339-346.

Hutt, G., A. Molodkov, H. Kessel, and A. Raukas 1985. ESR dating of subfossil Holocene shells in Estonia. Nucl. Tracks 10, Nos. 4-6: 891-898.

Ikeya, M. and K. Ohmura 1981. Dating of fossil shells with electron spin resonance. J. Geol. 89:247-256.

Johansson, H.G. 1976. A report on a Scandinavian questionnaire on till, 1974. In W. Stankowski (ed.), Till, Its Genesis and Diagenesis. Univ. A. Mickiewicza, Poznan. Ser. Geogr. 12:198-202.

Jurgaitis, A. 1982. Criteria for recogni-tion and methods of investigation of genetic types of glaciofluvial deposits. In Ch. Schlüchter (ed.), Commission on Genesis and Lithology of Quaternary Deposits, Report on Activities 1977-1982, p. 38-41. Zurich, ETH.

Jurgaitis, A. 1984. Litogenez fluvioglia-cialika otdozhenii oblasti poshdnego materikovogo oledeneniia. Moscow, Nedia.

Kaerlein, F. 1985. Bibliographie der Geschie e des pleistozanen Vereisungsgebietes Nord-Europas. Teil II, Mitt. Geol. Palaont. Inst. Univ. Hamburg, 59.

Kinzelbach, W. 1986. Groundwater modelling. Amsterdam, Elsevier.

Mark, D.M. 1973. Analysis of axial orien-tation data, including till fabric. Geol. Soc. Amer. Bull. 84:1369-1374.

Orviku, K. 1958. Lithological investiga-tions of the till of the last glaciation in Estonia with quantitative methods (Russian, German summary). In ENSV Teaduste Akadeemia Geoloogia Instituudi Uurimused (Tallinn) 3:213-252.

Powers, M.C. 1953. A new roundness scale for sedimentary particles, J. Sed. Petrol. 23:117-119.

Raukas, A. 1962. Regularities in the distribution of pebbles in the tills of Estonia (Russian, English summary). In Eesti NSV Teaduste Akadeemia Toimetised, XI. Fuusikalis-matemaatiliste ja teh-niliste teaduste seeria 2:140-153.

Raukas, A. 1976. Procedures of till investigations in European countries and possibilities of their unification. In W. Stankowski (ed.), Till, its genesis

and diagenesis. Univ. A. Mickiewicza, Poznan. Ser. Geogr. 12:186-198.

Raukas, A. (ed.) 1980. Methods of the field and laboratory investigations of glacial deposits. Abstracts of the Sym-posium (Russian), 184 p.: Tallinn.

Raukas, A. 1982. Field and Laboratory Methods of Investigation of Tills. In Ch. Schlüchter (ed.), Commission on Genesis and Lithology of Quaternary Deposits, Report on Activities 1977-1982. p. 32-37. Zurich, ETH.

Raukas, A., D.M. Mickelson, and A. Dreimanis 1978. Methods of till investi-gation in Europe and North America. J. Sed. Petrol. 48:285-294.

Raukas, A., D.M. Mickelson, and A. Dreimanis 1979. Methods of the labora-tory investigations of glacial deposits in Europe and North America (Russian). In Eesti NSV Teaduste Akadeemia Toimetised, 28. Geoloogia 2:60-67.

Raukas, A. and L. Reintam 1965. The mechan-ical composition and some physico-chemical properties of the ground moraines of the last continental gla-ciation on Estonian territory (Russian, English summary). In Lithology and Stratigraphy of Quaternary Deposits in Estonia, 31-44. Tallinn.

Scott, J.S. 1976. Geology of Canadian Tills. In Leggett, R. (ed.), Glacial till, p. 50-66. Ottawa, Royal Society of Canada.

Scott, J.S. and D.A. St.-Onge 1969. Guide to description of till. Geol. Surv. Canada, Paper 68-6.

Serebrianniy, I.P. 1980. Laboratornii ana-liz V geomorfologic i chetvertichnoi paleogeografii. Moscow, Viniti.

Serebryanny, L.R. and A.V. Orlov 1982. Genesis of marginal moraines in the Caucasus. Boreas 11:279-289.

Shea, J.H. 1973. Proposal for a particle-size grade scale based on 10. Geology 1:3-8.

Stankowski, W. (ed.) 1976. Till, its gene-sis and diagenesis. Poznan: Univ. A. Mickiewicza, Ser. Geogr. 12.

Tavast, E. and A. Raukas 1982. The bedrock relief of Estonia (Russian, English sum-mary). Tallinn, Valgus.

216

*Genetic Classification of Glacigenic Deposits, Goldthwait & Matsch (eds)*
© *1988 Balkema, Rotterdam. ISBN 90 6191 694 1*

# Late glacial ice lobes and glacial landforms in Scandinavia

Jan Lundqvist
*Department of Quaternary Research, University of Stockholm, Sweden*

ABSTRACT: An early Younger Dryas readvance of the Scandinavian ice margin across
southern Sweden-Finland has been demonstrated. It is obvious from the distribution of
marginal deposits that the readvancing ice formed lobes. In connection with the readvance, the Baltic Ice Lake in front of the ice was reponded after having been temporarily lowered to sea level.
   Earlier the author has expressed the idea that the readvance was part of a rapid
downdraw of the central ice dome over the Gulf of Bothnia according to the model of
Denton and Hughes (1981). The ponding resulted in an increased hydrostatic pressure
below the ice and a heaving of its marginal parts. A connection between these processes
and glacial landforms was discussed.
   It is suggested that radial crevasses were formed in a marginal zone of divergent,
only partly compressive flow. In them glaciofluvial sediments were deposited as eskers.
In upslope positions compressive flow resulted in a thicker till cover. Otherwise the
till is very thin. Behind this marginal zone there was extending flow, in which transverse basal crevasses were formed. Changes in the water level caused break-up of the
crevasses, and De Geer moraines formed in them. Tabular icebergs were broken off and De
Geer moraines also formed along the ice margin. Behind the crevasse zone drumlinoid
features developed.
   De Geer moraines in Scandinavia also occur in other areas where there could have been
a floating, locally advancing ice margin affected by water-level changes, for instance
on the Swedish West Coast and in Finland south of the Salpausselkäs. Along the coasts of
the Gulf of Bothnia readjustment of the ice after collapse of the ice dome in the Gulf
had a similar effect.
   In interlobe areas there were radial crevasses or crossing crevasse systems where
radial moraines of different types formed. Meltwater deposited large amounts of gravel
and sand as deltas in these crevasse systems.
   Some of the conditions discussed may be relevant to areas with advancing and rapidly
disintegrating glaciers, such as Glacier Bay, Alaska.

## 1 INTRODUCTION

It is well known that the Younger Dryas
cool stage in Scandinavia caused a halt in
the retreat of the ice-sheet margin. This
event is marked across central Scandinavia
by the Ra moraines in Norway, the Central
Swedish marginal zone, and the
Salpausselkäs in Finland (Fig. 1). In
Finland and Norway as well as in part of
western Sweden these deposits form true
moraine ridges, whereas eastern Sweden has
only scattered but large marginal deltas.
All these features are portrayed on a
number of earlier published maps (e.g., G.
Lundqvist 1961).

The following paper discusses the
features of glacial geology related to the
Younger Dryas Zone, mainly in Sweden but
also in adjacent areas. These features are
characteristic and may be relevant in
other, formerly glaciated areas.

## 2 YOUNGER DRYAS GEOLOGICAL EVENTS

The following aspects of the glaciogeological development in the Younger Dryas
substage are essential. In the time and
region under discussion the ice terminated
almost entirely in open water. This water
was primarily the Baltic Ice Lake, ponded
up between the ice and the high ground

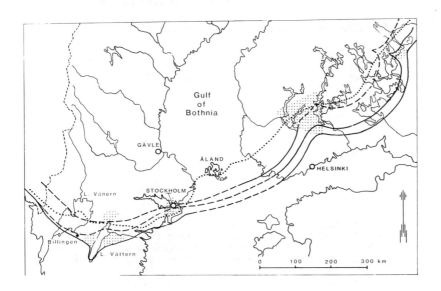

Fig. 1. The Younger Dryas ice-marginal lines across Sweden and Finland (solid and broken lines) and the tentative pre-Younger Dryas line of recession (dotted line) in Finland, according to Rainio (1985). Dotted areas are assumed interlobe regions.

around the southern part of the present-day Baltic Sea.

In the Alleröd interstade, preceding the Younger Dryas substage, the ice margin receded northward through the central Swedish lowland. When it passed the northern end of the Billingen hill, south-western Sweden, the Baltic Ice Lake lowered to sea level and a direct connection opened. How far this recession proceeded is unknown, but in Finland Rainio (1985) has shown a tentative position north of the Salpausselkäs. In Sweden we may also assume that it proceeded at least to north of the so-called Central Swedish marginal zone.

The Younger Dryas cooling implied a readvance of the ice margin. This is shown very clearly in Finland by the Salpausselkäs, far south of the preceding recession line (cf. Kujansuu and Niemelä 1984). In Sweden a readvance has been demonstrated by Johansson (1982), Svantesson (1981) and Strömberg (1985). The readvance caused a new damming of the water in the Baltic basin, as has been shown in the Billingen area by Björck and Digerfeldt (1984). The corresponding rise in water level is known from all over the southern Baltic basin (see Björck 1979, and Donner 1982, with references).

The pattern of the moraines, especially the Salpausselkä arcs (Kujansuu and Niemelä 1984) and the striae and glacial morphology behind them (also Punkari 1980)

clearly show that the readvancing ice formed large lobes (Brenner 1944). The author has suggested (lecture at the 17th Nordic Meeting in Finland 1986) that the advance was part of a rapid drawdown of the central ice dome over the Gulf of Bothnia, according to the model of Denton and Hughes (1981), and that it took place rapidly, possibly even as surges, with the advance of one lobe triggering the next. It was suggested that there may have been a close relationship between the advances and the rise of the Baltic Ice Lake. Actually it seems most logical that a climatically caused advance in the west blocked the outlet of the lake, and the rising water level triggered the advance of the eastern lobes. This hypothesis, however, is contradicted by the relation between the Salpausselkä I and shorelines of the lake found by Donner (1969). Nevertheless, the water-level rise of 26 m must have had a considerable effect upon the marginal parts of the ice.

After the initial readvance, the general recession of the ice sheet slowed down, and repeated minor oscillations of the ice margin took place (see Glückert 1981, Persson 1983, Kristiansson 1986). However, during the Younger Dryas substage there was only one cooling (Donner 1978) and one major ice advance, followed mainly by recession northward, during which time the Baltic Ice Lake was finally drained (Strömberg 1977).

218

Interaction between the ice margin and the water level in the Baltic basin is shown in Fig. 2. It is inferred that this interaction had a considerable effect upon the formation of glacial deposits. Some of these deposits are discussed below.

## 3 GLACIAL DEPOSITS IN THE YOUNGER DRYAS ZONE

### 3.1 Regional till cover

The regional till cover of the Younger Dryas zone in the Baltic area is generally thin, often absent, as seen from the geological maps of Sweden and Finland. In outline the surface of the pre-Quaternary bedrock is flat, but in detail rather hummocky. The till cover is mainly restricted to the lee-sides of the hummocks, whereas their upper surfaces are often not covered by till. An exception is the area of soft sedimentary Paleozoic

rocks east of Lake Vättern. Another essential condition is that the till cover seems to be insignificant on the low ground between the hummocks, as well.

The steep sides of the bedrock hummocks, commonly fractured by numerous tectonic joint systems, offer good possibilities for glacial erosion, favoured by frequent changes of ice movement as indicated by numerous striae. The fact that the till cover is insignificant therefore must be caused by the mode of glacial erosion. The ice may have been cold-based. This may have been the case at the Alleröd deglaciation (cf. Lagerlund et al. 1983). During the Younger Dryas deglaciation, however, the ice was probably warm-based, as indicated by the fact that the bedrock surface is everywhere striated and covered by numerous subglacial glaciofluvial deposits.

It is inferred that the entrainment of such an insignificant volume of glacial debris and the subsequent deposition of so little till indicates extending flow, a

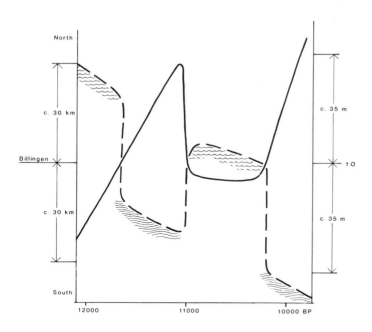

Fig. 2. Inferred interaction between the receding ice front (solid line, left scale) and the water level in the Baltic basin (broken line, right scale). Sparse wave-lining indicates the Baltic Ice Lake and close-spaced wave-lines the sea. When the receding ice margin passed the Billingen hill, the Baltic Ice Lake was lowered to sea level. When the ice readvanced, there was a new ponding, until it finally receded again at the end of Younger Dryas time. Distances and altitudes depend on the geographical situation -- this sketch shows an estimation for southcentral Sweden.

condition that allows very little transport of basal debris upward in the ice. It is also inferred that the flow was divergent toward the convex front of the ice lobes, resulting in the dispersal of debris over a larger area.

Only in a narrow marginal zone was there compressive flow sufficient to allow transport (upward) and concentration of debris to form a somewhat thicker till cover. This happened mainly in positions at the margin where the regional slope was toward the advancing ice lobes. Such zones are clearly visible on some geological maps (Svantesson 1981, Persson and Lundström 1973). In no place, however, did a moraine in the morphological sense, composed of till, form. The concentration of till took place only by the thrusting of thin slabs on top of each other, according to the process described from terrestrial ice margins (Boulton 1972).

## 3.2 De Geer moraines

A most characteristic feature at some distance (up to more than 50 km) behind the Younger Dryas margin is the sequence of small moraines, usually referred to as De Geer moraines (Hoppe 1959), and often very similar to the washboard moraines defined by Mawdsley (1936). They occur in several areas in Scandinavia, as will be further discussed below, but this is their main region (see map of G. Lundqvist 1961).

The mechanism of formation of De Geer moraines has been much debated. According to De Geer's (1940) original idea they are marginal, annual deposits. According to a different opinion (Elson 1957, Hoppe 1957, Strömberg 1965), they were formed in basal crevasses in the ice and have no chronological significance, even if there is sometimes a correlation between them and the annual recession of the ice margin (Möller 1962).

The conspicuous distribution behind the limit of the Younger Dryas readvance implies a close relationship. If we consider the idea of a very rapid readvance -- possibly surges -- and the extending flow discussed above, we may infer that the moraines represent a zone of ice with mainly transverse crevasses (Fig. 3) similar to that seen in surging glaciers (Robin and Barnes 1969). The pattern of the moraines and their combination with radial forms, demonstrated by Strömberg (1965) and seen even in De Geer's (1940) original works, clearly resembles a crevasse system. In some areas it is almost identical with the subglacially formed

pattern in Glacier Bay, Alaska (Haselton 1979).

It is inferred that the crevasses were formed in the basal part of the extending-flowing ice, and that these crevasses were opened further during changes in the water level. The heaving effect and hydrostatic pressure in the ice helped to open the crevasses. At the final lowering of the lake further break-up took place and subglacial debris was squeezed into the crevasses to form moraines. During the process of deglaciation, tabular icebergs broke off at the crevasses; thus, the moraines may also have coincided with the ice margin during retreat by calving. Such a process accords with the varying internal composition of the moraines (J.

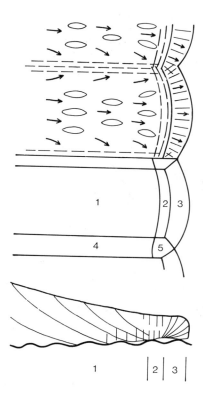

Fig. 3. Crevasse systems and flow directions in the outer part of two adjacent ice lobes and slip lines in a longitudinal section through a lobe. 1: Extending flow diverging toward the sides of the lobes. 2: Zone of transverse crevasses. 3: Marginal zone with diverging, compressive flow. 4: Interlobe area with radially fractured ice. 5: Area with complicated, intersecting crevasses.

Lundqvist 1977).

## 3.3 Radial moraines

Radial moraines -- crag-and-tail forms as well as more irregular ridges -- are common in one part of the Younger Dryas zone and north of it (e.g., Bergdahl 1953, Strömberg 1981). Some of these features are without any doubt subglacially formed (Möller 1960), whereas others are debatable. Among the latter are the moraines in the Gävle area, much discussed in the literature (see G. Lundqvist 1963). Recent investigations indicate that they are composed of basal meltout till, and thus formed subglacially and not in open crevasses.

All these features belong mainly to an area that has been interpreted as the interlobe zone between one lobe in the Baltic depression forming the western Salpausselkä arc in Finland and another, less well-defined lobe over eastern Sweden (Punkari 1984). This is indicated by the directions of the striae (cf. G. Lundqvist 1961). The zone extends northward between the areas covered by a Baltic lobe and the inland ice over Sweden (cf. Järnefors and Fromm 1960, G. Lundqvist 1963). The whole area was characterized by upbreaking of the ice margin with formation of deep, narrow calving bays (Strömberg 1981). It is suggested here that the radial moraines should be interpreted as interlobe subglacial formations.

## 3.4 Eskers and deltas

A conspicuous feature in the zone just north of the Younger Dryas limit is the large radial eskers. They are very regularly spaced (G. Lundqvist 1959), indicating a regular diverging flow towards the ice margin. In Finland the eskers form joint moraines of meltwater sediments along the ice margin, the Salpausselkäs (see Ignatius, Korpela, and Kujansuu 1980, Kujansuu and Niemelä1984). In that flat country the flow has taken place on a broad front as well as in regularly spaced channels. In the more broken Swedish country no continuous moraine is formed. The esker systems widen to form scattered large deltas along the ice margin.

In some areas, these delta deposits are more frequent and form a very irregular pattern. This applies to the Stockholm region, south of the region with radial moraines, and a similar area is found around the northern part of Lake Vättern. It may be possible to arrange the deposits

into "moraines" (e.g., Mörner 1970), but the significance of these "moraines" is highly debatable. Instead, some other features are probably more relevant.

In the Stockholm region some of the large eskers split to form several branches. Each of these is marked by very large deposits, similar to deltas but not reaching the corresponding sea level. The stratigraphy of these deposits indicates that deposition started subglacially, often on the lee-side of bedrock knobs. An "esker" formed on top of the subglacial coarse sediment, and finally a long delta covered it all.

In the Vättern region, kettle-holes and ice-contact features indicate that the deposits accumulated in contact with the ice. The shape of the deltas demonstrates accumulation in bays in the ice margin. A conspicuous feature in connection with the deltas is a moraine morphology trending NE-SW and NW-SE (see Fig. 4 and geological maps, e.g., Svantesson 1981). We get the impression that the glaciofluvial deposits accumulated in two crossing systems of crevasses, with both directions at roughly 45° angles to the ice margin.

The deposits under discussion were probably formed within strongly fractured interlobe zones close behind the ice margin. In the Stockholm region this is the marginal part of the interlobe area marked by radial moraines. At Lake Vättern conditions were more complicated. Lobes on the western and eastern side of the lake interfered to form a local lobe in the lake basin. This interpretation of the glaciofluvial deposits in the Swedish sector is supported by conditions in Finland, where the interlobes are much more clearly developed. Examples are found above all in the Lahti area, in eastern Karelia and around Kankaanpää (see Punkari 1980, 1984; Kujansuu and Niemelä 1984).

## 4 COMPARISON WITH OTHER REGIONS

It has been suggested above that some specific glacial forms connected to the Younger Dryas ice margin have been caused by the extending flow of rapidly moving ice lobes, and the changes in water level/ hydrostatic pressure related to these movements. Because similar forms occur in other areas in Scandinavia it is necessary to look for comparable explanations in those instances.

## 4.1 De Geer moraines

Small moraines of the De Geer type are

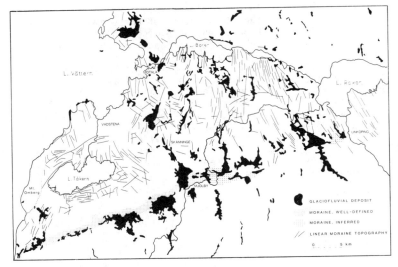

Fig. 4. Sketch map of the main glacio-morphological features at the Younger Dryas moraine east of Lake Vättern. The moraine topography implies the extension of till areas on the geological maps, mostly not true ridges. Redrawn from the geological map sheets Ae nos. 13, 19, 22, 24, 36, and 44 of the Geological Survey of Sweden.

frequent in some areas on both sides of the northern part of the Gulf of Bothnia (Hoppe 1948, Fromm 1965, Zilliacus 1981), and south of the Salpausselkäs in Finland (Zilliacus 1976, Kujansuu and Niemelä 1984). Along the Swedish west coast (Caldenius 1942, Hillefors 1979) and on the bottom of Hanö Bay, in the southern Baltic Sea (Tunander 1982), there are moraines that are sometimes described as De Geer moraines, although they are straighter and more regular.

In none of these cases can the Younger Dryas readvance be used as an explanation of the moraine formation. Only in one case may there have been a rise of an adjacent water level at the stage in question. However, if we consider the general glaciogeological environment, we find that in all these areas there was probably some readjustment of the balance in the ice when the ice margin was nearby.

On the Swedish west coast we can assume a certain outflow of ice from the South Swedish upland when the ice in the Kattegatt wasted away. Mörner (1969) has described how a wedge-like bay in the ice margin separated an ice lobe in the Kattegatt basin from the inland ice. We may assume that an outflow, a downdraw, of the inland ice took place when the buttressing effect of the Kattegatt ice ceased. This effect may have been increased by a contemporary sea level rise (see Lagerlund 1980).

In southern Finland the retreat of the ice front from the Palivere moraine in Estonia (e.g., Cebotareva 1972) and the northern scarp ("the Glint") of the Paleozoic rocks extending from Gothland to

northern Estonia (e.g., Martinsson 1979) would have had a similar effect. The buttressing effect of the steep scarp ceased and outflow against it followed.

In both the Gulf of Bothnia and the Hanö Bay the moraine formation may be related to the rapid break-up of large ice masses. In the Gulf of Bothnia this was related to the final collapse of the central dome of the ice sheet (Denton and Hughes 1981). The collapse of the waterbased dome implied a break-up with formation of basal crevasses and outflow of tabular icebergs. The remaining ice on both sides of the Gulf readjusted itself with further formation of basal crevasses in which the moraines could develop.

In the whole of the southern Baltic basin we may assume the existence of a vast, flat, down-wasting ice mass (cf. Mörner, Flodén, Beskow, Elhammer, and Haxner 1977). This ice seems to have disappeared more or less simultaneously from a vast region (Duphorn, Kögler, and Stay 1979). We may assume that this took place as an upflow and breaking off of tabular ice masses with formation of temporary basal crevasses.

It is essential that in all these cases the moraines are interpreted as formed in basal crevasses, although an ice margin might well develop along such a crevasse. The crevasses are related to a downdraw of the marginal part of the ice towards open water rather than by an increase of accumulation in an ice centre.

Outside Scandinavia, one area that may be compared to the De Geer moraine areas at the Younger Dryas Zone is the McBride Remnant Glacier, Glacier Bay, Alaska. The pattern of moraines described by Haselton (1979) is very similar to patterns of De Geer moraines and crag-and-tail forms at the Younger Dryas zone. One of the most important aspects of the glacial history of this region is the very rapid deglaciation, partly by downdraw at the break-up of tidal glaciers. Although the McBride Remnant Glacier had its base about 100 m above sea level it may have been affected by the rapid change of balance in a way that created basal crevasses.

Mickelson and Berkson (1974) have demonstrated how similar moraines continue down to the bottom of the adjacent Wachusett Inlet. The Plateau Glacier reaching into the inlet was affected by the tide. Most probably the changes of water level have contributed to an upbreaking similar to the one discussed for the Baltic area.


## 4.2 Interlobe eskers and deltas

In some areas in Scandinavia we find concentrations of large deltas and eskers similar to the ones discussed above as interlobate complexes. The location in all typical instances seems to be compatible with the discussed areas. Most of the good examples are, however, situated within the Younger Dryas zone itself, where the lobe pattern is best developed.

Outside the Younger Dryas zone, an area in southwestern Värmland, north of Lake Vänern (see J. Lundqvist 1958), is a case in point. There, several large deltas very similar to those of the Stockholm region occur. As in the Stockholm region, the pattern may be interpreted as a number of successive marginal positions. It would be possible to construct lines of recession, but there is no absolute correlation between different deposits. The general ice movement converged toward the area, although with rather small angles. Like the Stockholm deposits, these were formed far below (50-100 m) the corresponding sea level.

In Scania in southernmost Sweden, the separation of a Baltic ice stream moving westward in the extreme south, and a main ice moving from the northeast, has been recognized early. The details in this separation process have been very much debated, and are not completely understood (see summary by Lagerlund 1977).

Irrespective of the details in the process, the separation of the ice into two main units splitting up from the northwest (see Mörner 1969) seems beyond any doubt. The area of separation can be considered

a good example of an interlobe. It can be followed all through Scania as a system of very large glaciofluvial deposits extending from the Ängelholm region past the Söderåsen and Romeleåsen horsts. Large and thick deltas at Söderåsen are described by Sandgren (1983). In the basin of Lake Vombsjön the deposits are extremely large and complicated, glaciofluvial sediments alternating with glaciolacustrine. Toward the east, a broad zone of glaciofluvial deposits ends with the kame tract of Brösarps backar. This wide and long train of interlobate sediments is generalized to a broad zone of glaciofluvial deposits on small-scale maps (see G. Lundqvist 1959).

This Scanian interlobe area differs from those discussed above in its position in relation to a water level. Although there was a complicated system of glacial lakes (cf. Nilsson 1953) the deposits were formed in shallow water or in supra-aquatic position. Consequently, the morphology is much sharper and the individual deposits are not separated by deep-water sediments. This region is more compatible with the supra-aquatic interlobe areas of the Great Lakes region in North America, which are characterized by large volumes of hummocky glaciofluvial sediments and scattered, high kames (e.g., Alden 1918, Nelson and Mickelson 1977).

Thus we may distinguish between two types of interlobe areas: one belongs to debris-rich areas in supra-aquatic position, the other occurs in subaquatic areas with little debris. The difference may be expressed in terms of glacier flow. In the debris-rich areas there was compressive flow due partly to the supra-aquatic position with greater basal friction, partly to topographic conditions with upslope flow. In the areas with low content of debris there was extending flow due to downdraw at the break-up of a floating ice margin. Only in distinct upslope position was the flow locally compressive.


## 5 CONCLUSIONS

The rapid advance of ice lobes into deep water may account for a number of glaciomorphological features. We may assume that glacial advance was accomplished by extending flow caused by downdraw rather than increased accumulation. Some features, such as De Geer moraines and groups of large glaciofluvial accumulations, which have been difficult to interpret, find a natural explanation in this way, even though other aspects may have to be considered as well.

The process described may be explained in terms of the interaction between attenuated parts of the ice sheet and changes of water level in the Baltic basin. Thinning ice in combination with raised water level and increased hydrostatic pressure at the base of the ice may result in rapid calving and downdraw from the central ice dome. This is an interpretation that also has implications in other regions where we find similar glacial deposits and their combinations.

REFERENCES

Alden, W.C. 1918. The Quaternary geology of southeastern Wisconsin with a chapter on the older rock formations. U.S. Geol. Survey. Prof. Paper 106.

Bergdahl, A. 1953. Israndsbildningar i östra Syd-och Mellansverige med särskild hänsyn till åsarna. Medd. Lunds Univ. Geogr. Inst. Avhandl. 23.

Björck, S. 1979. Late Weichselian stratigraphy in Blekinge, SE Sweden, and water level changes in the Baltic Ice Lake. Univ. Lund Dep. Quat. Geol. Thesis 7.

Björck, S. and G. Digerfeldt 1984. Climatic changes at Pleistocene/Holocene boundary in the middle Swedish endmoraine zone, mainly inferred from stratigraphic indicators. In N.-A. Mörner and W. Karlén (eds.), Climatic Changes on a Yearly to Millennial Basis, p. 37-56. Dordrecht, D. Reidel Publ. Co.

Boulton, G.S. 1972. Modern Arctic glaciers as depositional models for former ice sheets. J. Geol. Soc. London 128:361-393.

Brenner, T. 1944. Finlands åsars vittnesbörd om ytgestaltningen hos landisen. Fennia 68:4.

Caldenius, C. 1942. Gotiglaciala israndsstadier och jökelbäddar i Halland. Geol. Fören. Stockh. Förhandl. 64:163-183.

Cebotareva, N.S. 1972. Die Gletscherströme der Valdaj-Vereisung. Petermanns Geogr. Mitteil. 116:247-254.

De Geer, G. 1940. Geochronologia Suecica Principles. Kungl. Sven. Vet.-Akad. Handl. Ser. 3, 18:6.

Denton, G.H. and T.J. Hughes 1981. The Last Great Ice Sheets. New York, J. Wiley and Sons.

Donner, J.J. 1969. Land/sea level changes in southern Finland during the formation of the Salpausselkä endmoraines. Bull. Geol. Soc. Finland 41:135-150.

Donner, J.J. 1978. The dating of the levels of the Baltic Ice Lake and the Salpausselkä moraines in South Finland. Comment. Phys.-Math. 48:11-38.

Donner, J.J. 1982. Fluctuations in water level of the Baltic Ice Lake. Ann. Acad. Sci. Fennicae A III 134:13-28.

Duphorn, K., F.-C. Kögler, and B. Stay. 1979. Late-glacial varved clays in the Bornholm Basin and Hanö Bay. Boreas 8:137-140.

Elson, J.A. 1957. Origin of washboard moraines. Geol. Soc. Am. Bull. 68:1721.

Fromm, E. 1965. Beskrivning till jordartskarta över Norrbottens län nedanför lappmarksgränsen. Sver. Geol. Unders. Ca 39.

Glückert, G. 1981. Salpausselkien rakenteesta ja synnystä Lohjalla. Publ. Dep. Quat. Geol. Univ. Turku 45.

Haselton, G.M. 1979. Some glaciogenic landforms in Glacier Bay National Monument, Southeastern Alaska. In C. Schlüchter (ed.), Moraines and Varves, p. 197-205. Rotterdam Balkema.

Hillefors, A. 1979. Deglaciation models from the Swedish West Coast. Boreas 8:153-169.

Hoppe, G. 1948. Isrecessionen fran Norrbottens kustland i belysning av de glaciala formelementen. Geographica 20.

Hoppe, G. 1957. Problems of glacial morphology and the Ice Age. Geogr. Ann. 39:1-18.

Hoppe, G. 1959. Glacial morphology and inland ice recession in northern Sweden. Geogr. Ann. 41:193-212.

Ignatius, H., K. Korpela, and R. Kujansuu. 1980. The deglaciation of Finland after 10,000 B.P. Boreas 9:217-228.

Järnefors, B. and E. Fromm. 1960. Chronology of the ice recession through Middle Sweden. Internat. Geol. Congr. Rep. XXI Session, Norden, 1960. Part IV, 93-97.

Johansson, B.T. 1982. Deglaciationen av norra Bohuslän och södra Dalsland. Chalmers Tekn. Högsk. Göteb. Univ. Geol. Inst. A 38.

Kristiansson, J. 1986. The ice recession in the south-eastern part of Sweden. A varve-chronological time scale for the latest part of the Late Weichselian. Univ. Stockh. Dep. Quat. Res. Rep. 7.

Kujansuu, R. and J. Niemelä (eds.) 1984. Suomen maaperä. Geol. Tutkimuskeskus, Helsinki. Map.

Lagerlund, E. 1977. Förutsättningar för moränstratigrafiska undersökningar på Kullen i Nordvästskåne - teoriutveckling och neotektonik. Univ. Lund Dep. Quat. Geol. Thesis 5.

Lagerlund, E. 1980. Litostratigrafisk indelning av Västskånes Pleistocen och en ny glaciationsmodell för Weichsel. Univ. Lund Dep. Quat. Geol. Rep. 21.

Lagerlund, E., K. Knutsson, M. Åmark, M. Hebrand, L.-O. Jönsson, B. Karlgren, J.

Kristiansson, P. Möller, J.M. Robison, P. Sandgren, T. Terne, and D. Waldemarsson 1983. The deglaciation pattern and dynamics in South Sweden, a preliminary report. Univ. Lund Dep. Quat. Geol. Rep. 24.

Lundqvist, G. 1959. Description to accompany the map of the Quaternary deposits of Sweden. Sver. Geol. Unders. Ba 17.

Lundqvist, G. 1961. Beskrivning till karta över landisens avsmältning och högsta kustlinjen i Sverige. Sver. Geol. Unders. Ba 18.

Lundqvist, G. 1963. Beskrivning till jordartskarta över Gävleborgs län. Sver. Geol. Unders. Ca 42.

Lundqvist, J. 1958. Beskrivning till jordartskarta över Värmlands län. Sver. Geol. Unders. Ca 38.

Lundqvist, J. 1977. Till in Sweden. Boreas 6:73-85.

Martinsson, A. 1979. The Pre-Quaternary Substratum of the Baltic. Acta Univ. Upsal. Symp. Univ. Upsal. Annum Quingentesimum Celebrantis 1, p. 77-86. Uppsala.

Mawdsley, J.B. 1936. The wash-board moraines of the Opawica-Chibougamau area, Quebec. Trans. Roy. Soc. Can. Ser. 3, 30(4):9-12.

Mickelson, D.M. and J.M. Berkson 1974. Till ridges presently forming above and below sea level in Wachusett Inlet, Glacier Bay, Alaska. Geogr. Ann. 56 A:111-119.

Möller, H. 1960. Moränavlagringar med linser av sorterat material i Stockholmstrakten. Geol. Fören. Stockh. Förhandl. 82:169-202.

Möller, H. 1962. Annuella och interannuella ändmoräner. Geol. Fören. Stockh. Förhandl. 84:134-143.

Mörner, N.-A. 1969. The Late Quaternary history of the Kattegatt Sea and the Swedish West Coast. Deglaciation, shore-level displacement, chronology, isotasy and eustasy. Sver. Geol. Unders. C 640.

Mörner, N.-A. 1970. The younger Dryas Stadial. Geol. Fören. Stockh. Förhandl. 92:5-20.

Mörner, N.-A. T. Flodén, B. Beskow, A. Elhammer, and H. Haxner 1977. Late Weichselian deglaciation of the Baltic. Baltica 6:33-51.

Nelson, A.R. and D.M. Mickelson 1977. Landform distribution and genesis in the Langlade and Green Bay glacial lobes, North-Central Wisconsin. Wisc. Acad. Sci. Arts and Letters 65:41-57.

Nilsson, E. 1953. Om Södra Sveriges senkvartära historia. Geol. Fören. Stockh. Förhandl. 75:155-246.

Persson, C. 1983. Glacial deposits and the Central Swedish end moraine zone in eastern Sweden. In J. Ehlers (ed.), Glacial Deposits in North-West Europe, p. 131-140. Rotterdam, Balkema.

Persson, C. and I. Lundström 1973. Beskrivning till geologiska kartbladet Nyköping SO. Sver. Geol. Unders. Ae 12.

Punkari, M. 1980. The ice lobes of the Scandinavian ice sheet during the deglaciation in Finland. Boreas 9:307-310.

Punkari, M. 1984. The relations between glacial dynamics and tills in the eastern part of the Baltic Shield. Striae 20:49-54.

Rainio, H. 1985. Första Salpausselkä utgör randzonen för en landis som avancerat på nytt. Geologi 37:70-77.

Robin, G. de Q. and P. Barnes 1969. Propagation of glacier surges. Can. J. Earth Sci. 6:969-977.

Sandgren, P. 1983. The deglaciation of the Klippan area, southern Sweden, a study of glaciofluvial and glaciomarine sediments. Lund Univ. Dep. Quat. Geol. Thesis 14.

Strömberg, B. 1965. Mappings and geochronological investigations in some moraine areas of south-central Sweden. Geogr. Ann. 47A:73-82.

Strömberg, B. 1977. Einige Bemerkungen zum Rückzug des Inlandeises am Billingen (Västergötland, Schweden) und dem Ausbruch des Baltischen Eisstausees. Z. Geomorph. N.F. Suppl. 27:89-111.

Strömberg, B. 1981. Calving bays, striae and moraines at Gysinge-Hedesunda, central Sweden. Geogr. Ann. 63A:149-154.

Strömberg, B. 1985. New varve measurements in Västergötland, Sweden. Boreas 14:111-115.

Svantesson, S.-I. 1981. Beskrivning till jordartskartan Hjo SO. Sver. Geol. Unders. Ae 44.

Tunander, P. 1982. Seismic stratigraphy of unconsolidated deposits in the Hanö Bay, southern Baltic. Stockh. Contr. Geol. 39.

Zilliacus, H.L. 1976. De Geer-moräner och isrecessionen i södra Finlands östra delar. Terra 88:176-184.

Zilliacus, H.L. 1981. De Geer moränerna på Replot och Björkön i Vasa skärgård. Terra 93:12-24.

*Genetic Classification of Glacigenic Deposits, Goldthwait & Matsch (eds)*
© *1988 Balkema, Rotterdam. ISBN 90 6191 694 1*

# Genetic classification of glaciofluvial deposits and criteria for their recognition

A.Jurgaitis
*Vilnius State University, Vilnius, USSR*

G.Juozapavičius
*Lithuanian Geological Research and Survey Institute, Vilnius, USSR*

ABSTRACT: A genetic classification of glaciofluvial deposits was developed according to work on Project 2C of the INQUA Commission on Genesis and Lithology of Quaternary Deposits. The proposed genetic classification is based upon a complex of different textural and structural properties and the composition of the deposits. This paper also presents the criteria for recognition of different types of glaciofluvial deposits.

## 1 GENETIC CLASSIFICATION

Glaciofluvial deposits are widely distributed in formerly glaciated areas. They are most commonly found as sandy-gravelly-pebbly deposits occurring as mounds or extensive plains, and are generally extremely variable in form or texture. They may form either inside or outside a glacier.

Investigation of glaciofluvial deposits began in the middle of the 19th century. Subdivisions of glaciofluvial deposits into englacial and proglacial are generally accepted (Shantser 1966; Price 1973; Embleton and King 1975), based on the relationship of the deposits to the glacier from which they derive. Furthermore, detailed analysis of data (Jurgaitis 1977, 1980; Jurgaitis and Juozapavičius 1979; J. Lundqvist 1979, 1985) distinguishes ice marginal deposits, which form immediately at the edge of a glacier.

According to work on Project 2C of the INQUA Commission on Genesis and Lithology of Quaternary Deposits, the genetic classification of glaciofluvial deposits was discussed and approved at the Xth INQUA Congress in Birmingham (England) in 1977 and at the XIth Congress in Moscow (USSR) in 1982.

On the basis of lithomorphogenetic study, three genetic types (Table 1) are distinguished among glaciofluvial deposits: englacial, ice-marginal, and proglacial.

The englacial glaciofluvial deposits, which have been formed by englacial, subglacial, and supraglacial streams of meltwater, are subdivided into kame and esker deposits.

Kames are formed in depressions on the glacier surface and other cavities; they appear as rounded or irregular hills. The best definition of kame is given by C.D. Holmes (1947): "a kame is a mound composed chiefly of sand and gravel, whose form has resulted from original deposition modified by any slumping incident to later melting of glacial ice against or upon which the deposit accumulated."

According to geological structure and genesis, kames (Niewiarowski 1963) are subdivided into glaciofluvial, limno-glacial, and mixed; and according to the place of their formation in the glacier, into englacial, subglacial, and supraglacial.

Kame terraces are distinguished as a variety of kame deposits. They are formed by streams of glacial meltwater between the glacier on one side, and a highland or valley slope on the other. A terrace is formed when the supporting ice is removed.

Esker deposits are formed by glacial meltwaters in tunnels and crevasses in stationary or retreating glaciers, and are finally deposited as long, narrow, and sinuous ridges.

Radial eskers are oriented perpendicular to the glacier margin. From detailed study of eskers in Canada, Banerjee and McDonald (1975) suggest three ways that eskers form: in open channels, in tunnels, and as deltas.

Glaciofluvial deposits of so-called end (terminal) moraines described by Flint (1971) are expressed as marginal ridges.

Table 1. Genetic classification of glaciofluvial deposits.

| Genetic type | Genetic subtype | Genetic species | Relationship to ice |
|---|---|---|---|
| Englacial | Kames | Glaciofluvial kames<br>Limnoglacial kames<br>Mixed kames<br>Kame terraces | Englacial<br>Subglacial<br>Supraglacial |
| | Eskers | Radial eskers | Englacial<br>Subglacial<br>Supraglacial |
| Ice-marginal | Marginal ridges | Piled marginal ridges<br>Piled and ice-pushed marginal<br>  ridges | Marginal |
| Proglacial | Sandurs | Elementary sandurs<br>Outwash plains<br>Interridge sandurs | Marginal<br>Proglacial |
| | Glaciofluvial deltas | Outwash deltas<br>Underdeveloped deltas<br>Deltas of glacial tunnels and<br>  channels<br>Deltas of valley mouths of<br>  meltwater discharge | Marginal<br>Proglacial<br>Lateral<br>Distal |
| | Glaciofluvial terraces | Terraces of large lateral valleys<br>Terraces of valleys of glacio-<br>  fluvial discharge | Lateral<br>Distal |

It is suggested that these can be formed at the frontal margins of continental glaciers from accumulations of stratified and unstratified glaciofluvial deposits. These form hilly rampart-like ridges, rows, or hills stretching along the frontal margin of the glacier. The ridges described by Parizek (1969) in Canada and the U.S.A. are examples. In characterizing stagnation deposits of glaciers, Embleton and King (1975) distinguish these as a separate group.

Glaciofluvial marginal ridges are basically divided into accumulated (piled) ridges that were formed at the stationary margins of glaciers or by retreating glaciers, and ice-pushed ridges of glaciofluvial material formed by the action of glaciers advancing over outwash.

Proglacial meltwater stream deposits comprise sandurs, glaciofluvial deltas, and glaciofluvial terraces.

Sandur deposits, gently sloping fanlike plains, originate as a result of accumulation and erosion processes of freely wandering glacial meltwater streams at the front of continental ice sheets. Detailed morphological and structural investigations have been published by Jewtuchowicz (1955) and Mikalauskas (1963). Three types of sandurs are distinguished: elementary or onecone sandurs, composite or outwash plains, and interridge sandurs.

Glaciofluvial delta deposits are formed in subaerial and subaquatic positions at the mouths of glaciofluvial rivers entering proglacial water bodies. They are represented by flat alluvial proglacial plains. These deposits differ little from ordinary deltas; therefore they are rarely distinguished in the literature on glaciofluvial formations. Four varieties of glaciofluvial deltas are distinguished: outwash deltas, underdeveloped deltas, deltas of glacial tunnels and channels, and deltas of valley mouths of meltwater discharge.

Glaciofluvial terraces represent the deposition of clastic material by meltwater streams, and are confined to valley sides in the proglacial zone. Terraces and deposits of large lateral glaciofluvial rivers are described in detail by Woldstedt (1950) and Galon (1961). Two species of glaciofluvial terraces are distinguished: terraces of large lateral valleys and those of glaciofluvial discharge valleys.

In finishing the review of definitions and terminology of glaciofluvial deposits, it should be noted that certain conditions must exist for inclusion of a particular variety into a specific genetic type. For example, esker deposits formed as deltas can be considered marginal glacial or even proglacial deposits. Conversely, the deposits of pitted sandur originally formed on the surface of underlying ice are included by R.J. Price (1973) in a group of englacial deposits.

## 2 METHODS OF INVESTIGATION

On the basis of 56 replies (from 7 countries) to a questionnaire, Table 2 was compiled, emphasizing the importance of methods used for investigation of the genetic varieties of glaciofluvial deposits.

Replies to the questionnaire suggest that at present geomorphologic and lithologic criteria are used almost equally in northern hemisphere countries to determine the genesis of glaciofluvial deposits. Among the lithologic criteria, structural peculiarities and indices of granulometric composition are distinguished by a greater importance and reliability. Sometimes the orientation of dip of cross-bedded laminae is used as well. About 70% of all criteria used yield results under field conditions.

Because glaciofluvial deposits are characterized by complicated composition and primary structure and texture, special methods are needed for their study. Our studies are based on a detailed investigation of deposits -- texture, structure, granulometric, mineralogic, petrographic and chemical composition, and in the western regions of the U.S.S.R.: in

Karelia, Estonia, Latvia, Lithuania, Byelorussia, and the northwestern Ukraine. Results of these investigations have been presented in several papers, for instance Jurgaitis, Mikalauskas, and Juozapavičius 1982; Juozapavičius and Ekman 1984; Jurgaitis 1984.

Mean size of grains in mm ($M_\alpha$) and sorting coefficient ($\sigma$) were determined by using the following formulas:

$$\log \quad M_\alpha = h + w v_1,$$

$$\sigma = w\sqrt{v_2 - v_1},$$

where $M_\alpha$ is the value of a mean weight dimension of grains expressed in millimetres, $\sigma$ is a weight coefficient of sorting, h - a half sum of logarithms of the most distributed fraction, w - difference of logarithms of finite dimensions of the most distributed fraction in (1), w - a mean interval between the logarithms of finite dimensions of fractions in (2), $v_1$ and $v_2$ - the first and second moments (Jurgaitis 1980:92).

A detailed field study of deposits and analysis of laboratory results are necessary to differentiate rather similar hydrodynamic situations. Figure 1 is a schematic suggestion of investigation methods for glaciofluvial deposits.

Before studying any lithologic body it is necessary to determine in general outline at least its geomorphology to obtain an initial interpretation of the paleogeographic situation of deposition of fragmental material. It is known that deposits of different genesis are expressed as different landforms: plains are formed by the deposition of proglacial glaciofluvial deposits, a hilly relief commonly is composed of englacial and marginal deposits, and valley areas are occupied by the deposits of glaciofluvial

Table 2. Importance of the parameters studied for investigation of glaciofluvial deposits, in percentages.

| Criteria used | Percentage |
|---|---|
| I. Geomorphologic | 48.3 |
| II. Lithologic | 51.7 |
|    1. Structural | 18.8 |
|    2. Granulometric | 13.9 |
|    3. Orientation of dip of cross-bedding | 5.7 |
|    4. Mineralogic-petrographic | 3.8 |
|    5. Morphologic peculiarities of fragments | 3.3 |
|    6. Orientation of long axes of pebbles | 3.1 |
|    7. Others | 3.1 |

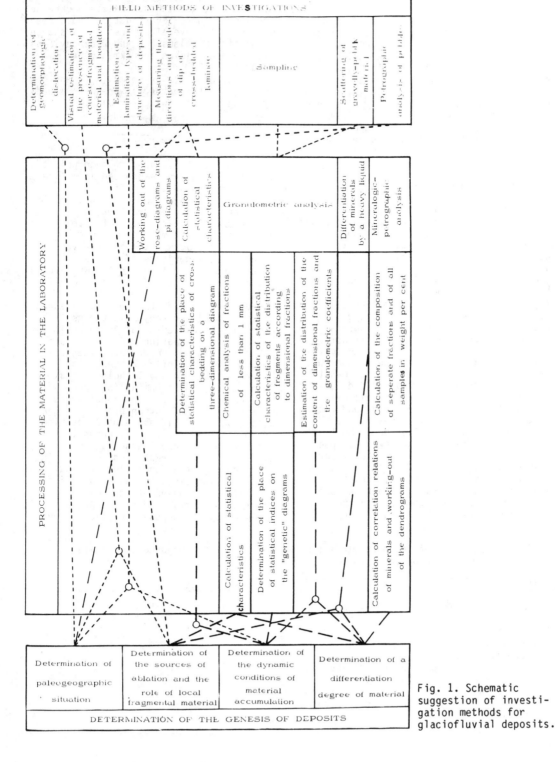

Fig. 1. Schematic suggestion of investigation methods for glaciofluvial deposits.

terraces and deltas.

## 2.1 Bedding and structure

In visual study of the stratigraphic sections of deposits, attention should always be given to the presence of beds of sandy-gravel material, thickness, quantity, petrographic composition and sorting of boulders, the occurrence of which eliminates the possibility of the formation of deposits in basins of low hydrodynamic activity. A considerable quantity of large boulders, or sometimes of thick beds of rounded but unsorted small boulders might prove proximity of the ice margin during the period of accumulation of sediments (Fig. 2). Beds of almost equal rounded boulders are usually formed under high hydraulic pressure in the streams forming eskers (Fig. 3).

A detailed study of the lamination and other structural peculiarities of material provides valuable information about the

Fig. 3. Nonstratified coarse-fragmental deposits (cobble-rounded stone with small admixture of pebbles) in the esker of Iizaku (Estonia).

sedimentation conditions. It has been found that every type of bedding corresponds to a certain phase of sedimentation created by a certain velocity of stream and its bed-load. The post-depositional dislocations of bedding (disjunctive and by pressure) indicate former contact of the depositing medium with the glacier (Fig. 4).

Detailed measurements on the orientation and dips of cross-bedded laminae allow us to reconstruct the direction of transportation of material and the strength and degree of variation of stream flow. The statistical characteristics of elements reflecting the variability of a spatial distribution of cross-bedded laminae made it possible to construct a three-component diagram (mean dip of cross-bedded laminae, variation coefficient, variation coefficient of the directions of dip). The diagram (Fig. 5) helps to distinguish between the deposits of anastomosing streams (the deposits of kames, marginal glaciofluvial ridges, and sandurs) and the deposits of single-channel streams (eskers, glaciofluvial deltas, and terraces).

Eskers and marginal glaciofluvial ridges have high dip angles of crossbeds. Taking into account the relief forms of glaciofluvial deposits, the given diagram makes it possible in the end to recognize esker deposits, marginal glaciofluvial ridges, kames, sandurs, and glaciofluvial deltas and terraces combined.

Horizontal bedding with frequent alternation of gravel and pebble beds with beds of sand, presence of faults and absence of boulders are typical of kame-forming depo-

Fig. 2. Nonstratified sandy-gravelly-pebbly-boulder deposits of the marginal glaciofluvial ridge in the gravel pit of Malkovo (Pskov district).

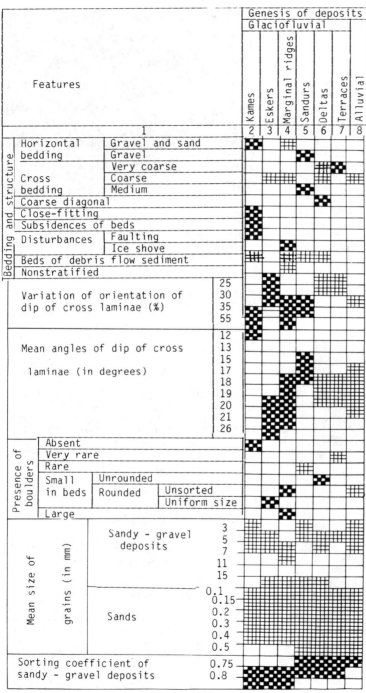

Fig. 4. Lithologic criteria for recognition of the genesis of glacio-fluvial and alluvial sandy-gravel and sandy deposits in the eastern Baltic area and northern Byelorussia (After Jurgaitis, Mikalauskas, and Juozapavičius 1982: Table 4).

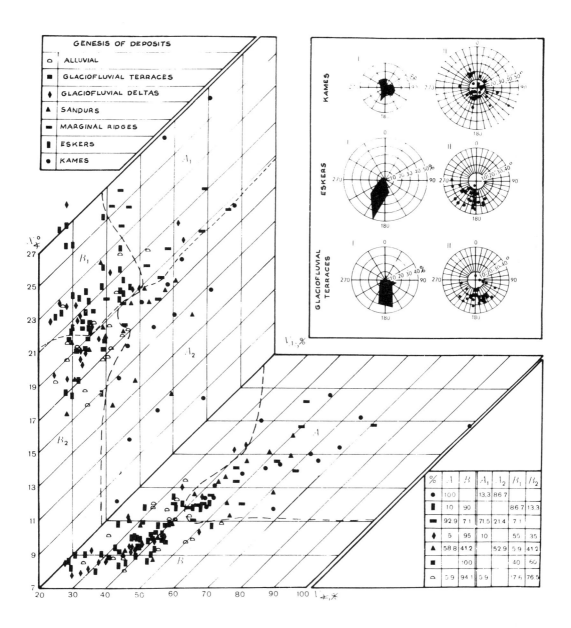

Fig. 5. Diagram of mean values of dip angles ($X_{\nleftrightarrow}$), variations of orientation ($V_{Az}$) and angles of gradient ($V_{\nleftrightarrow}$), rose diagrams (I), diagrams (II) of orientation, and angles of gradient of cross laminae of glaciofluvial and alluvial deposits.
Fields of deposits: A - anastomosing streams, B - single-channel streams; $A_1$ - marginal glaciofluvial ridges, $A_2$ - kames and sandurs; $B_1$ - eskers, $B_2$ - glaciofluvial deltas, glaciofluvial terraces, and alluvium.

sits (Fig. 6). The coarseness of fragments changes more quickly vertically through a section and less quickly horizontally. Esker deposits differ by thicker cross beds noticeable in cross sections, by strict orientation of the direction of cross-bedding along the ridge, and by large mean of their dip angles (Fig. 5). In eskers, a regularity of change of coarseness of the fragments is observed from "upper reaches" to "lower parts." In the "upper reaches" of eskers, boulders are found and pebbly fractions prevail, whereas in the "lower parts" finer gravelly and sandy fractions prevail. The presence of indistinct horizontal- and in some places cross-bedding and of many large boulders is typical of the deposits of the marginal glaciofluvial ridges. The orientation of the cross-bedding is not unidirectional and values of their dip angles are large. Sometimes dislocations by ice thrust and lenses of till are found.

Sandur or outwash deposits are characterized by a cross-bedding typical of them, with small lenses (0.5-5 m) and a frequent change of bedding types (Fig. 7). A wide variation in the orientation of cross-bedding, but consistently low dip angles is also typical for them (Fig. 5). The deposits of glaciofluvial deltas are distinguished by large inclined beds (Fig. 8). In proximal parts of partially developed deltas, one may observe beds of debris-flow sediments containing a large quantity of small single-dimensional non-rounded boulders of local rocks. The deposits of glaciofluvial terraces are characterized (Fig. 9) by large well-preserved cross-beds (length of lenses 5-15 m).

2.2 Granulometric composition

Subsequent studies are carried on by a detailed analysis of selected samples. Sampling should be done by the channel method in every bed. Under field conditions the determination of petrographic composition of gravel is carried out by dispersion of a large amount of sandy-gravel material on a sieve with openings of 2.5 mm. The degree of the influence of local material during the formation of deposits can be estimated in the field. Regional mapping of gravel composition makes it possible to determine their provenance.

The peculiarities of the source rocks and glacier transport have a considerable influence upon the composition of Quaternary deposits. Therefore, comparative analysis of granulometric and lithologic composition of deposits of different genesis is appropriate only for regions of similar geological structure and composition. Typical examples of such regions within Scandinavian Continental Glaciation might be: 1. the region of the Baltic Shield, where the glacier had flowed over igneous and metamorphic rocks and old soils; 2. the East-European Plain, where thin Quaternary deposits lie over comparatively hard carbonate rocks of Paleozoic age and as a result have been enriched by them; 3. another part of the territory where, due to thick deposits of older glaciations, the last glaciation Quaternary deposits do not have a direct influence upon the relief-forming deposits. Analysis of granulometric and lithologic composition revealed considerable interregional variation even for one genetic material type.

Detailed analysis of granulometric composition of the fines of tills showed the unsorted nature of their granulometric composition (Fig. 10). The general character of the distribution curves of their average granulometric composition and coefficients of variation point to a common factor that has determined these characteristics. Higher clay contents occur in till deposits covering sedimentary rocks. Smoothing out of maxima and minima occurs in adjacent fractions. Grain-size distribution curves are more displaced to the negative asymmetry. The given features of average granulometric composition and its variation for tills are significant and reliable since the mentioned differences depend upon the law of distribution. Hence, the glacier meltwater, which had formed glaciofluvial deposits in relation to the distribution of tills, reworked poorly sorted material but of homogeneous composition.

The grain-size composition of the material imprinted in the till by the glacier can be observed in all the sandy deposits of glaciofluvial genesis and can be traced more distinctly in englacial and marginal sands and less distinctly in proglacial sands. During their formation clayey and aleurite (silty) particles had been washed out by streams and meltwater to the basins of uniform hydrodynamic state and, as a result, in glaciofluvial deposits the sandy fractions became dominant. However, the position of maxima and minima in the distribution curves are completely predetermined by the parent till or glacial debris (Fig. 10). According to the distribution character, content of aleurite-clayey particles, and

Fig. 6. Character of bedding and fault disturbances in the deposits of a mixed kame; gravel pit of Villas (Latvia).

Fig. 8. Sandy-gravelly-pebbly diagonally-bedded deposits of the glaciofluvial delta in the gravel pit of Gramzda (Latvia).

Fig. 7. Cross-laminated inequigranular gravelly sand of sandur in gravel pit of Kaniukai (Lithuania).

Fig. 9. Cross-bedded sandy-gravelly deposits of the glaciofluvial terrace in the gravel pit of Rikliskes (Lithuania).

some granulometric coefficients, the sands of marginal glaciofluvial ridges in any region are most similar to parent rocks. They are the finest and poorest-sorted of all glaciofluvial sands.

There are more fine sandy, aleurite and clayey particles in the sands of mixed kames but less in the sands of marginal glaciofluvial ridges (Fig. 10). Composition of sands is more homogeneous and a little better sorted. The sands of eskers are sorted much better. Those that developed in the plains underlain by carbonate rocks are slightly enriched with clayey and aleurite particles as a result of abrasion of relatively soft carbonate

rocks or their weathering after sedimentation.

All englacial glaciofluvial and alluvial sands are characterized by a significant quantity of aleurite particles, and in some cases -- with better sorting of sand grains in glaciofluvial deltas -- by a strong prevalence of some mixed fractions and by rather peaked curves of distribution (Fig. 10). The sands of glaciofluvial deltas are better sorted than the sands of

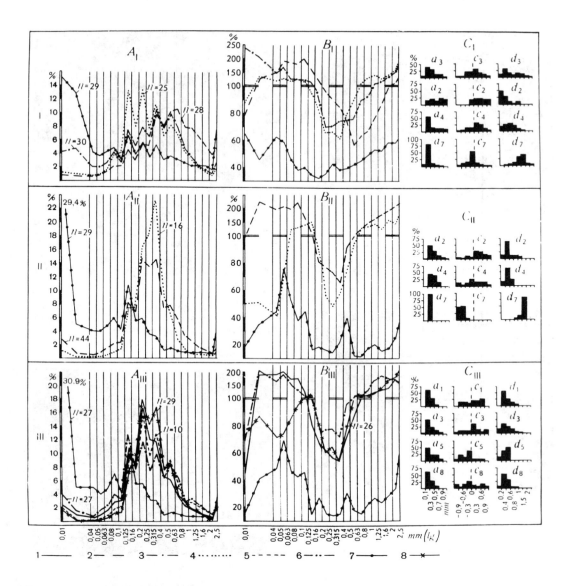

Fig. 10. Distribution of average granulometric composition (A), coefficients of variation of the content of separate fractions (B), and histograms of granulometric coefficients (C) of glaciofluvial, alluvial sands, and tills. I - area of the Baltic Shield; II - area of thin Quaternary deposits underlain by hard Paleozoic carbonate rocks; III - area with thick Quaternary cover. Genesis of deposits: 1. kames; 2. eskers; 3. marginal glaciofluvial ridges; 4. glaciofluvial deltas; 5. sandurs; 6. glaciofluvial terraces; 7. till; 8. alluvium. Granulometric coefficients: a. mean size of fragments; b. sorting coefficient; c. coefficient of asymmetry; d. coefficient of variation.

different genesis. This is caused by the more homogeneous hydrodynamic environment of sedimentation. On the whole, variations of content of separate size fractions in glaciofluvial sands are rather great as a result of an extremely variable melting of glacier. Only in individual cases (kames, glaciofluvial deltas) is the given process uniform, and that shows in the curves of distribution of variation of content of

236

separate granulometric fractions (Fig. 10).

Alluvial sands that were formed by the reworking of older glacial sediments do not differ from englacial glaciofluvial sands according to their granulometric composition (Fig. 10). Morainic sands do differ if redeposited. When reworked, the alluvial sands are a little more poorly sorted than the sands of glaciofluvial deltas; however, there are fewer silty and clayey particles in them than in the rest of stream deposits.

Another part of industrially more important glaciofluvial deposits is composed of sandy-gravelly-pebbly-boulder material. The influence of local hard rocks is observed more distinctly in granulometric composition of such glaciofluvial deposits than in sands. Thus, among the deposits on the Baltic Shield and in the plains over shallow Paleozoic carbonate bedrock fragments of more than 70 cm (boulders) prevail. An exception is the sandy-gravel-pebbly deposits of glaciofluvial deltas (Fig. 11). For genetic interpretation the sorting coefficient is very important. It is often less than 0.75 for proglacial glaciofluvial deposits and usually more than 0.75 for englacial and marginal glaciofluvial deposits. Distinctive features of granulometric composition of bouldery fragmental deposits of marginal glaciofluvial ridges are as follows: general prevalence of boulder material, simultaneous increase of the smallest particles content, and comparatively high coefficients of variation of content of gravelly-pebbly fractions. A large amount of small fragmental particles is also typical of sandy-gravelly deposits of mixed kames in which, contrary to the above-given formations, there is little boulder material (large boulders are completely absent), and the content of gravelly-pebbly fractions is the most common if compared to the rest of rudaceous glaciofluvial deposits.

An increase of open and coarse sandy grains and a small admixture of fine material are typical of proglacial glaciofluvial sandy-gravelly deposits (Fig. 11). However, rudaceous deposits of glaciofluvial deltas are characterized by a peaked curve of distribution and by sorted and more constant composition.

The character of distribution of fragments according to sandy-gravelly-pebbly fractions for alluvium is very similar to that of glaciofluvial deposits. Uniform content of gravelly-pebbly fractions is a very significant feature (Fig. 11). According to the degree of sorting

and to content of aleurite-clayey particles, two groups of rudaceous alluvial deposits are distinguished. Well-washed and sorted rudaceous deposits were formed by large rivers, but the same deposits are clayey and more poorly sorted in the valleys of small rivers. If the quantity of boulders was not considered it would correspond to granulometric composition of marginal formations. This confirms the thesis that even in the postglacial period the degree of sorting and reworking of the parent material have a direct influence upon the structure of rudaceous deposits.

## 2.3 Mineralogic-petrographic composition

Petrographic composition of rudaceous fractions does not carry information about the genesis of glaciofluvial deposits and only shows the composition of parent rocks. Minerals react more sensitively to hydrodynamic conditions of transportation and to grain deposition. Individual mineral grains are more receptive to the hydrodynamic conditions of transport and deposition. Their distribution in the various size fractions depends upon such physical features as density, cleavability, crystal form, hardness, and buoyancy, among others. Ultimately, the mineralogic distribution is influenced by the genesis and composition of the bedrock from which they were derived.

Tills (Fig. 12:I) show the original mineralogic composition that subsequently becomes redistributed by glaciofluvial processes according to the laws of hydraulic equivalence. Having observed the correspondence of mineral distribution to the given regularities, we can estimate the degree of reworking of the parent material for deposits of any genesis, and then we can proceed to restore the conditions of formation.

Strong and moderate positive correlation exists among quartz, mica, pyroxene, amphibole, and epidote of all investigated fractions of tills over the Baltic Shield, suggesting mixed parent material of variable origin.

In the sands of mixed kames and eskers the minerals are grouped according to their specific weights (Fig. 12). Nevertheless, in the sand of mixed kames a strong correlation appears among the minerals of different density and form (mica-limonite, mica-ilmenite-magnetite, mica-leucoxene). In the sands of eskers the minerals of the fractions of 0.1-0.05 and 0.2-0.1 mm are better isolated.

In the sands of glaciofluvial deltas

there are a great number of related minerals (Fig. 12). However, a number of minerals of different sizes that form separate groups show an average degree of reworking of parent material. In the sands of sandurs a large number of homogeneous minerals form separate groups in which minerals of similar physical features show the strongest correlation. Among deposits of glaciofluvial genesis the process of differentiation of mineral composition in the sands of glaciofluvial terraces has advanced. Here separate groups of different sizes are formed for one mineral. In the correlation line heavy minerals of all fractions are well isolated, and they are gradually formed into separate groups according to their density, mass, and form

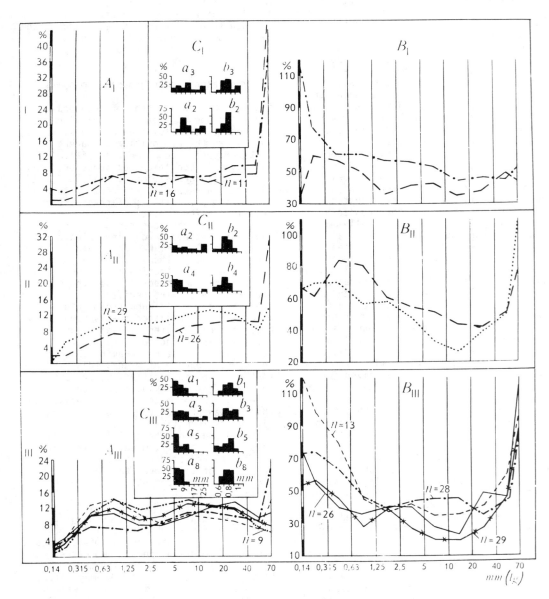

Fig. 11. Distribution of mean granulometric composition: A - coefficients of variation of the content of separate fractions; B - histograms of granulometric coefficients; C - of glaciofluvial and alluvial sandy-gravelly deposits. For legend, see Fig. 10.

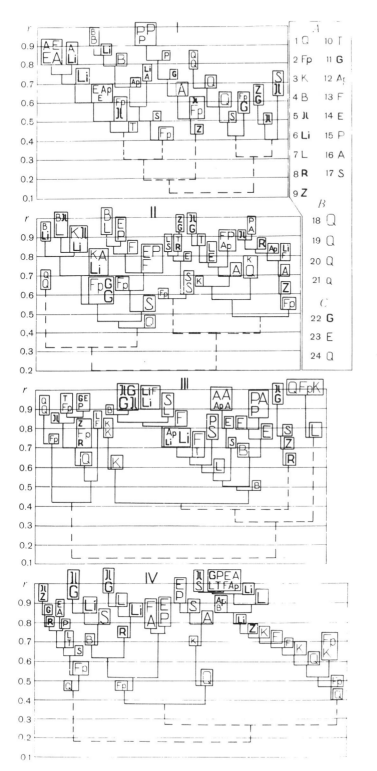

Fig. 12. Dendograms of correlation relations of the content of essential minerals in the tills of the Baltic Shield (I), in the sands of mixed kames (II), of glaciofluvial deltas (III), and glaciofluvial terraces (IV). A - minerals: 1. quartz; 2. feldspar; 3. mica; 4. grains of carbonate rocks; 5. ilmenite-magnetite; 6. limonite; 7. leucoxene; 8. rutile; 9. zircon; 10. turmaline; 11. garnet; 12. apatites; 13. phosphates; 14. group of epidotes; 15. pyroxene; 16. amphiboles; 17. staurolites. B - size fractions, mm: 18. 0.4-0.315; 19. 0.315-0.2; 20. 0.2-0.1; 21. 0.1-0.05. C - density of minerals, kg/m$^3$: 22.>4; 23. 3-4; 24.<3. r - correlation coefficient. Minerals with same correlation coefficient are grouped by rectangles.

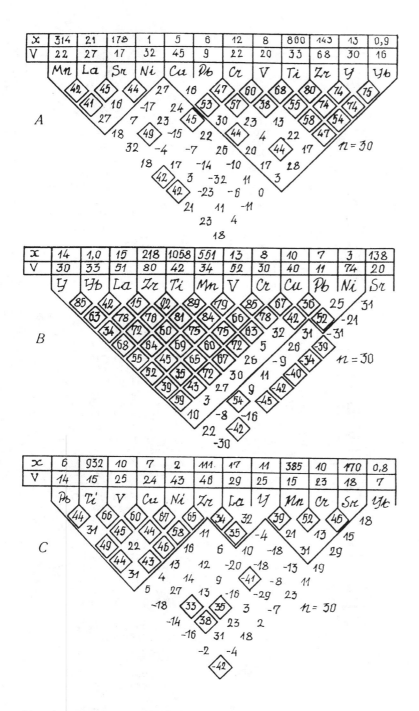

Fig. 13. Correlation (%) of the content of (A) minor elements in the deposits of a mixed kame of Marjamagi Otepaa Hills (Estonia); (B) of the sandur of Borovka (Byelorussia); and (C) of alluvium of the river Daugava near the village of Ellerne (Latvia). x = mean content, $10^{-4}$%; v = coefficient of variation, %; n = number of samples. Rhombuses represent relations that are significant when q = 0.01.

(Fig. 12). The material available suggests
the following sequence of increasing differentiation of their mineral composition:
mixed kames -- eskers -- glaciofluvial
deltas -- sandurs -- marginal glaciofluvial ridges -- glaciofluvial terraces.

Even in carbonate terrain during long
transportation by meltwater streams, the
sands are enriched with more stable
quartz. Thus, in the sandy fractions of
glaciofluvial deposits and of alluvial
terraces there is always about 5-10% more
quartz and 2-3 times less carbonate in
comparison with the regional background
under which deposits of different genesis
are formed. Some of the quartz may have
derived, however, from the Devonian
sandstone bedrock.

## 2.4 Chemical composition

Associations of chemical elements in the
Quaternary deposits of the Baltic Shield
prove that variations of chemical composition are determined mainly by a mechanical redistribution of fragments of
different genesis, not by geochemical
processes. In the material of the southern
fringe of the Baltic Shield, fragments of
carbonate rocks are admixed and migration
conditions of elements are changed.

However, in the glaciofluvial deposits
of englacial and marginal genesis slightly
reworked by water streams, the chemical
peculiarities of parent material are
distinctly observed. For example, in the
deposits of mixed kame in Marjamagi Otepaa
Hills (Estonia), the association Mn-La-Sr
and interrelation Sr-Ni (Fig. 13) show
that some part of the material is represented by chemically disintegrated
material deposited in a marine environment
(Fig. 13) and characteristic of carbonate
formations of the Paleozoic in Estonia. At
the same place, a single association of
rare earths having Ti, V, Cr, positive
correlation of Ni with Cr and Mn with V
and Ti shows the influence of unweathered
material from the Baltic Shield.

Past geochemical processes are hidden by
a comparatively short-term process of
sedimentation in glaciofluvial streams,
low temperatures, distinct mechanical differentiation and sorting of fragmental
material. Therefore, in the better
reworked glaciofluvial deposits, "false"
single statistical correlation of
microelements arises (Fig. 13).

In the postglacial period under formation of alluvial deposits, chemical
migration of matter is more active as a
result of increased temperatures and organic matter present in geochemical processes. Therefore, in spite of
comparatively good mechanical sorting of
alluvial sands, the single ("false") statistical association becomes disintegrated
(Fig. 13), as in the given conditions such
minerals as Mn, Ni, Cu and oxides Ca, Mg,
and carbon dioxide combined with them
become mobile.

## 3 CONCLUSIONS

The genetic classification of glaciofluvial deposits presented in Table 1 is confirmed by a complex of different
structural and textural indications and
features of material composition.

We recommend investigations to determine
the genesis of glaciofluvial deposits
according to the following scheme: morphologic determination of bedding type,
orientation and dip of cross-bedding laminae, sampling of sandy-gravel and sand
deposits, their detailed granulometric
analysis with calculation of statistical
parameters of grain distribution, determination of variation of the content of
different size fractions according to
geological body, study of mineralogic composition of some fractions, determination
of the weight contents of minerals and
their interpretation by methods of mathematical statistics, and systematic analysis of all results obtained.

## REFERENCES

Banerjee, J. and B.C. McDonald 1975.
    Nature of esker sedimentation, Soc.
    Econ. Paleont. and Miner., spec. publ.
    23:132-154.
Embleton, C. and C.A. King 1975. Glacial
    geomorphology. London, E. Arnold.
Flint, R.F. 1971. Glacial and quaternary
    geology. New York, John Wiley and Sons.
Galon, R. 1961. Morphology of the Notec-
    Warta (or Torun-Eberswalde) ice marginal
    streamway, Warsaw.
Holmes, C.D. 1947. Kames. American Journal
    of Science 245:240-249.
Jewtuchowicz, S. 1955. Struktura sandru,
    Lodz.
Juozapavičius, G.A. 1984. Geochemical dif-
    ferentiation of fragmental deposits as
    an index of paleogeographic situations
    of Quaternary lithogenesis of glacial
    regions, p. 187-196 (in Russian). Paleo-
    geography and stratigraphy of Quaternary
    of the Baltic area and adjacent areas,
    Vilnius.
Juozapavičius, G.A., A.A. Jurgaitis, J.Z.

Zimkute 1981. Peculiarities of lithogenesis of the deposits of marginal glaciofluvial ridges and their importance
for paleogeographic investigations, p.
23-30 (in Russian). Geology of
Pleistocene in the north-western part of
the USSR, Apatiti.

Juozapavičius, G.A., I.M. Ekman 1984.
Peculiarities of formation of structure
and composition of fragmental deposits
of the Baltic Shield, p. 14-25 (in
Russian). Nature and economy of North.
Murmansk Publishing House.

Jurgaitis, A.A. 1977. Genetic classification of glaciofluvial deposits and
methods of their investigation, p. 231.
10th INQUA Congress Abstracts,
Birmingham.

Jurgaitis, A.A. 1980. Genetic classification of glaciofluvial deposits and
methods of their investigation. Univ. A.
Mickiewicza Poznaniu, Ser. Geogr.
20:85-93.

Jurgaitis, A.A. 1984. Lithogenesis of glaciofluvial deposits of the region of the
last continental glaciation (in
Russian). Moscow, Nedra Publishers.

Jurgaitis, A.A., G.A. Juozapavičius 1979.
Lithologic criteria for the determination of genesis of Quaternary sandy
and sandy-gravelly deposits. Lithology
and Mineral Resources (in Russian)
5:142-146.

Jurgaitis, A., A. Mikalauskas, G.
Juozapavičius 1982. Bedded structures of
glaciofluvial deposits in the Baltic
Area (in Russian). Vilnius, Mokslas
Publishers.

Lundqvist, J. 1979. Morphogenetic classification of glaciofluvial deposits.
Sver. Geol. Unders, C 767.

Lundqvist, J. 1985. What should be called
glaciofluvium? In Glaciofluvium, Striae
22:5-8.

Mikalauskas, A. 1963. On the differences
between sandurs and glaciofluvial terraces (in Lithuanian). Geographical Annual
6-7:109-118.

Niewiarowski, W. 1963. Types of kames
occurring within the area of the last
glaciation in Poland as compared with
kames known from other regions. Report
of the 6th INQUA Congress, Warsaw
3:475-485.

Parizek, R.R. 1969. Glacial ice-contact
rings and ridges, Geol. Soc. Amer. spec.
publ. 123:49-102.

Price, R.J. 1973. Glacial and
Fluvioglacial Landforms. Edinburgh,
Oliver and Boyd.

Shantser, E.V. 1966. Studies of genetic
types of continental sedimentary formations (in Russian). Moscow, Nauka
Publishers.

Woldstedt, P. 1950. Norddeutschland und
angrenzende Gebiete im Eiszeitalter,
Stuttgart.

*Genetic Classification of Glacigenic Deposits, Goldthwait & Matsch (eds)*
© 1988 Balkema, Rotterdam. ISBN 90 6191 694 1

# Classification of glaciolacustrine sediments

Gail M. Ashley
*Department of Geological Sciences, Rutgers University, New Brunswick, N.J., USA*

ABSTRACT: Glacier-fed lakes receive a substantial proportion of their annual water and sediment budgets from meltwater. A classification of these lakes includes water bodies developed under glaciers, on or surrounded by active or stagnating ice, adjacent to ice (ice-contact lakes), and separate from ice (distal lakes). The great variety of lithofacies deposited in lakes is due to several interrelated factors: proximity to ice (i.e., lake type), lake basin geometry, slope stability, nature of sediment source, position within the lake water column of the meltwater inflow, relative densities of stream and lake water, nature of lake stratification, and seasonal and nonseasonal factors influencing both runoff to the lake and ice cover. Sediment dispersal and sedimentation driven by ice, gravity, or fluvial processes decrease in importance with distance from the glacier, whereas effects of wind, limnological circulation, and organisms increase. Individual lithofacies units are identified and related to the sediment dispersal processes. The units are grouped into four commonly occurring lithofacies groups: I Proximal (subaqueous outwash), II Proximal (glaciolacustrine delta), III Intermediate (lake basin), and IV Distal (lake basin).

## 1 INTRODUCTION

During the last several decades a considerable database has accrued from modern and ancient glaciolacustrine studies (Forel 1892; Antevs 1922; DeGeer 1940; Arnborg 1955; Mathews 1956; Jopling and Walker 1968; Howarth and Price 1969; Ashley 1972; Churski 1973; Gilbert 1973; Gustavson 1975a; Shaw 1977; Sturm and Matter 1978; Carmack et al. 1979; Cohen 1979; Wright and Nydegger 1980; Smith, Vendl, and Kennedy 1982; Pickrill and Irwin 1983; Reimer 1984). Although often logistically difficult, studies of contemporary glacial environments provide an excellent opportunity to relate physical processes to the resulting deposit. Quaternary (ancient) deposits provide thick accumulations that can be examined for vertical and lateral variations; however, sedimentation rates and physical processes such as transport and deposition can only be interpreted. Synthesizing data from both modern and ancient studies allows a link to be made between specific types of glaciolacustrine deposits and their genesis (physical process involved in sedimentation) (Eyles and Miall 1984;

Ashley, Shaw, and Smith 1985; Drewry 1986; Edwards 1986).

The damming of meltwater by a general isostatic depression of the crust due to the weight of an ice sheet or simple impounding of drainage by glacial deposits provide a variety of glacier-fed lake types in terms of size (area and depth), longevity, and limnological characteristics. All are fed by meltwater, however, and can be classified according to their position with respect to ice as either ice-contact lakes or lakes separate from ice (distal) (Table 1).

This paper reviews recent concepts and studies of the physical processes occurring in glacier-fed lakes and discusses sediment dispersal mechanisms in terms of physical limnology. The wide variety of known lithofacies (each characterized by texture, sorting, sedimentary structures, and biologic component) is a product of a spectrum of sediment dispersal and deposition processes. The purpose of this paper is to present a descriptive classification of glaciolacustrine sediments that can be used to interpret their genesis. To do this, individual lithofacies (units) will be linked with glaciola-

Table 1. Classification of lakes: comparison of terminology to INQUA Commission Working Group 4.

| Ashley 1987 | Elson 1980 |
|---|---|
| Glacier-fed lakes | Glacio-lacustrine lakes |
|  | A. Glacigenic lakes |
| A. Ice-contact lakes | B. Glacier lakes |
|   1. Subglacial lakes |   1) Subglacial |
|  |   2) Englacial |
|   2. Supraglacial lakes |   3) Supraglacial |
|     a) Isolated thermokarst lakes |     a) Ice-walled |
|     b) Ice-stagnation lake network |     b) Ice-basin |
|  | C. Glacial lakes |
|   3. Ice-marginal lakes |   1) Ice marginal lakes |
|     a) Dammed by ice |   2) Ice dammed lakes |
|     b) Dammed by topography |  |
|       1) River lakes |  |
|       2) Semi-permanent lakes |  |
| B. Distal lakes |   3) Proglacial lakes |

custrine processes and distinct suites of lithofacies (groups) will be related to environments of deposition within the various lake types.

## 2 GLACIER-FED LAKES

### 2.1 Ice-contact lakes

An ice-contact lake is a ponded body of meltwater with at least a portion of the lake in direct contact with glacial ice, either active or stagnating. Subglacial lakes are known from airborne radar records (Drewry 1981) to exist under an Icelandic ice cap (Bjornsson 1974) and beneath the East Antarctic Ice Sheet. Supraglacial lakes develop as ablation occurs, creating isolated depressions in the ice (ice-stagnation kettle lakes) (Clayton 1964) (Fig. 1A) and complex interconnected lakes (i.e., stagnation lake networks) (Fig. 1B). Supraglacial lakes are relatively short-lived and change configuration continuously as the ice melts. Ice-marginal lakes that form adjacent to ice may be dammed by ice or topography. Lakes dammed locally by ice are likely to be ephemeral (Thorarinsson 1939; Marcus 1960; Clement 1984), and may drain and refill more than once during the melt season. Other ice-contact water bodies include river lakes, which are characterized by short residence times (Fig. 2A). Glacial discharge moves slowly but continuously away from the ice, and the impounded meltwater may be thought of as a low-velocity river or high-velocity lake (Ashley, Shaw, and Smith 1985). Ice-marginal lakes that occupy pre-existing natural basins, dammed river valleys, or isostatic depressions may be semi-permanent and tend to enlarge as the glacier recedes (Fig. 2B).

### 2.2 Distal (non ice-contact) lakes

A distal glacier-fed lake is a water body located at a distance from the ice, but fed primarily by glacial meltwater via outwash streams. Distal lakes may occupy pre-existing natural basins, and thus have the potential to persist for hundreds to thousands of years. In contrast, distal lakes impounded by sediment dams are likely to be shorter-lived (tens to hundreds of years).

## 3 FACTORS THAT AFFECT GLACIOLACUSTRINE SEDIMENTATION

The dominant factors affecting sedimentation are: 1. proximity to ice (i.e., lake type); 2. number and directions of sediment sources; 3. lake basin geometry; 4. slope stability; 5. position in lake (areally); 6. position in the lake water column of the meltwater inflow; 7. the relative densities of lake water and influent meltwater; 8. nature of density stratification in the lake; and 9. extent of ice cover and ice rafting. Each of these factors is discussed below.

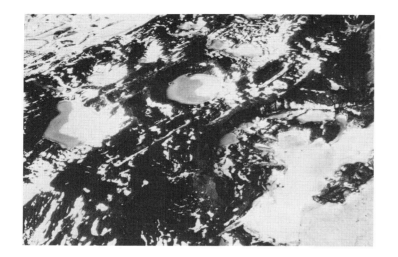

A

B

Fig. 1. Ice-contact, supraglacial lakes, Malaspina Glacier, Alaska: (A) ice-stagnation kettle lakes; (B) ice-stagnation lake network. The mounds in the foreground are ice-cored sediment, which supports growth of spruce and alders.

## 3.1 Lake basin characteristics

1. Glacier-fed lakes include water bodies developed under glaciers, on or surrounded by active or stagnating ice, adjacent to ice, and separate from ice. Table 1 outlines a classification of lakes with only minor differences from the classification originally proposed by Elson (1980). It is expected that processes involved in sediment transportation and deposition vary systematically with distance from the glacier, and thus vary with lake type.

2. Lakes are efficient traps, retaining the bulk of sediment transported into them. The number of sediment sources supplying the lake, as well as variation in transport directions, creates local proximal-distal trends within the larger lacustrine depositional setting. Figure 3 depicts examples of single and multiple source directions existing in a variety of lake types. Occasionally, a lake dammed by ice or topography may burst and drain into another, transporting eroded lake bottom sediment in the flood waters. Resedimenta-

Fig. 2. Ice-marginal lakes, dammed by topography: (A) river lake, Matanuska Glacier, Alaska; (B) semi-permanent lake, Bering Glacier, Alaska.

tion of this material adds to the complexity of the record.

3. The geometry (depth, area, relief, etc.) of the lake basin is a critical factor determining the relative importance of each physical process, particularly mass movement and the manifestation of the density-driven sediment dispersal mechanisms (Ashley, Shaw, and Smith 1985).

4. Slope instability is common where sedimentation is rapid and/or episodic. Mass movement of unstable deltaic and lake margin sediments is known to be an important mechanism of sediment dispersal (Mathews 1956; Fulton and Pullen 1969; Gilbert 1972; Bryan 1974; Gustavson 1975b; Smith 1978; Pharo and Carmack 1979; Pickrill and Irwin 1983). Rapid sedimentation of proximal glaciolacustrine deposits triggers foundering, slides, and slumps (Eyles, Clark, and Clague 1987). Debris flows may generate surge-currents (short-lived density currents) that continue along the bottom (Shaw 1977; Shaw, Gilbert, and Archer 1978; Pharo and Carmack 1979). In supraglacial lakes (Figs. 1 and 3), gravity-driven mass movements are likely to be the most important mechanism of sediment transport. Glacial debris continuously sloughs off the steep-sided ice-cored topographic highs that occur throughout the ice-stagnation lake networks.

5, 6, 7. Sorting and dispersal of sediment within the lake are a function of relative densities of the lake water and the influent stream water (Bates 1953), and the level at which the influent stream discharges into the lake water column. The existence and nature of density stratification in the lake plays a critical role in controlling the dispersal of sediment as it enters the lake. The following is a discussion of these interrelated factors.

## 3.2 Lake stratification

Thermal Stratification: Water density is a function of temperature; however, the relationship is a nonlinear one with the maximum density (near 1.00 cm$^3$) at 4°C and lower densities at temperatures both colder and warmer than 4°C (Fig. 4A). Solar radiation to the lake surface warms the water from the top down, creating a lower-density layer on the surface (Fig. 4B). Hypothetical thermal profiles produced range from winter (e) through mid-summer (a) with a complete mixing of water column occurring when the lake is isothermal (d), generally in spring and fall. The warmer the surface water, the greater the density contrast with the bottom water and the less likely that mixing will occur. Solar heating, then, produces a thermal (density) stratification of the warmer, less dense water mass (epilimnion) superimposed on a cooler, more dense mass (hypolimnion) separated by a layer (metalimnion) with a pronounced thermal (density) gradient (Fig. 4C). Circulation within the epilimnion is commonly enhanced by wind.

Sediment Stratification: Water density is also a function of suspended sediment concentration. Lakes being fed by melt-water streams with high sediment loads (>1 g/l) may develop a density stratification that is a function of a gradual increase of suspended sediment concentration with depth. Gustavson (1975b) measured nearly 0.2 g/l at the surface of Malaspina Lake (an ice-contact semi-permanent lake) and 0.8 g/l at mid-depth of the 80-meter deep lake. As compared to the sharper more persistent thermal stratification (Fig. 5A), the density stratification produced by continuously settling fine sediment is likely to be vague and diffuse and to lack temporal and spatial persistence (Fig. 5B).

No Stratification: Water bodies do not necessarily develop any density stratification. If they are shallow lakes with a short residence time, or are well mixed by wind, the water column may have an equal density throughout (Fig. 5C) (Smith, Vendl, and Kennedy 1982).

## 3.3 Position of influent stream

Several detailed studies of modern glacier-fed lakes have provided insight into the physical processes related to sediment dispersal that occur when a melt-water stream enters a lake (Nydegger 1967; Gilbert 1972 and 1973; Gustavson 1975b; Carmack et al. 1979; Wright and Nydegger 1980; Gilbert and Shaw 1981; Pickrill and Irwin 1982; Smith, Vendl, and Kennedy 1982; Dominik, Burrus, and Vernet 1983; Weirich 1984). Although the specifics of each scenario vary from lake to lake, the overall theme is the same; both short-term fluctuations (hours and days) in stream discharge and sediment load and long-term fluctuations (weeks and months) in stream characteristics and lake density stratification occur. These temporal fluctuations in stream and lake properties control the various rhythms of sedimentation that are common to glaciolacustrine deposits.

Streams enter at the top (directly off the ice or via an overland stream) or at

## SINGLE SOURCE DIRECTION

### ICE-MARGINAL  ICE-CONTACT LAKE

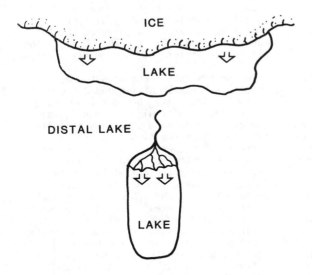

### MULTIPLE SOURCE DIRECTIONS

#### ICE-STAGNATION  KETTLE LAKE

#### ICE-STAGNATION LAKE NETWORK

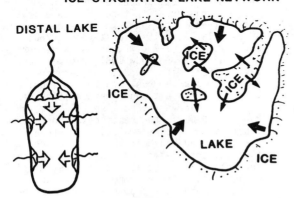

Fig. 3. Sediment sources for glacier-fed lakes. Single source direction: (1) ice-contact, ice-marginal lake and (2) distal lake with no side streams. Multiple source directions: (3) ice-stagnation kettle lake, (4) ice-stagnation lake network and (5) distal lake with side streams.

**A**

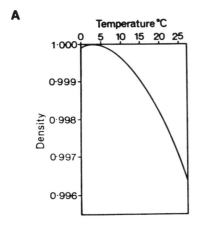

**B**

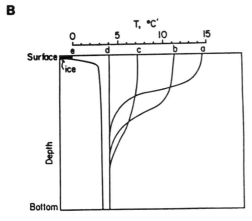

**C**

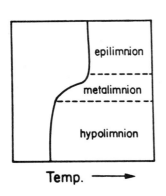

or near bottom (from englacial or subglacial streams). Streams entering at the top form deltas (Axelsson 1967), whereas streams entering at the bottom produce subaqueous outwash deposits (Rust and Romanelli 1975) or glaciolacustrine fans (Boothroyd 1984).

Stream Entering at Top: Channelized stream water expands upon entering the lake basin, and the density difference between the inflowing water and the lake water determines how the two water masses will mix (Bates 1953). The greater the density contrast, the less the mixing. Consequently, rivers with high sediment loads may "dive" to the bottom and flow as a quasi-continuous subaqueous density flow along the lake floor (Fig. 5A, 5B) (Gilbert 1973; Gustavson 1975a; Lambert, Kelts, and Marshall 1976; Carmack et al. 1979; Weirich 1984). Coarse grains (cobbles to sand) carried by the river are deposited on the pro-delta slope (Gilbert 1890; Mathews 1956; Axelsson 1967; Gilbert 1975; Gustavson, Ashley, and Boothroyd 1975; Bogen 1983; Pickrill and Irwin 1983).

If the density of the inflowing water is less than the bottom water (hypolimnion), the stream moves as a plume above the thermocline at a level determined by its density, yielding overflows at the surface and interflows at any level below the surface (Fig. 5A).

Equal density mixing occurs when the inflowing river water is of equal density to that of the lake and the lake is either unstratified or weakly stratified. Rather than following a specific level within the lake, the flow expands in three dimensions and the river and lake water gradually mix as the sediment moves by diffusion throughout the entire water column (Fig. 5C).

Stream Entering at Bottom: The likelihood is great of meltwater streams entering at or near the bottom of ice-contact lakes. Englacial or subglacial streams typically carry high sediment loads, and this sediment is discharged directly into the lake. Flow exits from the tunnel as an axial jet. Coarse clasts are deposited proximally and the sediment-charged melt-water splays out onto the lake bottom (Albertson et al. 1950; Aario 1972; Rust and Romanelli 1975; Thomas 1984), building up thick accumulations in

Fig. 4. (A) Density of fresh water as a function of temperature. (B) Seasonal variation in hypothetical thermal profiles range from mid-summer (a) to winter (e). Mixing occurs when lake is isothermal (d). (C) Temperature profile resulting from solar radiation effectively dividing the lake water column into three layers (from Ashley, Shaw, and Smith 1985).

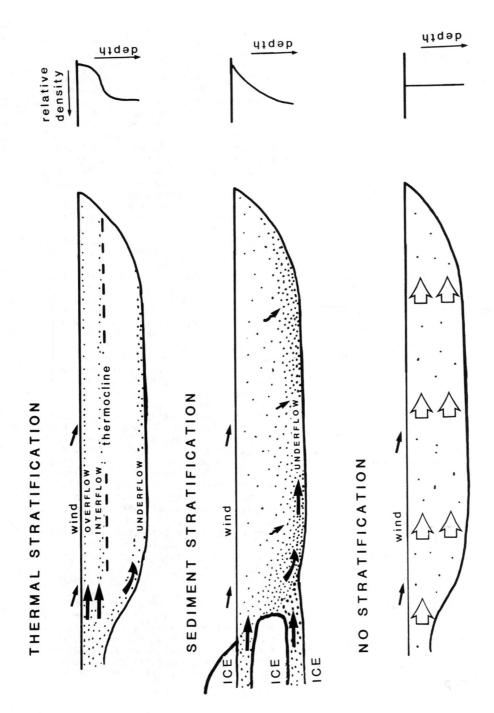

Fig. 5. Density stratification in glacier-fed lakes: (A) thermal stratification, most common in distal lakes; (B) sediment stratification, most common in ice-contact lakes; (C) no stratification, typical of ice-contact river lakes. Arrows indicate direction of moving flows.

a fan-shaped geometry of coarse sediment called "subaqueous outwash" or "glaciolacustrine fans" (Boothroyd 1984).

# 4 SEDIMENTARY PROCESSES

Sedimentation in glacier-fed lakes is dominated by: 1. mass movement processes that vary with slope angle, water content, and clast size; and 2. current-related processes that are driven by glacier and proglacial hydrology and vary with proximity to ice, slope angle, and temporal variability of meltwater discharge. Additional minor processes include melting of sediment-rich icebergs and biogenic processes. These sedimentary processes are discussed only briefly, as an in depth treatment is beyond the scope of this paper.

## 4.1 Mass movement

Downslope movement of sediment under the influence of gravity includes a continuum of sediment flow types ranging from high strength, continuous, slow moving creep to dynamic, highly fluidized, episodic surge currents. The initial grain-size distribution, bulk density, porosity, and water content all contribute to shear strength of deposit and susceptibility to mass movement (Rust 1977; Lawson 1981; Eyles, Eyles, and Miall 1983; Eyles, Clark, and Clague 1987).

## 4.2 Current-driven processes

Meltwater streams may discharge into ice-contact lakes at any level of the water column, whereas flow enters only at the top in distal lakes. Although velocities fluctuate, flow is generally maintained during the melt season, and these currents are identified as quasi-continuous density currents to distinguish them from short-lived surge currents generated from slope failure. Current strength is generally a reflection of the velocity of the melt-water stream. Overflow-interflow velocity is commonly enhanced by katabatic winds (Fig. 5).

## 4.3 Additional processes

Melting of sediment-rich icebergs and the release of debris into the lake water column is an important sedimentation process in ice-contact lakes. Bergs may be moved tens of kilometers by wind. Finally, organisms spending their entire life cycle or only a portion of it in glacial lakes enhance sedimentation rates (Smith and Syvitski 1982) and leave records of their presence in the form of a variety of trace fossils on bedding surfaces as well as body fossils (Ashley 1972; Gibbard and Dreimanis 1978).

# 5 LITHOFACIES

Sediment dispersal and sedimentation controlled by ice, gravity, or currents decrease in importance with distance from the glacier, whereas effects of wind, limnological circulation, and organisms increase. Table 2 outlines the sedimentary processes operating in glacier-fed lakes and the types of expected lithofacies units produced by each process. Few lithofacies units are unique to a specific process. Sedimentary structure descriptions follow Blatt, Middleton, and Murray 1980 and Walker 1984.

## 5.1 Lithofacies units

Mass movement processes (Table 2) produce diamicts with a wide spectrum of possible bedding characteristics ranging from massive to crudely stratified and from matrix-supported to clast-supported (Eyles, Eyles, and Miall 1983). The characteristics of the diamicts are both a function of the original sediment composition and the processes of resedimentation. Increased water content leads to liquefaction and fluidization and the production of gravity flows in which sediment is transported by water rather than water by sediment (Fig. 6). Surge currents and grain flows produce moderately sorted sediments ranging from massive to parallel and graded bedding. Surge currents may create cross bedding produced by both ripples and larger bedforms, as well as draped lamination during waning flow.

Quasi-continuous density currents produced by the inflowing meltwater stream produce structures similar to surge currents: large-scale planar cross beds, parallel and graded bedding, climbing ripples, draped lamination, and load structures (Fig. 7). Wind-driven currents disperse fine silts and clay throughout the lake. Deposition from overflow-interflow produces parallel bedding and multiple graded beds. Wind-generated waves winnow fines from bottom sediments along shallow lake margins. Longshore currents

A

B

Fig. 6. I Proximal: (A) ice-marginal river lake with glacial debris flows, Matanuska Glacier, Alaska; (B) lithofacies Group I matrix-supported diamict with minor silty-clay parallel bedding (glacial Lake Noatak, Alaska).

Table 2. Sedimentary processes and lithofacies units in glacier-fed lakes.

| Sedimentary processes | Lithofacies units |
| --- | --- |
| Mass movement<br>  Creep<br>  Slump<br>  Debris flow | Matrix supported and clast supported, massive to stratified diamicts |
|   Surge current | Cross bedding, load structures, fluidization structures, massive beds, parallel and graded bedding, climbing ripples, draped lamination |
| Currents<br>  Quasi-continuous | Parallel bedding, graded bedding, large-scale tabular bedding, climbing ripples, multiple graded beds, load structures, fluidization structures |
|   Wind-driven<br>    Longshore<br>    Lake circulation | Small-scale cross bedding, parallel lamination<br>Parallel bedding, multiple graded beds |
|   Combined | Rhythmites |
| Ice rafting | Debris blebs; dropstones |
| Biogenic processes | Body or trace fossils |

commonly produce rippled and parallel bedded sands making up a variety of coastal landforms such as beaches, spits, and longshore bars.

Ice rafting produces isolated dropstones with deformed bedding and blebs of partially sorted debris.

## 5.2 Lithofacies groups

The record of sedimentation in glacier-fed lakes is complex, but is composed of individual building blocks (lithofacies units). Many of the units frequently occur together and these lithofacies associations are found in particular depositional environments.

A lithofacies classification scheme for glacier-fed lakes utilizing common facies associations (groups) is proposed (Table 3). The classification organizes lithofacies units into four groups: I Proximal (related to ice-contact subaqueous outwash), II Proximal (related to ice-contact lake delta or distal glacier-fed lake delta), III Intermediate (lake basin), and IV Distal (lake basin).

I Proximal glaciolacustrine facies accumulate at and lakeward of subaqueous tunnels as meltwater discharges into the water column. The facies assemblage consists of intercalated diamicts, poorly stratified gravel, parallel and cross bedded sands, ice-rafted debris, and laminated silt and clay (Fig. 6). Glacial sediment transported by mass movement from the ice into the lake is interbedded with sediment moved by ice rafting and density currents. Ice margin fluctuations commonly bulldoze these deposits into "push moraines."

II Proximal glaciolacustrine facies are deposited as subaqueous extensions of Gilbert-type deltas and may consist of an occasional diamict, poorly sorted stratified sands and gravels, or well-stratified coarse to fine sediments composed of climbing ripple sequences and draped lamination (Fig. 7). Fluidization structures, rhythmic sediments, and biogenic structures are common. Sediments are transported to the delta by outwash streams and into the lake by density flows, grain flows, and mass movement processes. This lithofacies assemblage is

Table 3. Lithofacies groups with commonly occurring lithofacies units.

I Proximal

II Proximal

Landform: subaqueous outwash or push moraine

Landform: ice-contact delta or distal lake delta

| Massive matrix-supported diamicts | Massive to crudely stratified diamicts |
| Crudely stratified diamicts | Poorly-sorted stratified gravel |
| Open framework, clast-supported gravel | Well-stratified sand and gravel |
| Poorly-sorted, stratified gravel | Parallel bedded sand |
| Parallel bedded sand | Graded bedding, normal and reverse |
| Cross-bedded sand | Climbing ripple sequences |
| Climbing ripple sequences | Draped lamination |
| Draped lamination | Ice-rafted debris |
| Ice-rafted debris | Laminated silt and clay |
| Laminated silt and clay | Load and fluidization structures |
| Load and fluidization structures | Rhythmites |
| Collapse structures | |

III Intermediate

Landform: lake basin

Diamict lenses
Ice-rafted debris
Ripple drift sequences and draped lamination
Laminated sands and silts with multiple graded beds
Rhythmites
Trace and body fossils

IV Distal

Landform: lake basin

Laminated silts with multiple graded beds
Rhythmites
Trace and body fossils
Rare ice-rafted debris

254

A

B

Fig. 7. II Proximal: (A) coarse gravel foreset beds in an ice-marginal delta built into a semi-permanent lake; (B) climbing ripple sequence overlain with draped lamination (Gustavson, Ashley, and Boothroyd 1975).

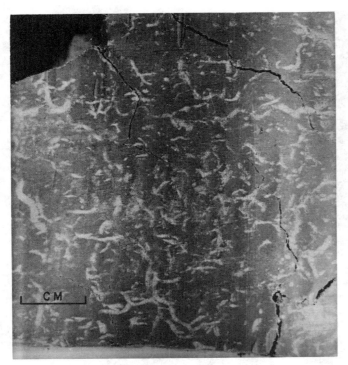

Fig. 8. III Intermediate and IV Distal: (A) biogenic structures found on a bedding plane within a rhythmite; (B) rhythmic bedding (varves) produced by seasonally controlled alternation of sediment influx.

deposited in the lake beyond the margin of the glacier, and is separated from the ice by a broad to narrow subaerial outwash plain, fan, or delta, which tends to intercept part of the coarser sediment fraction before it reaches the lake. A I Proximal facies may evolve into a lake surface if the subaqueous splay grows upward to the lake surface or if the level of the lake falls.

III Intermediate and IV Distal glaciolacustrine facies represent a continuum of lake basin sediments dominated by density current and suspension deposits (Fig. 8) although either may be absent in proglacial lake deposits. In the III Intermediate facies, sediments transported and deposited exclusively by underflow are likely to contain current bedding structures in proximal areas and to consist of multiple graded beds in the distal portions of the lake. Presumably, each micrograded bed represents a pulse in the underflow or the "tail-end" of an underflow. Clay brought in with each underflow is kept in suspension by underflow activity as a "nepheloid layer." The difference in settling time of fine silt (4-15 days) and coarse clay (2-7.7 months) is conducive to efficient separation of clay from silt. The clay settles out after cessation of inflow activity as a thick blanket in the topographic lows of the lake basin, although weak thermally-driven lake water circulation may extend the settling process for months.

In the IV Distal facies, sediments are transported predominantly by overflow-interflow and tend to be fine-grained (silt and clay), consisting of a summer layer that is relatively uniform in grain size and gives an appearance of being faintly laminated to massive. The winter layer is a fining-upward clay unit. Any fluctuations in sediment size or concentration of the inflowing stream are "smoothed out" during transport into the lake in the epilimnion and by settling through the thermocline to the lake bottom. Fine silt should settle out within a few weeks of influx in the summer. This leaves only clay in the water column to settle during the winter, completing the seasonally controlled sedimentation cycle that produces rhythmites (Fig. 8B).

## 6 DISCUSSION

It is clear from the spectrum of sedimentary processes and lithofacies (Table 2) and their grouping with respect to depositional environment (Table 3) that there are three encompassing factors that dominate sedimentation in glacier-fed lakes: proximity to ice, seasonal variation in glacier melting, and seasonal freezing and thawing that controls both the wind energy on the lake surface and the ice rafting of detritus. Thus, lake deposits are expected to vary with type of lake (Table 1). Specifically, the deposits reflect the relative importance of each of the physical factors that operate in each lake.

Ice-contact lake sediments as a group are highly variable in lithofacies with a mixture of mass movement and current related deposits. Except for deposits of the large semi-permanent lakes, ice-contact lakes are not likely to demonstrate any spatial or temporal trends with respect to grain size or sorting. All lithofacies groups can be found in ice-contact lakes; however I and II dominate in small lakes.

Distal lakes by definition are depositional environments far removed from the highly variable proximal zone of the glacier. Widely fluctuating magnitudes of discharge are buffered by the lake; distal lake sediments are thus characterized by both temporal and spatial trends in sediment type. Lithofacies groups II, III, and IV are characteristic of distal glacier-fed lakes.

A secondary control of glaciolacustrine sedimentation is seasonal variation in processes. Seasonally controlled alternation between rapid sedimentation during a short glacial-melt season and deposition during the balance of the year produces an overprint of rhythmicity. Within this two-part cyclic pattern, however, other forms of rhythmic sedimentation occur that reflect temporal variations in sediment influx and the sediment dispersal mechanisms within the lake. Temporal variations range from slump-generated surge currents (minutes) and fluctuations in sediment load of the inflowing meltwater stream (minutes and hours) to local weather variations affecting ablation (days). Sediment dispersal mechanisms vary throughout the year depending upon the nature of mixing of the inflowing stream and the lake water. Thus, there are many potential causes of rhythmic sedimentation in glacier-fed lakes, each representing different time scales. Rhythmic sedimentation related strictly to the annual cycle (varves) is best developed at a distance from the glacier where the lake acts as a buffer to input variations (Drewry 1986).

# 7 CONCLUSIONS

1. The wide variety of lithofacies types in glacier-fed lakes is a function of proximity to ice (i.e., lake type), nature of sediment source, lake basin geometry, slope stability, position in the lake (areally), vertical position of inflow within the lake water column, relative densities of stream and lake water, lake stratification, and seasonal and non-seasonal factors influencing both runoff to the lake and ice cover.

2. Glacier-fed lakes can be classified according to their position with respect to the ice: ice-contact lakes (including subglacial, supraglacial and ice-marginal lakes); and distal (non ice-contact) lakes.

3. Sedimentary processes are driven primarily by gravity (mass movement) and glacially controlled hydraulics (quasi-continuous currents), although wind-generated currents can be important along shallow lake margins.

4. Physical processes of sediment dispersal are a direct function of slope stability and the interaction of the melt-water stream and the lake water. The nature of these interactions is related to the relative densities of the two water masses, the presence and nature of lake stratification, vertical position within the lake water column of meltwater inflow, and wind energy available to the lake surface.

5. Physical processes of transport and deposition lead to a wide variety of lithofacies units, which have been grouped into four commonly occurring lithofacies groups: I Proximal subaqueous outwash, II Proximal delta front, III Intermediate lake basin, and IV Distal lake basin.

6. The major control of sedimentation with respect to lithofacies distribution is proximity to ice. Ice-contact lakes may contain lithofacies groups I, II, III, and IV, whereas distal lakes will have only II, III, and IV.

7. A secondary control of sedimentation is the seasonally controlled melting of ice. Rhythmicity of sedimentation, the signature of glacier-fed lakes, is dampened near the ice and is best developed at a distance from the ice. Nonglacial lakes also display rhythmicity, especially when deep and/or covered by winter ice.

# ACKNOWLEDGMENTS

Bill Mathews' pioneering work on a modern glacier-fed lake in the early 1950s sparked my interest in glaciolacustrine processes. His well-documented study of Lake Garibaldi answered many questions regarding possible mechanisms of sediment dispersal and deposition in ancient (Quaternary-age) lakes. I am grateful for the example he set.

Many of the ideas presented in this paper have developed through numerous discussions. I would particularly like to thank the following: J. Boothroyd, J. Caggiano, M. Church, C. Eyles, N. Eyles, T. Gustavson, T. Hamilton, J. Hartshorn, W. Mathews, G. Reimer, J. Shaw, N.D. Smith, J. Southard, B. Stone, and J. Teller.

# REFERENCES

Aario, R. 1972. Associations of bed forms and paleocurrent patterns in an esker delta, Haapajarvi, Finland. Ann. Acad. Scientiarum Fennicae, ser. A, pt. 3, no. 111.

Albertson, M.L., Y.B. Dai, R.A. Jensen, and H. Rouse 1950. Diffusion of submerged jets. Trans. Amer. Soc. Civil Engrs. 115:639-664.

Antevs, E. 1922. The recession of the last ice sheet in New England. Am. Geog. Soc. Research Ser., no. 11.

Arnborg, L. 1955. Ice-marginal lakes at Hoffellsjokull. Geogr. Annaler 37:202-228.

Ashley, G.M. 1972. Rhythmic sedimentation in glacial Lake Hitchcock, Massachusetts-Connecticut. Geology Pub. #10, Amherst, Univ. Mass.

Ashley, G.M., J. Shaw, and N.D. Smith 1985. Glacial Sedimentary Environments, SEPM Short Course No. 16, Tulsa, OK.

Axelsson, V. 1967. The Laitaure delta: A study of deltaic morphology and processes. Geogr. Annaler 49A:1-127.

Bates, C.C. 1953. Rational theory of delta formation. Amer. Assoc. Petrol. Geol. Bull. 37:2119-2162.

Bjornsson, H. 1974. Explanation of jokulhlaups from Grimsvotn, Vatnajokull, Iceland. Jokull 24:1-26.

Blatt, H., G.V. Middleton, and R.C. Murray 1980. Origin of sedimentary rocks, 2nd ed. Englewood Cliffs, NJ, Prentice Hall.

Bogen, J. 1983. Morphology and sedimentology of deltas in fjord and fjord valley lakes. Sediment. Geol. 36:245-267.

Boothroyd, J.C. 1984. Glaciolacustrine and glaciomarine fans: A review. Geol. Soc. America, Abs. Programs 16:4.

Bryan, M.L. 1974. Sublacustrine morphology and deposition, Kluane Lake, Yukon Territory. Amer. Geog. Soc., Icefields Ranges Research Project, Scientific

Results 4:171-187.

Carmack, E.C., C.B.J. Gray, C.H. Pharo, and R.J. Daley 1979. Importance of lake-river interaction on the physical limnology of the Kamloops Lake/Thompson River system. Limnol. Oceanogr. 24:634-644.

Churski, Z. 1973. Hydrographic features on the proglacial area of Skeidararjokull. Geographia Polonica 26:209-254.

Clayton, L. 1964. Karst topography on stagnant glaciers. Jour. Glaciol. 3:107-112.

Clement, P. 1984. The drainage of a marginal ice-dammed lake at Nordbogletscher, Johan Dahl Land, South Greenland. Arctic Alpine Research 16:209-216.

Cohen, J.M. 1979. Deltaic sedimentation in glacial Lake Blessington, County Wicklow, Ireland. In Ch. Schlüchter (ed.), Moraines and varves, p. 357-367. Rotterdam, Balkema.

DeGeer, G. 1940. Geochronologia Suecica Principles: K. Svenska Vetensk. Handl., ser. 3, 18(6). Stockholm, Almquist and Wiksells.

Dominik, J., D. Burrus, and J.P. Vernet 1983. A preliminary investigation of the Rhone River plume in eastern Lake Geneva. Jour. Sed. Petrol. 53:159-163.

Drewry, D.J. 1981. Radio echo sounding of ice masses: Principles and applications. In A.P. Crackell (ed.), Remote sensing in meteorology, oceanography and hydrology, p. 270-284. Chichester, Ellis Horwood.

Drewry, D.J. 1986. Glacial geologic processes. London, Edward Arnold Publishers.

Edwards, M. 1986. Glacial environments. In H. Reading (ed.), Sedimentary environments and facies, p. 445-470. Blackwell Scientific Pub.

Elson, J.A. 1980. Project B-3: Glaciolacustrine. In W. Stankowski (ed.), Tills and glacigenic deposits, p. 77-83. Uniwersytet im. Adama Mickiewicza W. Poznaniv, Seria Geografia NR 20.

Eyles, N., B.M. Clark, and J.J. Clague 1987. Coarse-grained sediment gravity flow facies in a large supraglacial lake. Sedimentology 34:193-216.

Eyles, N., C.H. Eyles, and A.D. Miall 1983. Lithofacies types and vertical profile models; an alternative approach to the description and environmental interpretation of glacial diamict and diamictite sequences. Sedimentology 30:393-410.

Eyles, N. and A.D. Miall 1984. Glacial facies. In R.G. Walker (ed.), Facies models, 2nd ed., p. 15-38. Geoscience Canada Reprint Series 1.

Forel, F.A. 1892. Le Leman, Monographie Limnoligique 1:1-543. Lausanne, F. Rouge.

Fulton, R.J. and M.J.L.T. Pullen 1969. Sedimentation in Upper Arrow Lake, B.C. Can. Jour. Earth Sci. 6:785-791.

Gibbard, P.L. and A. Dreimanis 1978. Trace fossils from late Pleistocene glacial lake sediments in southwestern Ontario, Canada. Can. Jour. Earth Sci. 15:1967-1976.

Gilbert, G.K. 1890. Lake Bonneville. U.S. Geol. Surv. Monogr. 1:1-438.

Gilbert, R. 1972. Observations on sedimentation at Lillooet delta, British Columbia. In O. Slaymaker and H.J. McPherson (eds.), Mountain Geomorphology, p. 187-194. Vancouver, Tantalus Press.

Gilbert, R. 1973. Processes of underflow and sediment transport in a British Columbia mountain lake, p. 493-507. Proc. 9th Canad. Hydrology Symposium, Natl. Res. Council Canada.

Gilbert, R. 1975. Sedimentation in Lillooet Lake, British Columbia. Canad. Jour. Earth Sci. 12:1697-1711.

Gilbert, R. and J. Shaw 1981. Sedimentation in proglacial Sunwapta Lake, Alberta. Can. Jour. Earth Sci. 18:81-93.

Gustavson, T.C. 1975a. Sedimentation and physical limnology in proglacial Malaspina Lake, southeastern Alaska. In A.V. Jopling and B.C. McDonald (eds.), Glaciofluvial and glaciolacustrine sedimentation, p. 249-263. Tulsa OK, S.E.P.M. Special Pub. 23.

Gustavson, T.C. 1975b. Bathymetry and sediment distribution in proglacial Malaspina Lake, Alaska. J. Sed. Petrol. 45:450-461.

Gustavson, T.C., G.M. Ashley, and J.C. Boothroyd 1975. Depositional sequences in glaciolacustrine deltas. In A.V. Jopling and B.C. McDonald (eds.), Glaciofluvial and glaciolacustrine sedimentation, p. 264-280. Tulsa OK, S.E.P.M. Special Pub. No. 23.

Howarth, P.J. and R.J. Price 1969. The proglacial lakes of Breidamerkurjokull and Fjallsjokull, Iceland. Geog. Jour. 135:573-581.

Jopling, A.V. and R.G. Walker 1968. Morphology and origin of ripple-drift cross-laminations, with examples from the Pleistocene of Massachusetts. Jour. Sed. Petrology 38:971-984.

Lambert, A.M., K.R. Kelts, and N.F. Marshall 1976. Measurement of density underflows from Walensee, Switzerland. Sedimentology 23:87-105.

Lawson, D.E. 1981. Sedimentological characteristics and classification of depositional processes and deposits in the glacial environment. CRREL, Report 81-27.

Marcus, M.G. 1960. Periodic drainage of glacier-dammed Tulsequah Lake, British Columbia. Geogr. Rev. 50:89-106.

Mathews, W.H. 1956. Physical limnology and sedimentation in a glacial lake. Geol. Soc. Amer. Bull. 67:537-552.

Nydegger, P. 1967. Investigation of fine-sediment transport in rivers and lakes; concerning the origin of turbidity layers; and river-induced inflow currents in Lake Brienz and other lakes. Beitr. Geol. Schweiz. Hydrol. 16:1-92.

Pharo, C.H. and E.C. Carmack 1979. Sedimentation processes in a short residence-time intermontane lake, Kamloops, British Columbia. Sedimentology 26:523-541.

Pickrill, R.A. and J. Irwin 1982. Predominant headwater inflow and its control of lake-river interactions in Lake Wakatipu. New Zealand J. Freshwat. Res. 16:201-213.

Pickrill, R.A. and J. Irwin 1983. Sedimentation in a deep glacier-fed lake --Lake Tekapo, New Zealand. Sedimentology: 30:63-75.

Reimer, G.E. 1984. The sedimentology and stratigraphy of the southern basin of glacial Lake Passaic, New Jersey. Unpublished M.S. thesis, Rutgers University, New Brunswick, N.J.

Rust, B.R. 1977. Mass flow deposits in a Quaternary succession near Ottawa, Canada: Diagnostic criteria for suba-queous outwash. Canad. Jour. Earth Sci. 14:175-184.

Rust, B.R. and R. Romanelli 1975. Late Quaternary subaqueous outwash deposits near Ottawa, Canada. In A.V. Jopling and B.C. McDonald (eds.), Glaciofluvial and glaciolacustrine sedimentation, p. 177-192. Tulsa OK, S.E.P.M. Special Pub. 23.

Shaw, J. 1977. Sedimentation in an alpine lake during deglaciation, Okanagan Valley, British Columbia, Canada. Geografiska Annaler 59A:221-240.

Shaw, J., R. Gilbert, and J.J.J. Archer 1978. Proglacial lacustrine sedimen-tation during winter. Arctic and Alpine Research 10:689-699.

Smith, N.D. 1978. Sedimentation processes and patterns in a glacier-fed lake with low sediment input. Canadian Jour. Earth Sci. 15:741-756.

Smith, N.D. and J.P.M. Syvitski 1982. Sedimentation in a glacier-fed lake: The role of pelletization on deposition of fine-grained suspensates. Journal of Sed. Petrology 52:503-513.

Smith, N.D., M.A. Vendl, and S.K. Kennedy 1982. Comparison of sedimentation re-gimes in four glacier-fed lakes of western Alberta. In R. Davidson-Arnott, W. Nickling, and B.D. Fahey (eds.), Proc. 6th Guelph Symposium on Geomorphology, Research in glacial, glaciofluvial, and glaciolacustrine systems, p. 203-238.

Sturm, M. and A. Matter 1978. Turbidites and varves in Lake Brienz (Switzerland): Deposition of clastic detritus by den-sity currents. In A. Matter and M.E. Tucker (eds.), Modern and ancient lake sediments, p. 147-168. Internat. Assoc. Sedimentologists Special Pub. 2.

Thomas, G.S.P. 1984. Sedimentation of a sub-aqueous esker-delta at Strabathis, Aberdeenshire. Scott. J. Geol. 20:9-20.

Thorarinsson, S. 1939. The ice-dammed lakes of Iceland, with particular reference to their value as indicators of glacier oscillations. Geog. Annaler 21:216-242.

Walker, R.G. (ed.) 1984. Facies models, 2nd edition. Geoscience Canada, Reprint Series 1.

Weirich, F.H. 1984. Turbidity currents: Monitoring their occurrence and movement with a three-dimensional sensor network. Science 224:384-387.

Wright, R.F. and P. Nydegger 1980. Sedimentation of detrital particulate matter in lakes: Influence of currents produced by inflowing rivers. Water Resources Res. 16:597-601.

*Genetic Classification of Glacigenic Deposits, Goldthwait & Matsch (eds)*
© *1988 Balkema, Rotterdam. ISBN 90 6191 694 1*

# A provisional genetic classification of glaciomarine environments, processes, and sediments

Harold W.Borns, Jr.
*Department of Geological Sciences and Institute for Quaternary Studies, University of Maine, Orono, Maine, USA*

Charles L.Matsch
*Department of Geology, University of Minnesota, Duluth, Minn., USA*

ABSTRACT: A provisional genetic classification of glaciomarine processes and sediments is presented to provide a framework for discussions. The objective is to refine the classification and expand its scope so as to be useful to future research. The classification includes type of glacier sediment source, position of sediment release and deposition with respect to grounded and floating glacier ice, mechanisms of sediment release, and an inventory of delivery and depositional processes. Genetic varieties of glaciomarine sediment (GMS) are grouped according to marine environment of deposition, position of accumulation with respect to relevant glacier zones, and the processes of deposition.

## 1 INTRODUCTION

During the last decade there has been a significant increase in publications relating to glaciomarine processes and sediments. This activity is generally coincident with the growing realization that large portions of the late Pleistocene, as well as earlier ice sheets, were grounded below sea level. This realization in turn has created a focus on the possible influence of marine-based glaciers on global events of the Quaternary (Denton and Hughes 1983; Denton et al. 1985) and on the causes of ice ages in general. However, not all glaciomarine sediments are derived from marine-based glaciers, and it is recognized that land-based glaciers contribute sediments to the marine environment, mainly by meltwater streams. Marine sediments, in turn, can be incorporated into terrestrial glacigenic sequences.

This expanding interest in glaciomarine sediments and processes has led to a rapid increase in the number of glacial geologists, marine geologists and glaciologists researching glacier dynamics and the geological work of marine-based glaciers, both past and present. In the field of sedimentology this research has resulted in local descriptions of glaciomarine sediments and their morphology and, hence, to the development of a growing number of predictive sedimentary models derived in part from the study of modern glaciers in various settings (e.g., Armstrong and Brown 1954; Rust and Romanelli 1975; Boulton and Deynoux 1981; Powell 1981; Elverhøi, Lonne, and Seland 1983; Eyles, Eyles, and Miall 1985).

The INQUA Commission on Genesis and Lithology of Glacial Deposits recognized this new trend and that the developing body of knowledge includes overlapping and often confusing nomenclature involving many styles of sedimentary descriptions, a wide variety of process descriptions and interpretations, and many models. Some of these models, based upon modern glacial environments, are predominantly deductive because modern sedimentary processes and the resultant sediments are not adequately observable offshore on a three dimensional or regional basis through either acoustic stratigraphy or coring. It is also apparent that, as valid as many of these locally-derived models may be, no single model yet developed can be considered representative of all past or present glaciomarine settings. Given the growing interest in glaciomarine processes and sediments and their importance to understanding the role of marine-based glaciers in the geological record and in global events, Work Group 5 of the Commission was formed. The Working Group concluded that it would be desirable to develop a fundamental genetic classification of glaciomarine environments, associated processes, and sediments. A first

Table 1. A provisional genetic classification of glaciomarine processes and sediments.

| Sediment source | Position | Debris release mechanism | Delivery/depositional process | Depositional genetic varieties of glaciomarine sediment (GMS) | | |
| --- | --- | --- | --- | --- | --- | --- |
| | | | | By marine environment | By position | By process |
| Marine-based glaciers, ice shelves, and glacier tongues | Grounding line<br>-Ice cliff<br>-Buoyancy line<br><br>Seaward of the grounding line<br><br>-Ice shelf<br> Base<br> Edge<br>-Open marine<br> Floating ice<br><br>Supraglacial<br><br>Englacial<br>-Cavity<br><br>Subglacial<br>-Cavity | Melt-out above the water line<br><br>Melt-out at or below the water<br><br>Iceberg calving and fragmentation<br><br>Sublimation | Initial sedimentation<br><br>-Settling<br>-Meltwater flow<br>-Mass movements<br>-Extrusion flow<br>-Floating ice<br>-Marine currents<br>-Compound<br><br>Resedimentation<br><br>-Mass movements<br>-Marine currents<br>-Sea ice<br>-Compound | Proximal GMS<br><br>-Intertidal<br>-Estuarine (fjord)<br>-Continental shelf<br>-Continental slope<br><br>Distal GMS<br><br>-Continental shelf<br>-Continental slope<br>-Deep ocean basin | Grounding line GMS<br><br>Sub-ice shelf GMS<br><br>Calving line GMS<br><br>Open marine GMS<br><br>-Iceberg zone GMS | Direct melt-out and settled<br><br>Extruded<br><br>Mass-moved<br><br>Suspension-deposited<br><br>Traction-current deposited<br><br>Ice-rafted<br><br>Mixed |
| Land-based glaciers | Proglacial | Melt-out<br><br>Sublimation<br><br>Fragmentation | Initial sedimentation<br><br>-Meltwater streams<br>-Density flows<br>-Mass movements<br><br>Resedimentation<br><br>-Mass movements<br>-Marine currents<br>-Sea ice | Proximal<br><br>-Intertidal<br>-Estuarine (fjord)<br>-Continental shelf<br><br>Distal<br><br>-Continental shelf<br>-Continental slope<br>-Deep ocean basin | Proglacial<br><br>-Shoreline<br>-Nearshore<br>-Offshore | Mass-moved<br><br>Suspension-deposited<br><br>Traction-current deposited<br><br>Ice-rafted<br><br>Mixed |

NOTE: Each vertical column is independent, and no horizontal correlation is implied.

262

approximation of such a basic classification is presented in Table 1 to provide a framework for discussions between researchers in the field, with the ultimate objective that it be refined and expanded in scope to be useful to future research.

## 2 DEFINITION

Definitions of the term "glaciomarine (or glacial marine)" have been most recently reviewed by Andrews and Matsch (1983) in a volume encouraged by the INQUA Commission on Genesis and Lithology of Quaternary Deposits and its Work Group 5. Some would restrict the use of the term to sediment transported by and dropped from floating ice (Conolly 1978; Anderson et al. 1980). That definition would exclude a whole variety of deposits associated with non-floating glaciers whose termini lie in contact with, or close to, marine water. Others would restrict the glaciomarine environment to that marine zone in which glacial meltwater is a dominant influence on the water structure (Boulton and Deynoux 1981). In this view, sediments delivered to the deep sea, even by ice-rafting, would be excluded because of the profound separation of the site of deposition from other glacial influences. We endorse the broad, inclusive definition proposed by Andrews and Matsch (1983:2):

"Glacial-marine sediment includes a mixture of glacial detritus and marine sediment deposited more or less contemporaneously. The glacier component may be released directly from glaciers and ice shelves or delivered to the marine depositional site from those sources by gravity, moving fluids, or iceberg rafting. The marine component...is comprised mainly of terrigenous and biogenous sediments. Glacial-marine sediments vary laterally from ice-proximal diamicton, gravel, and sand facies, to an intermediate pebbly silt and mud facies, to distal marine environments where the glacial imprint is seen in ice-rafted debris (IRD) particles usually in the -1 to 4 o fraction." For other recent discussions on terminology and definitions of glaciomarine sediments and environments of deposition, see Boulton and Deynoux (1981), Molnia (1983), Eyles, Eyles, and Miall (1985), Drewry (1986), and Edwards (1986).

## 3 CONTROLS ON DEPOSITION

The primary factors controlling glaciomarine sedimentation and facies patterns

include the climatic setting, the nature of the glacier interface with the marine environment (grounded ice, ice wall, ice shelf), and the water depth of the basin of deposition (Anderson 1983). In temperate climates, glaciers produce meltwater, which is a major transporting mechanism for delivering sediment to the marine environment. This regime is dominant today in the area draining into the Gulf of Alaska and its fjords (Molnia 1983). In contrast, polar climates support glacier systems that do not lose significant mass by melting. Such regimes result in the delivery of most sediment to the sea by the glaciers themselves, as is the case on the Antarctic Shelf. The contrasts in depositional style, rates of deposition, and characteristics of the sediments have resulted in the development of several models. Subarctic (Molnia 1983) and polar (Anderson et al. 1983) paradigms are based on direct observations of those modern glaciomarine settings. Models have also been presented for tidewater glaciers (Powell 1981, 1984) and ice shelves (Drewry and Cooper 1981; Orheim and Elverhøi 1981; Powell 1984). For a detailed depositional model of lithofacies deposited in an isostatically depressed basin, see Domack (1983). More general glaciomarine sedimentary associations related to distance from glacier margins and water depth are modeled in Boulton and Deynoux (1981) and Andrews and Matsch (1983). Generalized lithofacies profiles for a variety of depositional environments are presented in Eyles, Eyles, and Miall (1985).

## 4 ENVIRONMENTS OF DEPOSITION

For marine-based glaciers, the major glaciomarine zones are: 1) subglacial, 2) grounding line, 3) ice shelf, and 4) open ocean (iceberg). For glaciers that terminate in the sea as ice cliffs, the calving zone coincides with its grounding line (Powell 1984). For ice shelves, the grounding line marks the place where a glacier begins to float (Hollin 1962). Shoreward of the grounding line, subglacial processes prevail for both terrestrial and marine-based glaciers. Seaward, a floating ice shelf may extend the ice margin hundreds of kilometers from the grounding line. Beyond the ice shelf margin, floating icebergs may carry glacial debris thousands of kilometers from the calving zone.

All sediments deposited in subglacial environments not directly influenced by the presence of marine water are properly considered simply as glacial deposits. Typical facies are sequences of basal tills produced by lodgment and melt out with associated subglacial meltwater sediments. Therefore, even though they may be deposited by marine-based glaciers, these sediments are terrestrial rather than marine. The only subglacial deposits included in our definition are those accumulated in cavities filled with seawater at the time of deposition, and having a direct connection to the sea.

The zone proximal to the contact between grounded glacier ice and the marine environment can be a highly energetic one, depending upon the stability of the ice margin or grounding line, the style of ablation, and the nature of the ice front. Where steep ice cliffs meet the sea, the interplay among depositional processes, including meltwater, gravity flows, iceberg calving, and marine currents makes a complex sedimentary facies predictable. At submarine grounding lines, substantial morainal banks can accumulate (Powell 1981). If significant melting occurs, accompanied by the concentration of meltwater into streams, glacial outwash is a dominant sediment facies.

The extension of glacier ice beyond the grounding line as a floating shelf considerably modifies the marine environment beneath, especially in the way of damping wind-driven oceanic circulation. Ice shelves transport glacial sediment seaward, potentially hundreds of kilometers from the line of flotation. Depending upon thermal conditions, bottom melting may release debris within a few kilometers of the grounding line, or seawater may freeze on, allowing sediment to be transported to the shelf edge. The former condition eventually produces clean shelf ice, whereas the latter results in sediment-laden ice reaching the calving edge (Powell 1984). Large-scale grounding of ice shelves during periods of lowered sea level may recharge the basal sediment load by freeze-on; subsequent lift-off may result in renewed sediment fall-out (Orheim and Elverhøi 1981). Such a flotation event may also bare glacially abraded pavements to fall-out of debris from floating ice, producing a cover of glaciomarine sediments directly upon the glacial pavement. Mass wasting in the form of sediment flows seaward of the grounding line or at local submarine topographic highs can be a major depositional process (Kurtz and Anderson 1979).

The width of iceberg zone sedimentation ranges considerably, from tens to a few hundred kilometers beyond ice shelf margins (Powell 1984). Sedimentation rates are highly variable, depending mainly on the original debris content of the icebergs, the size, melting rates, and the iceberg flux. Glaciomarine sediments accumulating in this zone may owe their dominant character to contributions from other sediment sources, especially in environments close to the sediment source. This is especially true in subarctic fjords where ice-rafted input is diluted by the much greater volume of finer grain sizes settling out from suspension (Molnia 1983). Generally, higher rates of accumulation are associated with proximity to the grounding line. The wider the ice shelf, the cleaner are the icebergs produced from it. The highest volume of debris-rich icebergs is produced by outlet glaciers and ice streams because they drain broad areas of interior ice at high rates, and in some cases move at rates greater than 1 km/yr (Anderson et al. 1983); this source accounts for the major portion of ice-rafted detritus (IRD) in today's oceans (Drewry and Cooper 1981).

Glaciers that terminate on land (land-based) can make significant sediment contributions to the marine environment only if they generate abundant meltwater and the glaciofluvial systems drain to the sea. Coarse-grained sediment is deposited in the shoreline and near shore zones, and fine-grained detritus is carried in suspension as sediment plumes offshore. Both mass-wasting and marine processes can result in resedimentation. According to Molnia (1983), glaciofluvial input is the most important mechanism for the introduction of present-day glacially-derived material into the Gulf of Alaska.

## 5 CLASSIFICATION

The classification (Table 1) includes type of glacier sediment source, position of sediment release and deposition with respect to grounded and floating glacier ice, mechanisms of sediment release, and an inventory of delivery and depositional processes involved in the accumulation of glaciomarine sequences. Depositional genetic varieties of glaciomarine sediments are grouped according to: 1) the marine environment of deposition; 2) the position of their accumulation with respect to relevant glacier zones; and 3) the process of deposition. The design of the classification was influenced by the genetic classification of till presented as Table 14 in Dreimanis (this volume).

## 5.1 Glacier source

Marine-based (or marine) glaciers are grounded below sea level. If a marine-based glacier terminates in the sea as an ice cliff, it is called a tidewater glacier. Commonly, marine glaciers have floating extensions in the form of small tongues or larger ice shelves. It is practical to group all of these sources in one category because they all contribute sediment directly into the marine environment. In contrast, sediment from land-based glaciers is delivered by meltwater streams.

## 5.2 Position

The grounding line, rather than the glacier margin, is chosen as the reference line for position of deposition because it includes both the position of the ice terminus for a tidewater glacier and the line of buoyancy (boundary between grounded and floating ice) for an ice tongue or ice shelf. Sedimentation can occur below the water line onto the surface of grounded ice (supraglacial) and in englacial and subglacial cavities. For land-based glaciers the marine environment is some distance beyond the ice margin (proglacial).

## 5.3 Debris release mechanisms

All of the debris release mechanisms are related to processes of ablation. They include subaerial melt-out and sublimation, melt-out at or below the water line, and physical separation such as occurs in iceberg calving. The calving process may release sediment along the initial plane of separation or as a consequence of subsequent fragmentation. Sediment may be released from land-based glaciers by ice fall along steep margins or by other mass movements, as well as by melt-out and sublimation.

## 5.4 Delivery/depositional process

Sediments finally accumulate in the glaciomarine environment as a result of delivery by and deposition from a wide range of transporting processes and agents. As a rule, the systems are dynamic, with complex interactions. Individual categories range from specific (settling) to broad. Mass movements, for example, include many different kinds of translocations, such as falls, slides, and flows. Meltwater flow encompasses subaerial

sheetflow, channeled flow, and submarine density flow. Icebergs and sea ice are the major kinds of floating ice, and a variety of processes are implied (i.e., dumping and rain-out). Waves are included with marine currents. Meltwater streams commonly have seaward extensions in the form of sediment-laden density flows.

## 5.5 Depositional genetic varieties

Within the context of the foregoing framework, a variety of genetic groupings is made and categorized by marine environment of accumulation, by position with respect to the major zones within which ice interfaces with marine water, and by process of deposition. Obviously, not all combinations are feasible.

## 6 CONCLUSION

The next stage in this work will be to expand and refine this genetic classification with the purposes of increasing our understanding of this complicated sedimentary environment and assuring the proper interpretation of glaciomarine sediments, in both ancient and modern sequences.

## REFERENCES

Anderson, J.B. 1983. Ancient glacial-marine deposits. In B.F. Molnia (ed.), Glacial-marine sedimentation, p. 3-92. New York, Plenum.

Anderson, J.B., C. Brake, E. Domack, N. Myers, and R. Wright 1983. Development of a polar glacial-marine sedimentation model from Antarctic Quaternary deposits and glaciological information. In B.F. Molnia (ed.), Glacial-marine sedimentation, p. 233-264. New York, Plenum.

Anderson, J.B., D.D. Kurtz, E.W. Domack, and K.M. Balshaw 1980. Glacial and glacial marine sediments of the Antarctic Continental Shelf. Journal of Geology 88:399-414.

Andrews, J.T. and C.L. Matsch 1983. Glacial marine sediments and sedimentation: An annotated bibliography. Bibliography No. 11. Norwich, Geo Abstracts, Ltd.

Armstrong, J.E. and W.L. Brown 1954. Late Wisconsin marine drift and associated sediments of the Lower Fraser Valley, British Columbia, Canada. Geol. Soc. America Bull. 65:349-364.

Boulton, G.S. and M. Deynoux 1981. Sedimentation in glacial environments and the identification of tills and tillites in ancient sedimentary sequences: Precambrian Res. 15:397-422.

Conolly, J.R. 1978. Glacial marine sediments. In R.W. Fairbridge and J. Bourgeois (eds.), The encyclopedia of sedimentology, p. 355-357. Stroudsburg, Dowden, Hutchinson and Ross, Inc.

Denton, G.H. and T.J. Hughes 1983. Milankovitch theory of ice ages: Hypothesis of ice-sheet linkage between regional insolation and global climate. Quaternary Research 20:125-144.

Denton, G.H., T.J. Hughes, and W. Karlen 1985. Global ice sheet system interlocked by sea level. Quaternary Research 26:3-26.

Domack, E.W. 1983. Facies of Late Pleistocene glacial-marine sediments on Whidby Island, Washington: An isostatic glacial-marine sequence. In B.F. Molnia (ed.), Glacial-marine sedimentation, p. 495-533. New York, Plenum.

Drewrý, D. 1986. Glacial geologic processes. London, Edward Arnold.

Drewry, D.J. and A.P.R. Cooper 1981. Processes and models of Antarctic glaciomarine sedimentation. Annals of Glaciology 2:117-122.

Edwards, M.B. 1986. Glacial environments. In H.B. Reading (ed.), Sedimentary environments and facies, p. 445-470. Oxford, Blackwell Scientific Publications.

Elverhøi, A., O. Lonne, and R. Seland 1983. Glaciomarine sedimentation in a modern fjord environment, Spitzbergen. Polar Res. 1:127-149.

Eyles, C.H., N. Eyles, and A.D. Miall 1985. Models of glaciomarine sedimentation and their applications to the interpretation of ancient glacial sequences. Palaeogeography, Palaeoclimatology, Palaeoecology 51:15-84.

Hollin, J.T. 1962. On the glacial history of Antarctica. Jour. Glaciology 4:173-195.

Kurtz, D.D. and J.B. Anderson 1979. Recognition and sedimentologic description of recent debris flow deposits from the Ross and Weddell Seas, Antarctica. Jour. Sed. Pet. 49:1159-1170.

Molnia, B.F. 1983. Sub-arctic glacial-marine sedimentation. In B.F. Molnia (ed.), Glacial-marine sedimentation, p. 95-144. New York, Plenum.

Orheim, O. and A. Elverhøi 1981. Model for submarine glacial deposition. Annals of Glaciology 2:123-128.

Powell, R.D. 1981. A model for sedimentation by tidewater glaciers. Annals of Glaciology 2:134.

Powell, R.D. 1984. Glaciomarine processes and inductive lithofacies modeling of ice shelf and tidewater glacier sediments based on Quaternary examples. Marine Geology 57:1-52.

Rust, B.R. and R. Romanelli 1975. Late Quaternary subaquatic outwash deposits near Ottawa, Canada. In A.V. Jopling and B.C. McDonald (eds.), Glaciofluvial and glaciolacustrine sedimentation, p. 177-192. Soc. Econ. Paleontol. Mineral, Spec. Publ. No. 23.

*Genetic Classification of Glacigenic Deposits, Goldthwait & Matsch (eds)*
*© 1988 Balkema, Rotterdam. ISBN 90 6191 694 1*

# Classification of glacial morphologic features

Richard P.Goldthwait
*Department of Geology and Mineralogy, Ohio State University, Columbus, Ohio, USA*

ABSTRACT: The primary criterion for this classification of glacial deposits and eroded features is morphology. A second useful criterion is association of forms because these imply origin. While the nature of the deposits (till versus meltwater-sorted sediments) is a partial basis for its construction, the classification is not a sedimentological one. Eolian features and forms measured in millimeters or centimeters are not classified, whereas those in meters or kilometers are. Two broad genetic categories are defined: (I) those landforms caused directly by ice push, scraping, melting, and evaporation (top, base, or margin), and (II) those caused by meltwater. The second order of division is location with respect to the glacier: (A) well back beneath ice of significant depth (100 m+), (B) reasonably near the ice margin, or (C) proglacial, to the farthest dominant effects of meltwater. The third order of division is based on the difference in terrains as it relates to dominant process (erosion or deposition), and on relation to the pattern of glacier ice or meltwater movement.

## 1 PURPOSE

This classification of glacial deposits and eroded features was developed from the most common ideas of professional field research geologists and geographers. The objects are 1) to afford the best and most recognized terminology in English possible at this time, and 2) to indicate the most prevalent ideas of origin. Of course we hope the classification may be used by those making surficial maps, and by students and trainees using glacial maps. It is the product of INQUA Commission II, Work Group 2F (formerly 6).

## 2 HISTORY

Classification is not new. In the early nineteenth century subdivisions started when glaciers were studied around the Alps. Most authors dealt with the ice itself, but deposits were described in French and German (Chorley, Dunn, and Beckinsale 1964; Charpentier 1841; and Martins 1842). The divisions were primarily: end moraine, erratic blocks, and outwash (valley train). As it was realized that similar shapes and materials extended across the plains of northern Germany and Poland, former continental glaciation

became an accepted idea in Europe and then America (Agassiz 1840). Near the close of the nineteenth century, chapters or articles in several languages set the basic pattern for most modern subdivisions of glacigenic landforms (Heim 1885; Woodworth 1899). In English the most notable was Chamberlin (1894). He described the principal ideas and relationships that dominate twentieth-century texts and this classification.

Glacial geology texts became numerous in the twentieth century. Most concentrate in the decades 1947 to 1957 and 1968 to 1979. Looking at twenty of these books we find that the chapter groupings were solely morphologic up until about 1955. All authors considered glacially eroded bedrock in a chapter separate from deposits, and many completely separate mountain (Alpine) forms from continental glacial deposits (Thwaites 1935). Features on live glaciers, like medial moraines (and glaciology in general) are likewise grouped in a separate chapter. The main subdivisions of till-dominant vs. aqueous (Okko 1955), or ice-made vs. water-made (Woldstedt 1954), or till drift vs. sorted drift (Sugden and John 1979) have so many exceptions as to be difficult categories. Nevertheless, shape and general content are the best field guides to mapping and similar origins.

# 3 CRITERIA AND BASIS

The primary criterion or basis here is one of shape, that is, morphology. What do you see when walking the surface, or studying stereo airphotos or detailed contour maps? A second criterion proved to be groups or associations of forms as these imply how the features originated: e.g., under fairly deep ice, under fast moving or stagnant ice, very near the glacier margin in tubes or crevasses, sliding into perforated ice surface pits, pushed up at a slow-moving ice edge, squeezed up from a deeper/earlier surface below? Equally important is the material of which the glacigenic feature is made: is it plucked or abraded in bedrock? Is it glacial drift -- unsorted, mixed, or deformed as in most tills? Or is it sorted, that is, handled by meltwater, as in glacial outwash, deltas, and lacustrine silts? Structures are sometimes recognizable after detailed field work and are the fourth means of verification. Some would make these, particularly deformation structures, the entire basis for classification, but they do not cover morphology at all, and in some large glaciated areas nothing would be served. Anyway, the parent Commission 2 specified "morphology."

# 4 LIMITATIONS

Our classification makes no pretext of being a sedimentological one, important as that is (see Lawson, this volume). It does have to be based in part upon whether till, in its broadest sense (Dreimanis, this volume) is a critical part of the deposit, or whether meltwater-sorted sediments are dominant. Sometimes these must be identified from animal burrows, vegetation, or auger holes, but preferably from a river-cut, road-cut, or pit seen while mapping. Bedrock may be evident on photos from patterns, color, or vegetation. Slightly earlier drifts may be evident from vegetation, soil, color, loess cover, and degrees of postglacial gully erosion. However, in no detailed sense is this an age-of-glaciation classification either. To some degree gross erosion features like cirques, or multilayered deposits like ground moraine may imply time or duration of erosion. To be certain of age one must have radioactive material, growth rings, indicator fossils, chemical change, or impact dating of some contents. Thus very

old buried drifts are not served much by this classification, nor are any of the paleoglaciations (e.g., Permo-Carboniferous). The classification is not fundamentally stratigraphic.

Some have insisted that, with "glacigene" in the title, the classification should be entirely genetic. Would any two professionals agree on just how a "drumlin" (one shape) formed? The main thing about a drumlin is that it does demonstrate the presence of a former glacier, and circumstances indicate a significant thickness of ice during shaping.

# 5 NECESSARY EXCLUSIONS

Many features often associated with glacial deposits are eliminated from this classification. For example, eolian features like a loess blanket or a chain of dunes may accompany and cover glacial deposits owing to ice-intensified winds, but they may also occur without any glacier. Where very common, as covering old ground moraine or dotting a glaciolacustrine shoreline, they may be mentioned in passing. In any case, INQUA has a Loess Commission, so that classification is not our prerogative. And wave action is critical even on a glaciolacustrine shoreline. How far from the glacier is it glaciolacustrine? The Work Group 2 says "so far as meltwater is a dominant and effective water source." There are indeed some "gray areas"!

Still another exclusion is permafrost and frozen-ground patterns. Inasmuch as these require temperatures below 0°C, and vary with ground-moisture accumulation, they frequently appear where a glacier expanded over cold-based ground. But many hundreds of hectares also have frost-wedge patterns and solifluction lobes, far from any glacier. These matters are left to the INQUA Permafrost Commission and the International Permafrost Association.

Many features on present-day glaciers, such as medial moraines or crevasse systems, are not finished or deposited in any particular fashion. They may be mentioned as sources of pattern but they are not classified here; they are not features of the postglacial landscape. Certainly they belong in any classification of temporary active glaciological landscapes. Glacial debris in transport is important, but material belongs in this classification only the day the ice disappears from the site.

Many small markings made during

transport, such as striae or crescentic markings, are omitted. Such tertiary features are not generally part of the gross landscape, although they are important marks of genesis. They are not contourable. Items measured in millimeters or centimeters are not classified; those in meters or kilometers are.

Postglacial surface modifications, such as landslides into a cirque, or gullying of glacial deposits like kame terraces, are not classified. Material moved by such processes must be mentally replaced in order to recognize what the glacier really left. Of course such processes carried on long enough in old or very old drift may alter the drift beyond recognition; in most climates this takes less than a half million years of exposure. However, in very arid Antarctica complete modification may take several million years. A sedimentological approach may be the only one possible for old or buried deposits; these may only indicate former shape at best.

## 6 METHOD OF CONSTRUCTION

This is the product of at least one hundred critics who suggested specific changes over a span of 15 years. It started with my suggested classification at the Christchurch, New Zealand VIII INQUA, 1973 (see Goldthwait 1975) and it has been presented to and modified by small groups at each INQUA Congress thereafter (including XII, 1987). Actually it was edited and reprinted eight times as it improved with each of 14 presentations:

1973 Questionnaire discussed at VIII INQUA Congress mailed to 120 glacial geologists.
1975 Key to glacial deposits based on shape and content mailed out.
1976 Stockholm, Sweden: "Till Sweden," Geological Survey of Sweden, eliminated that "key to morphology."
1980 Geilo, Norway: International Glaciological Society Symposium.
1981 Mail-out to 90 correspondents for answers.
1982 Moscow, USSR: INQUA XI, Commission 2 accepted simplified major categories.
1984 Alberta, Canada: Commission 2 Field Conference, some reordering and elimination.
1985 Helsinki, Finland: Commission 2 Field Conference, foreign equivalents attempted.
1985 Tallin, Estonia SSR: Commission 2 meeting, with foreign equivalents.
1985 Manchester, England: First Geographi-

cal Congress, very few additions.
1986 Illinois, USA: Geological Society of America, Northeast Section.
1986 Ohio, USA: Ohio Academy of Science, general approval.
1986 Glacier Bay, Alaska, USA: Commission 2 Field Conference, critique of moraine grouping.
1987 Ottawa, Canada: INQUA XII, sessions of Commission 2, final revisions, additions.

All reprinted versions were mailed to 60 to 80 persons at their request (Work Group 6) as "associate Commission members," as well as to all full members. Each mailing received 15 to 40 responses. The most notable contributors, critics, and translators are listed below; of course no one of these is responsible for the final result:

J.T. Andrews (USA)
M.N. Alekseev (USSR)
H.W. Borns (USA)
G.S. Boulton (UK)
T.   Chinn (NZ)
L.   Clayton (USA)
E.A. Colhoun (Australia)
I.D. Danilov (USSR)
S.   Daveau (Portugal)
G.   DeMoor (Belgium)
J-C. Dionne (Canada)
L.A. Dredge (Canada)
A.   Dreimanis (Canada)
E.   Drozdowski (Poland)
J.A. Elson (Canada)
M.M. Fenton (Canada)
P.L. Gibbard (UK)
G.   Gillberg (Sweden)
E.F. Grube (GFR)
     and group
K.A. Habbe (GFR)
D.   vanHusen (Austria)
S.   Jelgersma (Netherlands)
A.A. Jurgaitis (USSR)
W.   Karlen (Finland)
P.F. Karrow (Canada)
A.   Karukäpp (USSR)
T.J. Kemmis (USA)
R.   Kujansuu (Finland)
     and group
Y.A. Lavrushin (USSR)
D.E. Lawson (USA)
J.   Lundqvist (Sweden)
J.   Mangerud (Norway)
C.L. Matsch (USA)
W.H. Mathews (Canada)
G.D. McKenzie (USA)
J.   Menzies (Canada)
K.D. Meyer (GRF)
D.M. Mickelson (USA)
E.H. Muller (USA)

S. Occhietti (Canada)
R.D. Powell (USA)
V.K. Prest (Canada)
A. Raukas (USSR)
H. Ruszczyńska-Szenajch (Poland)
V-P. Salonen (Finland)
C. Schubert (Venezuela)
G. Seret (Belgium)
J. Shaw (Canada)
W.W. Shilts (Canada)
J.L. Sollid (Norway)
D.A. St.-Onge (Canada)
D. Sugden (Scotland)
J.T. Teller (Canada)
K.N. Thome (GFR)
W.B. Thompson (USA)
E. Virkkala (Finland)
A. Weidick (Denmark)
H.L. Zilliacus (Finland)

Ever since the first questionnaire it has been certain that no one final product could satisfy every research person in glacial geology/geography. It has been equally evident that many, if not most, are very tolerant, and that the product would be very useful. A few suggested new terms, not yet in the literature; they will be unhappy that their new terms do not appear. Others misuse terms already in the literature. I had to go to the original definition of "washboard moraine," and mark such equivalents used (in parentheses) with "#" where they are not correct.

An often misused term is "DeGeer moraine," but since DeGeer originally suggested "annual," and that is shown incorrect, these uses are marked "#" also. There are still others, most of which elicited great arguments. Only a few respondents said "this is no good," or "prejudiced," or "utterly incorrect." One or two of them have become converted! Our principal objective is understanding with brevity, not long descriptive phrases.

## 7 THIS CLASSIFICATION

Originally only the main titles or categories of terrain were planned. These were most welcomed at the 1982 (Moscow) INQUA XI Congress. Now revised (Table 1), they consist of two broad genetic categories: (I) those landforms caused directly by ice push, scraping, melting and evaporation (top, base, or margin), and (II) those caused by the icy meltwater racing, tunneling, standing and dropping its load. It might seem the former are marked by till deposits and the latter by sorted, layered deposits -- but not exclusively so. Massive till (balls, lenses, or layers) and sorted silt-sand lenses are found in both. The second order of division is location with respect to the glacier: (A) either well back underneath ice of significant depth (100 m+), or (B) reasonably near the ice margin in ragged thin ice or within a kilometer of a discrete "dying"

Table 1. Analytical classification of glacigene landforms (short form).

I. Direct glacial landforms
  A. Subglacially formed, parallel to ice motion
    1. LARGE PLUCKED BEDROCK FORMS
    2. STREAMLINED BEDROCK
    3. STREAMLINED DRIFT
  B. Subglacially formed, transverse to ice motion
    1. SUBGLACIAL MORAINE RIDGES
    2. GROUND MORAINE
  C. Ice-marginal landforms
    1. END MORAINES
    2. LATERAL MORAINES
    3. OTHER MORAINES, MOSTLY ABLATION

II. Indirect glacially induced landforms
  A. Subglacial meltwater features
    1. BASAL WATER EROSION
    2. ESKER SYSTEMS
  B. Near-ice-margin meltwater forms
    1. HILLSIDE CHANNELS
    2. KAME FIELDS
  C. Proglacial meltwater landforms
    1. CHANNELED TOPOGRAPHY
    2. OUTWASH FEATURES
    3. GLACIOLACUSTRINE FEATURES
    4. GLACIOMARINE FEATURES

edge, or (C) proglacial, to the farthest dominant effects of meltwater. The third order is that of two to four (1, 2, 3, 4) differing terrains, depending on whether erosion or deposition is dominant and on its relation to the pattern of glacier ice or meltwater movement. These 16 main categories of agreement are in capital letters.

So many queries and suggestions were received and argued that a fourth order of eroded or deposited features was developed and repeatedly revised. Some entries were even switched from terrain to terrain, but all are believed by most of us to belong in the terrain shown on Table 2 (the long table at the end of this chapter), and constitute examples of what may be found in that terrain. These are not all the forms you may find and wish to list, but they are prime examples. They are lettered with small letters (a, b, c, d, e, f) -- 63 in all. Indeed, you the worker may wish to add more, and you should for your own use.

## 8 FOREIGN EQUIVALENTS

Many names now used in English were adopted in the last century from the French (cirque, moraine), from Scandinavian (fjord, sandur) or modified from Scottish and Irish (kaim, druim). Nevertheless, meanings broaden or change with time. Perhaps the most bothersome is "moraine" (Moräne) used for the material (till) in northern Europe as discussed by Dreimanis, (this volume). Some entries in the lists below were declared utterly wrong by many using those languages. Unfortunately the alternates suggested by them did not agree one with the other. At INQUA XII in 1987, conflicting lists were circulated to the other speakers of that tongue and some compromise is now possible. It is presented, with duplicates or equivalents, as follows:

## GLACIGENE LANDFORMS

| English | French | German |
|---|---|---|
| N. America, N.Z., and U.K. | S. Europe, Alps, and Quebec | N. Europe, Alps, and Austria |
| I.Direct glacial landforms | Formes glaciaires de contact | Glaziale Formen, direkte |
| A.Subglacially formed, parallel | Sous-glaciaire, alignée parallèle | Exarationsformen, parallel |
| 1.Plucked bedrock | Roche affouillé | Lineare Hohlformen |
| 2.Streamlined bedrock | Roche fuselé au profile | Rundhöcker (Strömlinienförmig) |
| 3.Streamlined drift | Depôt fuselée au profile (Ridée longitudinal) | Drumlin (Strömlinienförmig) |
| B.Subglacially formed, transverse | Sous-glaciaire alignée transversale | Exarationsformen quer (transversal) |
| 1.Subglacial moraine | Crête sous-glaciaire (Rides sous-glaciaires) | Rogenmoräne (unter Eis Moränen) |
| 2.Ground moraine | Moraine de fond | Grundmoräne |
| C.Ice-marginal landforms | Formes marginales | Eisrand-Formen |
| 1.End moraines | Moraines frontales | Endmoränen |
| 2.Lateral moraines | Moraines laterales | Seitenmoränen |
| 3.Other moraines (Mostly ablation moraine) | Moraines d'ablation | Satzmoräne (Abschmelzmoräne) |
| II.Indirect glacially induced landforms | Formes glaciaire de l'eau (Fluvioglaciaire) | Indirekte glaziale Formen (Schmelzwasser-einfluss) |
| A.Subglacial meltwater features | Eaux de fonte (sous-glaciaire) | Subglazial Schmelz-wasserformen |

| 1.Basal water erosion | Chenal d'eau de fonte (erosion des eaux sous-glaciaire) | Subglazial Erosions-formen |
|---|---|---|
| 2.Esker systems glaciofluvial | Complexes d'esker | Wallberge (Osersysteme) |
| B.Near-ice-marginal forms | Formes marginales jusqu'á glaciaire | Glazifluviale Eisrandnahe Formen |
| 1.Hillside channels 2.Kame fields | Rigoles de versant Kames | Flankentälchen Kame-felder (Kamelandschaft) |
| C.Proglacial meltwater landforms | Formes d'eaux proglaciaires (justaglaciaires) | Proglazialer Formen-bereich, Schmelzwasser |
| 1.Channeled topography | Chanaux proglaciaires (topographie ravinée) | Abflussrinnen (Erosionfläche) |
| 2.Outwash systems | Formes d'épandages | Sandersystem |
| 3.Glaciolacustrine features | Formes glaciolacustres | Becken (Glazila-kustrine Formen) |
| 4.Glaciomarine features | Formes glaciomarines | Glazimarine Formen |

Of course English-speaking authors have adopted many useful foreign terms (here emphasized in bold face) and vice versa. Nevertheless there are no exact translations for roche moutonnée and many others. An Urströmaler has no short English equivalent, as pointed out here by Niewiarowski (this volume). Some fine meanings are lost, as in "push moraine" -- acting like a bulldozer blade; but there is "thrust moraine": meaning forcing up nonglacial material from below. Stauchmoränen conveys this in German.

I like to make a distinction between dead ice (Tôteis or glace mort) and stagnant ice (just barely moving, up to 1 m or 2 m per year, due to surface slope). These are hardly discriminated in North America, let alone in all the polar-alpine world or by glaciologists. Other personal preferences exist, such as differentiating between "esker" and "kame" by defining an esker as more than 10x as long as it is wide, and kame as less than 10x. These distinctions were not accepted by the majority; too many believe these forms differ in origin as indicated by structure and sediment (see A. Jurgaitis, this volume).

REFERENCES

Agassiz, L.J. 1840. Glaciers, and the evidence of their once having existed in Scotland, Ireland, and England. Proceedings Geological Society 3:327-332.

Chamberlin, T.C. 1894. Proposed genetic classification of Pleistocene glacial formations. Jour. of Geology 2:517-538.

Charpentier, J. de 1841. Essai sur les glaciers et le terrain érratique du Bassin du Rhône. Lausanne.

Chorley, R.J., A.J. Dunn, and R.P. Beckinsale 1964. The glacial theory: In The history of the study of landforms or the development of geomorphology 1:191-297. London: Methuen.

Goldthwait, R.P. 1975. Glacial deposits. Benchmark Papers/21. Stroudsburg, Dowden, Hutchison and Ross.

Heim, A. 1885. Handbuch der Gletscherkunde. Stuttgart, J. Englehorn.

Martins, Ch. 1842. Sur les formes regulieres du terrain de transport des vallées du Rhin anterieur et du Rhin posterieur. Bull. Societeé Geologique de France, Ser. 1, 13:322-345.

Okko, V. 1955. Glacial drift in Iceland: Its origin and morphology. Commission Geologique de Finlande, Bull. 170.

Sugden, D.E. and B.S. John 1979. Glaciers and landscape. London, Edward Arnold and Son.

Thwaites, F.T. 1935. Outline of glacial geology. Ann Arbor, Edwards Bros.

Woldstedt, P. 1954. Das Eiszeitalter. 2nd ed., v. 1. Stuttgart, Ferdinand Enke.

Woodworth, J.B. 1899. The ice-contact in the classification of glacial deposits. American Geologist 23:80-86.

Table 2. Analytical classification of glacigene landforms (detailed list).

Main origins (I,II) are underlined: direct, ice-shaped, or indirect, ice-controlled.

Usual place of origin (A,B,C) is underlined: subglacial, ice-marginal, or proglacial.

Regional groups of forms, terrain types, are in capital letters, and numbered 1,2,3...
Examples are "STREAMLINED DRIFT," "LATERAL MORAINES," or "OUTWASH FEATURES."

Some common single units in each terrain type, or local examples, have small letters
(a,b,c,...) but no underline. Each researcher will map or identify more units or
varieties (g,h,i...) of his/her own.

In parentheses ( ) are some near-equivalents common in English literature. Those used
locally or only in one country have asterisks (*), and those now archaic, or believed
used incorrectly, are marked (#). If adopted recently from a non-English tongue, names
are in **bold type**. Those marked (+) are primarily or exclusively formed by mountain
(alpine) glaciers.

I. Direct glacial landforms -- ice contact, shaped by ice, primary; composed mostly of
   basal till, or eroded in much older bedrock.

   A. Subglacially formed, parallel to ice motion -- mostly active and warm-based ice;
      radial to ice source, features smoothed by basal and englacial lower-ice debris
      (lodgment, basal melt-out till); under thick, moving ice.
      1. LARGE PLUCKED BEDROCK FORMS -- large, mostly plucked (sapped?), deepened by
         ice erosion.
         a. Cirque (corrie+): large semicircular headwall, arêtes, rock basin, some
            tarns, bastions; presumably by sapping and linear extension of ice
            downvalley.
         b. Nivation basin+: shallow circular depression with steep but not cliffed
            sides -- incipient form of cirque mostly filled once by névé and snow.
         c. Glacial trough (U-valley): rounded cross profile, some with rock basins,
            lips, lake chains (paternoster lakes), hanging tributaries, finger lakes,
            beveled facets -- often floored with valley train.
         d. Fjord+: same as c. but flooded by sea or below sea level.
         e. Glacial lake basin: eroded in weak and/or deeply weathered bedrock;
            usually elongated parallel to ice motion or in hilly land; trapping
            surface waters.
      2. STREAMLINED BEDROCK -- smooth bedrock forms rounded by ice erosion, mostly
         striated bedrock with pockets of subglacial till cover. Including:
         a. Rock drumlin (whaleback, drumlinoid*): large elliptical or oval hill,
            bedrock except for patchy cover of till. Micro markings as for b. below.
         b. **Roche moutonnée** (sheepsback#): small oval or truncated elliptical
            outcrop, steep lee end, many cliffed: top may have striae, crag-and-tail,
            chattermarks, crescentic scars (marks), etc.
         c. Groove (p-forms#): a large linear depression in rock, containing multiple
            striae, smoothly curving, plastically molded in places.
      3. STREAMLINED DRIFT -- assemblage of oval or elliptical hills and hollows;
         molded by ice erosion and/or deposition; subglacial tills, many with cores
         of older bedrock, harder till, outwash, or even lacustrine sediments.
         Dominated by:
         a. Drumlin: "the ideal" inverted spoon-shaped hills, tend to be steeper on
            stoss end; in groups of 10 to 2000.
         b. Drumlinoid: including elongate, cigar-shaped ridges, and spindle forms;
            with elongate hollows, between very extended parallel drumlins.
         c. Flute (glacial scallop, radial moraine): sigmoid crest and long hollow
            cross-profile; parallel to basal ice motion; squeeze-up or subglacial till
            in some; others emanate from boulder crag-and-tail.
         d. Lee-side cone (lee-side ridge or till, large crag-and-tail*): down-glacier
            deposit behind protective bedrock knob; subglacial or squeeze-up till with
            gravel-sand-silt lenses and/or silty diamict and boulders interbedded; may
            show shear structures.

B. <u>Subglacially formed</u>, <u>transverse</u> to ice motion, or unoriented --
   1. SUBGLACIAL MORAINE RIDGES -- systematic ridge sets, may be crescentic
      (nesting) to straight (parallel); subglacial till (mostly deformed or
      squeeze-up); probably near ice margin. May involve bottom-drag; ice usually
      moving.
      a. Corrugated moraine (washboard moraine, DeGeer moraine#): very low ridges,
         like sea swells, in parallel sets, oblique to ice motion (seen on
         airphotos).
      b. Rogen moraine (ribbed moraine): irregular-width ridges, often high, sub-
         parallel to ice margin, stoss side or crest; sometimes fluted, may often
         grade to crescentic shapes and drumlins.
      c. Thrust moraine (overthrust moraine, pop-up): dragged older basement
         crescentic lead edge of slabs with till or till-on-bedrock blocks. See
         I.C.1.c. May be gentle hills of old rock slab and subglacial till; mostly
         in outer half of glaciated area.
   2. GROUND MORAINE -- non-linear, smooth to hummocky drift cover, mostly subgla-
      cial till (lodgment or basal melt-out), with minor subglacial outwash lenses
      and cover of supraglacial tills (melt-out, flow). From thin to thick:
      a. Cover moraine *(**Alvars**): patchy, thin till revealing full bedrock
         topography (veneer) or partly masking details of bedrock topography
         (blanket).
      b. Hummocky ground moraine: irregular sprawling area, rolling or rough without
         pattern, but may be endform of drumlin group.
      c. Till plain: nearly flat or slightly rolling and gently inclined plain;
         mostly thick cover, often with multiple till layers (varying compositions),
         completely masking bedrock undulations.
C. <u>Ice marginal landforms</u> -- from active but slow-moving ice -- depositional moraine
   ridges or belts, higher than adjacent ground moraine; produced by slowly moving
   upper ice, with surficial debris (melt-out, flowtill), plus many glaciofluvial
   and glaciolacustrine inserts. Some sand with boulders.
   1. MARGINAL MORAINE (RIDGES) -- perpendicular to ice motion; belt of higher, hum-
      mocky ground; mostly supraglacial till (melt-out, flowtill), some dominant in
      glaciofluvial content, boulders common; presume "conveyor belt" rise to ice
      margin. Includes such specific units as:
      a. End moraine (terminal or recessional moraine, sometimes): broad belt in
         extensive arcuate pattern, around broad ice lobe at significant maximum
         extent, hummocky, often pitted,# irregular short ridge crests; some conical
         high mounds of distinctive lithology (I.C.3.e).
      b. Fluted (drumlinized, readvance) moraine: similar to a fluted or drumlinized
         apron of subglacial till proximal to end moraine; but distinguishable as a
         retreatal position, or long halt, afterwards overridden; deformed till?
      c. Push moraine (**Stauchmoräne**): belt of low hills, generally crescentic,
         raised above ground moraine; any soft preglacial or glacial material;
         raised and deformed by ice-edge thrusting (subtly different from thrust
         moraine I.B.1.c).
      d. Boulder belt*: erratic boulders on or near surface; subparallel to other
         end moraines. May be at ground-moraine level in places.
      e. Kame moraine (see II.B.2.): sections of end moraine belt, especially
         reentrants between lobes; replete with bedded gravel-sand kames and
         kettles, suggesting dead ice, and including very large kame plateaus (see
         II.B.2.b.) with high local lacustrine surfaces.
   2. LATERAL MORAINES -- parallel to central ice stream; sharp ridges of till with
      boulders or benches on slope, often alternating till and coarse gravel; may
      represent several swellings of ice. Also, high hummocky ground between ice
      lobes. Besides sharp ridges parallel to the sides of ice lobes it includes
      some of the following:
      a. Perched moraine (valley-side or stranded lateral moraine)+: high on either
         valley side, marginal to an ice lobe, in sets like stairs; non-matching
         levels across or up valley.
      b. Looped marginal moraine: usually ice-pushed over outwash or projecting
         through enclosing outwash, at rounded end of former ice lobe. Generally
         applied to U-shaped moraine outlines including some end moraine (I.C.1).
      c. Interlobate moraine (radial moraine#): broad ridges with hummocky top area,

may be smoothed and truncated by overriding ice; some till and many gla-
    ciofluvial gravels.
  d. Glacial trimline+: on higher valley side; more or less horizontal break
     between vegetation, blockfields, or deep soil, and no vegetation or thin
     weak soil. Often old vegetation above and new younger vegetation below.
     Lower plants, lichen, or soils removed by recent glacial expansion and
     eroding.
3. OTHER MORAINES -- many subaqueous, some "ablation," a few minor, not exactly
   marginal, irregular ridges of surficial tills and erratic boulders or reworked
   sorted materials; from slow-moving or stagnant ice or from crevasses or shear
   zones or on downwasting ice masses. Secondary or reworked till. These include:
   a. Crevasse filling: subparallel, straight group of small ridges at angle to
      ice motion or margin, en echelon; sandy diamicton or sandy till (see also
      II.A.2.e. for well-sorted fillings).
   b. DeGeer moraine (annual moraine,# cross-valley moraine): low ridge, straight
      to gently arcuate, some branching; all may be subaquatic, parallel to ice-
      cliff edge, in crevasses transverse to ice motion; aquatic sand to clay or
      loose sandy diamicton; inferred retreating ice cliff in sea or lake.
   c. Disintegration ridge (ice-block ridge, doughnut mound): intersecting
      ridges, all orientations, kettled, surficial tills; may be small.
   d. Till plateau (moraine plateau): extensive sandy-gravelly moraine with
      superposed lacustrine silt-clay top (II.B.2.b. similar).
   e. Moraine dump (medial dump+): isolated high mound of angular debris, not a
      belt; accumulated at terminus of a medial moraine; often limited lithologic
      types.
   f. Perched boulder (balanced rock*): notably large, erratic, and transported
      -- landed or pushed alone onto other boulders or smoothed outcrop.

II. Indirect glacially induced landforms -- all meltwater-induced or fed; secondary gla-
    cial sediments; shaped by meltwater currents, with lake or sea ice, wind waves, ice-
    bergs, bottom currents; composed mostly of well-sorted sediments (glaciofluvial) or
    carved in earlier bedrock or till.

   A. Subglacial meltwater features -- slow-moving ice -- ice contact, but hidden,
      mostly under the ablation zone of outer/lower ice surface; including quantities
      of bottom melt from warm-based ice.
      1. BASAL WATER EROSION --
         a. Tunnel valley: long channel without floodplain; cut in drift or impinging
            bedrock from basin to basin so as to require ice-controlled escape; may
            connect to esker system.
         b. Glacial moulin (glacial pothole): circular or curved, smooth, either very
            big or very numerous; on high bedrock ridges or valley sides.
         c. Glacial chute: steep, diagonally downslope, may lead to kame field or
            esker; believed subglacial near ice-marginal region.
      2. ESKER SYSTEMS (OSE#) -- ridge forms generally on plains, in broad valleys, or
         across low saddles between basins, deposited in ice tunnels within or under
         ice; composed of sorted layers of gravel, sand, and in some places boulders or
         silt; rare boulders of till or interlayered till lenses; some with squeezed-up
         cores of till; lenticular bedding, often elongated downridge and arched over
         top, with backset and foreset laminae, often faulted or tilted, especially
         near sides. Includes one or more of:
         a. Beaded esker (hogback*, radial ose): a generally high, long, winding ridge,
            with a low trough on either side in places and a hummocky crest; some sec-
            tions anastomosing (doubling and rejoining); presumed subaquatic tube with
            uphill clast travel; end in lake or sea; may be ice-marginal (II.B.2.) in
            part.
         b. Engorged esker: down a notable slope, modest size, presumed subaerial,
            often a tributary to main esker.
         c. Squeeze-up esker*: ridge of reworked gravel, with depressed adjacent
            troughs; assumed thick ice.
         d. Esker chain (kame ridge): end-on-end segments of ridge, original breaks,
            and inserted broad kame groups; assumed subaerial and in thin stagnant ice.
         e. Crevasse filling: either (1) parallel small ridges, nearly straight, en

echelon side by side; believed oblique to ice motion; or (2) semi-circular
(subcircular esker), roof-collapse crevasses; mostly sand (also of till,
see I.C.3.a.); postulate thin dead ice; or (3) conjugate ridges (esker
nets) in rectilinear pattern.
B. <u>Near-ice-margin meltwater forms</u> -- ice contact but subaerial forms, from surface
summer melt into pools and streamlets on collapsing or retreating ice, dead or
slow-moving; or just under thin ice edge.
1. HILLSIDE CHANNELS -- incised by meltwater, restricted or contained by ice
walls (not exactly downhill), in rolling or mountainous topography; cut walls
of bedrock or till mostly. Includes some of the following:
a. Lateral drainage channel: shelf or sharp trench, following contour of
slope, gentle incline; often in subparallel sets downslope.
b. In-and-out channel: horseshoe pattern, hanging ends on hillsides; requiring
ice to conduct meltwater into and out of channel.
2. KAME FIELDS (glacial karst# deposits) -- patches of irregular hummocks, and
short ridges in any direction; generally on broad basin, valley floor or lower
slopes; mostly sand, some gravel and silt; often surficial till sheets or
masses; poorly to well bedded, from horizontal to rakish angles, often faulted
or collapsed. Some of these include:
a. Moulin kame*: isolated single conical mound, usually high; believed formed
by meltwater into moulin in still slowly moving ice near margin.
b. Kame plateau: flat-topped sand/gravel; usually lacustrine silts on top;
surrounded by steep ice-contact slopes and collapse structures; may be pitted; higher than adjacent deposits (I.C.3.d.).
c. Kame terrace (valley-side or lateral terrace): flat-topped but often
kettled; non-matching, step-like risers, generally on lower slopes; chaotic
ice-disturbed bedding, made against wasting ice lobe.
d. Kame delta (marginal **ose**): pitted but gently sloping top, apex channel
hanging atop ice-contact slope; mostly steep ice-contact sides; gravel top-
sets, long foresets, lobate into temporary pool or sea level.
C. <u>Proglacial meltwater landforms</u> -- (glaciofluvial, glaciolacustrine, and glaciomarine features) adapted to pre-existing land topography on bedrock or earlier
glacial drift; deposits composed of well-sorted and bedded sediments.
1. CHANNELED TOPOGRAPHY (scabland#) -- scoured by fluvial erosion, by
catastrophic and/or seasonal deluges of meltwater overflowing from marginal
meltwater lakes, commonly ice-dammed, shifting in level and locations;
catastrophic drops. These products include:
a. Meltwater gorge (coulée): cliffed bedrock walls, narrow floor, through
former divide, often reversing headwaters of preglacial stream, sometimes
extensive, with plunge pools, rapids and falls; quickly becomes misfit
today.
b. Meltwater spillway (glacial lake outlet, col channel): broad channel cut,
occasionally through ground or end moraine, often marshy with misfit stream
downvalley; occasional potholes.
c. Drainageway (Urstromtäler): broad shallow cut fluvialchannel, braided, with
islands; across till plains, thin outwash plains, lake bottom and sea bottom plains.
d. Giant bar (flood terrace): rounded gravel, cobbles; giant ripples of
boulders/cobbles common.
2. OUTWASH FEATURES (**SANDAR**) -- in front of, on, or even under thin marginal ice;
inwash included; fully loaded fluvial deposition, gently sloping plain,
steepest and highest near ice, slightly concave downvalley; with shallow
dividing or braided glacial channels; alternating sand and gravel beds
parallel to surface; lenticular channel-and-fill structure. May involve:
a. Outwash fan (sandur, sand plain): symmetrical gentle cone, distributary
dividing channels; coarsest gravel at head, finest sand at outer edge;
bedding parallel to surface. Kettle holes may be near apex, inwash
included.
b. Valley train[+]: belt of outwash, closely restricted by valley walls; braided
dry channels, concave and fining downvalley from ice; pitted especially
near head, matching terraces if eroded postglacially.
c. Pitted plain (disintegration plain): same fan or train with kettle holes,
slump structures; presume deposit on top of dead marginal ice sheet or

masses.
d. Outwash terrace (glaciofluvial terrace): series of step-like terraces of glaciofluvial sand and gravel, matching levels on opposite valley walls, representing decrease in surface (cut and fill series) during retreat of glacial ice or successive lesser ice advances; slope more gently the lower the terrace, so converge downvalley; beds parallel surface in each one; highest beds tend to be coarsest.
3. GLACIOLACUSTRINE FEATURES -- "glacial" if lake extended outward from glacier ice; largely meltwater fed; shore erosion and lacustrine deposition filling silt-clay into valleys, basins, and lower slopes. Thin over rises, coarser sand and mound shapes near shores and meltwater sources. These may include:
a. Outwash delta (ice-marginal delta): glaciofluvial with subaquatic flows wherever a loaded outwash river reaches an open lake; very gently dipping gravel topset (coarsest); foreset dipping structure steeper and finer: some hanging apex channels; top slope gently fanning out; lobate ice-free outer edge.
b. Subaquatic outwash (subwash): overlaid irregular lobate mounds, poorly sorted sand-to-pebbles diamicton, lenticular current beds arched; believed to stem out from underwater ice-cliff tunnel. Some eskers (see II.A.2) may be formed at underwater tunnel mouth below ice cliff.
c. Lacustrine strandline (glacial shoreline): bench on slope from any long-lasting glacially controlled water level; shore rises above bottom deposits, either (1) mostly erosional bench in till with lag boulders (if steep), may be cliffed; or (2) depositional, sand, bar-spit system (if gentle). Often later modified to dune belt. Many tilted by late postglacial rebound.
d. Lake-bottom plain (glaciolacustrine plain, varved plain#): undulating valley fill, bottom silt and clay in very well sorted layers (rhythmites, varves) filling deep areas; assume quiet water settling or massive mixing due to shallow bottom and density fluctuations; often with berg-rafted boulders or clusters of drop stones; cold fresh-water fossils if any.
4. GLACIOMARINE FEATURES -- "glacial" if glacier extends into sea; "glacial" as far as glacial sediments dominate; a, b, c similar to glaciolacustrine features: -- silt, clay, fine sand, often mixed in deformed and sheared layers; suggests ice grounding and shallow sea. Often a postglacially tilted shoreline.
a. Sea-bottom plain; massive silt-clay layers with lenses of subaquatic till; rhythmites rare, and rare or local marine fossils; dropstones and berg-marks suggest that ice floating, organic content increases near shore around large inland seas.

*Genetic Classification of Glacigenic Deposits, Goldthwait & Matsch (eds)*
© *1988 Balkema, Rotterdam. ISBN 90 6191 694 1*

# Some remarks on: Classification of glacial morphologic features, by R.P.Goldthwait

Władysław Niewiarowski
*Institute of Geography, Nicholas Copernicus University, Toruń, Poland*

Every attempt at classification of glacigene landforms is a complex problem because of the great diversity of these forms. Their genesis is not always well understood; hence descriptive terms are still in use. Apart from landforms shaped by one agent, there are landforms of more complicated genesis, for example, those created by a glacier and its meltwaters (tunnel valleys). There are transitional landforms as well, such as crevasse fillings and kame moraines. Well developed glacigene landforms exist in some glaciated areas, but are poorly developed or completely absent in others. Moreover, our knowledge about the origin of such landforms has been accumulating for at least the last hundred years or so, and the meaning of traditional terms has also changed continuously. Obviously, classification criteria may vary and every classification summarizes our developing knowledge at each stage. Thus, as knowledge about glacial processes improves, there is always the need to modify and improve classifications even more.

The classification by R.P. Goldthwait is a detailed one, based on genetic criteria that take into account the main agents and processes forming both direct glacial landforms and also indirect glacially-induced landforms, their morphologic expression, lithology, and relation to ice -- subglacial, icemarginal, or proglacial. However, despite the overall concern for detail in his classification, some glacigene landforms important in glacial morphology have been omitted.

1. The most striking omissions among ice marginal meltwater forms (II,B) or proglacial landforms (II,C) are ice-marginal streamways (German "Urstromtäler" and Polish "pradoliny"). They are huge valleys -- several hundred or even thousand kilometers long and several kilometers wide -- formed solely by glacial meltwater or with some contribution of discharge from extraglacial areas. Such streamways formed a complex network on the Polish-German Lowland along the periphery of Scandinavian inland ice, more or less parallel to the ice margin. They are also found on the Russian Plain and other areas and are a distinctive element among glacigene landforms. [Editor: This is now rectified by addition of II,C,1,C "Drainageway."]

2. "Glacial terminal basins" (German: Stamm-Zweig-Zungenbecken) should also be considered among "direct glacial landforms" (I), subglacial landforms (A). They are rather smooth, broad depressions, eroded in bedrock by glaciers or in the marginal part of inland ice lobes. Very commonly, they are surrounded from the proximal side by end moraine arcs. Their deepest parts are generally occupied by lakes. They can be found in mountain regions, for instance in the Alps, as well as in lowlands, e.g., the Polish-German Lowland.

3. Goldthwait's classification includes only a few of the so-called "ice-pressed forms" such as squeezed-up esker. In my opinion it is necessary to distinguish at least the following categories: a) "ice marginal landforms" (I,C): squeeze-up end moraines -- ridges and hummocks formed in front and along the ice margin as a result of ice pressure, built of deposits of different origin, mostly clays, tills, and silts, with folded and diapir structures. They differ in genesis and structure from push end moraines; b) "kame fields" (II,B,2): squeeze-up kames -- kames (hillocks, hills, and ridges) with squeeze-up internal structure; disturbed kame deposits deformed by the pressure of surrounding stagnant ice. This type of kame is known from areas of northern Germany, Denmark, and Poland.

4. In my view, each classification of glacigene landforms should distinguish the landforms on the basis of melting out of glacier dead ice (kettles). This has also been omitted.

*Genetic Classification of Glacigenic Deposits, Goldthwait & Matsch (eds)*
© *1988 Balkema, Rotterdam. ISBN 90 6191 694 1*

# Spectrum of constructional glaciotectonic landforms

James S. Aber
*Emporia State University, Kans., USA*

ABSTRACT: Constructional glaciotectonic landforms include a variety of hills and plains
that are composed wholly or partly of deformed bedrock or drift masses. These landforms
are classified into five ideal genetic types: 1) hill-hole pair, 2) large composite-
ridges, 3) small composite-ridges, 4) cupola-hill, and 5) flat-lying megablock.
Transitional or mixed landforms exist between these types; nonetheless, the five types
are distinctive and common enough to justify special recognition of each as an end-
member class. The typical characteristics of each type are illustrated with examples
from Canada and Denmark.

## 1 INTRODUCTION

Constructional glaciotectonic landforms
comprise a variety of types, including
hills, ridges, buttes, and plains, which
are composed wholly or partly of pre-
existing soft bedrock or drift masses that
were deformed or dislocated by glacier-ice
movement (Aber 1985). Although many kinds
of glaciotectonic landforms have been
described, classification and regional
mapping of these forms have only been
attempted recently (Moran et al. 1980).
Clayton, Moran, and Bluemle (1980)
classified ice-shoved hills of North
Dakota into three types: hill-depression
forms, transverse-ridge forms, and irregular forms. This classification is significant because the hill-depression form is
recognized as the fundamental type of glaciotectonic landform. Other distinctive
types of glaciotectonic hills and plains
are found both in North America and in
northern Europe.

An expanded classification for constructional glaciotectonic landforms includes
five types (Aber 1987): 1) hill-hole pair,
2) large composite-ridges, 3) small
composite-ridges, 4) cupola-hill, and 5)
flat-lying megablock. These classes represent ideal genetic types within a continuous spectrum of glaciotectonic landforms.
Intermediate, transitional, or mixed landforms exist between or among these types.
Nonetheless, the five types are common
enough to justify special recognition of
each as an end-member class. One case

example is selected to illustrate the
typical or salient characteristics of each
type. The examples come from western
Canada and southeastern Denmark.

## 2 HILL-HOLE PAIR

The hill-hole pair is perhaps the simplest
and most instructive type of glaciotectonic landform. It represents a basic combination of ice-scooped basin and
ice-pushed hill. Other types of ice-shoved
ridges and ice-scooped depressions are
variations of the basic hill-hole theme.
The association of individual ice-shoved
hills with discrete source depressions was
first described from Denmark by Jessen
(1931).

Pairing of anomalous hills and up-ice
depressions in central North Dakota was
noted more recently by Bluemle (1970) and
Clayton and Moran (1974), who correctly
recognized the glaciotectonic origin of
the hill-hole pairs. In the past, these
hills were often misidentified as kames or
outliers of bedrock, depending on their
internal composition. Hill-hole pairs are
now widely recognized.

Bluemle and Clayton (1984:284) described
the hill-hole pair as "... a discrete hill
of ice-thrust material, often slightly
crumpled, situated a short distance
downglacier from a depression of similar
size and shape." The hill and associated
depression are usually next to each other,
but may be separated in some instances by

as much as 5 km. Both pre-existing drift or bedrock may be involved in the dislocated hills.

The depression represents the source of material in the hill. Depressions today are often the sites of bogs, lakes, bays, estuaries, or simply low spots in the surrounding topography. In some cases, the source depression for a hill cannot be identified. The depression may actually exist but is hidden by younger sediment cover or is under a large lake or sea. It can be demonstrated in a few situations that the dislocated mass was a pre-existing hill that was simply moved to a new location (Moran et al. 1980). On the other hand, anomalous depressions without associated hills are also known in several regions (Ruszczyńska-Szenajch 1976, 1978). The sizes of hills and related depressions vary from <1 to >100 square km, and many hill-hole pairs of different sizes are often found in close proximity to each other.

Wolf Lake is situated in east-central Alberta, near the Saskatchewan border (Fig. 1). A large hill, here called "Wolf Hill," of ice-shoved material is located immediately south of the lake. This hill reaches a maximum elevation >150 m above the normal level of Wolf Lake. To the west of Wolf Lake, smaller, streamlined, ice-shoved hills, drumlins, and flutes are well developed. Wolf Lake and Wolf Hill in combination thus represent a large hill-hole pair in a complex setting (Fenton and Andriashek 1983). The distinct morphologic expression alone makes this an outstanding example, even though almost nothing is known concerning its internal structures or materials.

Wolf Lake and Wolf Hill are both shaped approximately as similar, aligned parallelograms, 7.5 km long east-west. The alignment of the lake and hill is emphasized by a conspicuous, southwest-trending lineament formed by the straight eastern edge of Wolf Lake and the straight eastern flank of Wolf Hill. This Wolf Lake lineament is more than 11 km long and displays 135 m of total topographic relief. Fenton and Andriashek (1983) have interpreted the eastern as well as western boundaries of Wolf Lake as major strike-slip faults. The boundaries of the ice-scooped depression are essentially tear faults along which material from Wolf Lake was laterally and vertically displaced into Wolf Hill.

Wolf Hill consists of a single, east-west trending mound with more-or-less symmetrical, steep northern and southern flanks and a rounded crest. A belt of narrow, parallel ridges occupies the eastern portion of the hill's northern flank. Such ridges are the most typical morphologic trait of ice-shoved hills. More subdued ridges are also present in western parts of the hill. The ridges indicate that the hill probably consists of imbricately stacked thrust blocks.

All the morphologic features in the Wolf Lake vicinity are consistent with south-westward ice movement during thrusting and streamlining of the terrain. Whether these features were all formed during a single ice advance or were formed and modified during more than one advance is not known. The northeasterly alignment at Wolf Lake is consistent with the regional pattern of ice movement emanating from the District of Keewatin. The physical conditions, such as permafrost development or ice thickness, that existed at the time of thrusting are unknown.

Moran et al. (1980) concluded that thrusting of this and other hills on the North American Prairies took place in a narrow frozen-bed zone (2-3 km wide) at the ice margin during glacier advance, while streamlining occurred farther back in a thawed-bed zone. Streamlined morphology was locally overprinted on thrust morphology when the thawed-bed zone migrated outward.

The model for glaciotectonic pushing by "block-movement" of the glacier front may also apply to the Wolf Lake situation (Stephan 1985). Pushing of large masses occurred in front of individual ice blocks. Conversely, smaller ridges were pushed parallel to block side-margins, which could also be the sites of strike-slip faulting. An en-echelon pattern of ice-shoved hills connected at right angles by tear faults or elongated drumlins is the ideal result.

3 LARGE COMPOSITE-RIDGES

The most typical and distinctive glacio-tectonic features are ice-shoved ridges found in many glaciated plains. Prest (1983:45) aptly described such ridges as "... a composite of great slices of up-thrust and commonly contorted sedimentary bedrock that is generally interlayered with and overlain by much glacial drift." A great many terms have been used in different languages for ice-shoved ridges; among the more popular are: Stauchmöranen, push-moraine, ice-thrust moraine, transverse ridges, ice-thrust ridges, etc. (Kupsch 1962). The terms "composite-ridges" or "ice-shoved ridges," which carry no structural connotation and do not include the often misused word moraine,

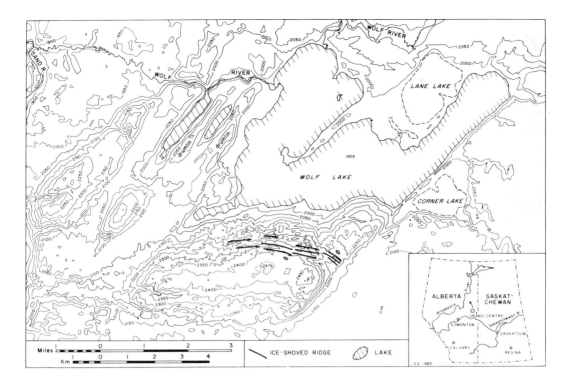

Fig. 1. Topographic map of Wolf Lake vicinity, Alberta. Position of ice-shoved ridges based on analysis of aerial photographs. Elevations in feet; contour interval = 50 feet (app. 15 m); perennial lakes shown by diagonal lining; ephemeral lakes shown by dashed outline; F = fire watch station.

are preferable.

Ice-shoved ridges are here divided into large (>100 m relief) and small (<100 m relief) categories. Large composite-ridges may be up to 200 m high, 5 km wide, and 50 km long. In map view, composite-ridges are often arcuate, concave up-ice with a radius of curvature of 2 to 10 km (Clayton, Moran, and Bluemle 1980). The arcuate pattern outlines the margin of the ice lobe or tongue that shoved up the ridges.

The ridges are developed on the crests of folds or the upturned ends of thrust blocks. A close correspondence typically exists between structural features and topography. Large composite-ridges usually involve considerable disruption of pre-Quaternary bedrock, which may comprise a substantial volume of the ridges. However, the depth of structural disturbance is generally not greater than 200 m (Kupsch 1962).

The folds and thrust blocks that form ridges have usually been detached, transported some distance, and stacked up in an imbricated structure. Composite-ridges are, thus, allochthonous in a glaciotectonic sense, and it may be possible to recognize the up-ice source from whence material in the ridges was derived.

Large composite-ridges are topographically and structurally similar to such thrust and folded mountain belts as the Canadian Rockies or Swiss Juras, which were formed by thin-skinned tectonics. The only real difference is size, ice-shoved ridges being an order of magnitude smaller than true mountains. Those geologists experienced in both structural and glacial geology have been impressed by the resemblance between orogenic hard-rock tectonics and glacial tectonics (Berthelsen 1979; van der Wateren 1985). Large composite-ridges may thus be viewed as natural scale-models of real mountains.

The large chalk cliffs of eastern Moen, southeastern Denmark, are justifiably famous for their scenic beauty and distinctive geological structure (Fig. 2). The cliffs are perhaps the finest example of glaciotectonic features in the world,

Fig. 2. Photograph looking north along Moens Klint. Sommerspiret is the vertical chalk mass, standing 102 m high, in the center. Photo by J.S. Aber, 1986.

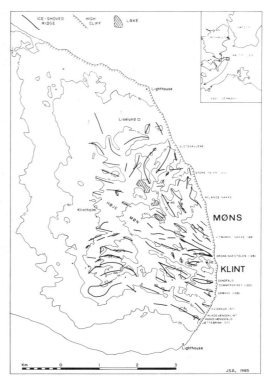

Fig. 3. Topographic map of eastern Moen showing ice-shoved ridges. Individual chalk cliffs and interchalk falls are indicated with heights of chalk cliffs given in parentheses. Contour interval = 25 m; Ht. = hotel. Position of ice-shoved ridges based on interpretation of aerial photographs. Taken from Aber (1985, Fig. 1).

and they have a long history of geological study (Agricola 1546; Puggaard 1851; Johnstrup 1874; Slater 1927; Hintze 1937; Haarsted 1956).

The high, rugged terrain known as Hoeje Moen, located at the eastern end of the island, is composed of several dozen chalk masses that were piled up during Weichselian ice advances 20,000 to 13,000 B.P. (Sjoerring 1981). Hoeje Moen generally exceeds 75 m elevation, reaching a high at 143 m (Fig. 3). Presumably undisturbed chalk bedrock is intersected in drill-holes at -20 to -40 m, indicating up to 180 m of structural relief.

The individual chalk bodies exposed in Moens Klint consist of upper Cretaceous (Maastrichtian) white "writing chalk" conformably overlain by two tills and stratified drift. The chalk and drift were thrust and deformed together as single, displaced masses. The upper dislocated till is dominated by erratics from the northeast, whereas the lower dislocated till has a Baltic composition (Hintze 1937). In places a younger, discordant drift is also deformed.

The chalk is very uniform in lithology; thus, stratigraphic correlation between chalk masses is difficult at best. Surlyk (1971) divided the Danish Maastrichtian into ten brachiopod biozones: zones 1 through 7 are lower Maastrichtian and 8 to 10 are upper Maastrichtian. These biozones are the best method for establishing correlation between chalk bodies.

Moens Klint and Hoeje Moen can be divided into three "morphostructural" regions on the basis of ridge morphology, cliff structures, and chalk stratigraphy. The southern region includes Jaettebrink through Sommerspiret (biozones 3, 4, and 5). The chalk masses form a series of imbricately thrust anticlines that dip southward and are increasingly deformed toward the north. These chalk masses continue inland as long, straight to arcuate ridges, concave toward the south. This region was thrust by ice movement directly from the south.

Dronningestolen along with Graederen and Maglevandspynten make up the central portion of the cliff. Dronningestolen is a huge composite of many lesser chalk floes (biozones 3 through 8) folded and stacked on top of each other in the overall form of a broad anticline. Dronningestolen continues inland as a massive ridge, beyond which the central region is marked by many short, offset ridges. The central region was apparently deformed by ice pushing from both north and south.

The northern region of Vitmunds Nakke through Slotsgavlene includes biozones 5 through 8 in chalk bodies oriented oblique to the coast. A regular shift is displayed in structural strike along the cliff from southeast in the south, to south in the center, to southwest in the north. This corresponds to the arcuate pattern of ridges inland from the cliff. Thrusting of the northern region was evidently brought about by ice advance from the east or northeast.

The Weichselian glaciation of Denmark occurred in four distinct phases separated by brief ice retreats (Berthelsen 1978; Houmark-Nielsen 1981): 1) Norwegian advance from the north, 2) Old Baltic advance from the southeast, 3) main Swedish advances from the northeast, and 4) Young Baltic advances from the southeast or south. All these advances covered Moen, except the Norwegian advance, and the main Swedish advance experienced two significant readvances during its retreat (Houmark-Nielsen 1986).

The upper and lower dislocated tills at Moens Klint are ascribed respectively to the main Swedish and Old Baltic advances. The overlying discordant drift probably relates to the Young Baltic advances. Thrusting of the southern region of Hoeje Moen was the final glaciotectonic disturbance and must be associated with the Young Baltic advances. Earlier thrusting of the northern region is related to either the initial Swedish ice advance or its first readvance.

## 4 SMALL COMPOSITE-RIDGES

Small composite-ridges are perhaps the most common type of glaciotectonic landform. They display the same morphologic traits and structural features as do large composite-ridges. Small composite-ridges are also found in similar circumstances, such as escarpments, islands, or other topographic obstacles to ice movement. A source depression is located a short distance up-ice in some cases. Finally, both small and large composite-ridges are usually associated with ice margins marking stillstands or readvances.

The size division between large and small composite-ridges is somewhat arbitrary, but one significant difference does exist. Large composite-ridges generally incorporate a considerable volume of pre-Quaternary bedrock that is consolidated to some degree. Small composite-ridges may or may not include such bedrock; in fact, many are composed largely of unconsolidated Quaternary strata. Being more susceptible to both glacial and nonglacial erosion, such ridges cannot maintain a high topographic relief. Thus the primary genetic difference between large and small composite-ridges is incorporation of consolidated bedrock rather than any difference in mode of glacier thrusting.

Small composite-ridges may be difficult to identify as glaciotectonic landforms, and have often been mapped as end moraines. Many so-called end moraines are now recognized to consist partly or wholly of ice-shoved material (Moran et al. 1980). Where disturbed bedrock is present, the glaciotectonic origin of such moraines is obvious. However, the absence of deformed bedrock does not preclude ice pushing as the primary means of constructing certain end moraines. Recognition of this situation may prove troublesome, however, as Moran (1971) pointed out, particularly if the displaced drift masses are similar in lithology to the enclosing drift.

Brandon Hills are a group of subparallel drift-cored ridges (Fig. 4), located at the northern end of Tiger Hills upland in southwestern Manitoba. Brandon Hills are part of the Manitoba Escarpment, located near the eastern edge of the Saskatchewan Plain. This escarpment rises abruptly above the lower Manitoba Plain to the east, and is marked by a series of uplands, which are cored by Upper Cretaceous shales of the Riding Mountain Formation. Bedrock of southwestern Manitoba is almost completely mantled by thick drift consisting mainly of till. Stratified drift is abundant on the uplands, where ice stagnated during deglaciation.

The glacial geomorphology and stratigraphy of southwestern Manitoba have been described in several recent investigations (Klassen 1975, 1979; Fenton et al. 1983; Fenton 1984). Brandon Hills have usually been mapped as part of the end moraine atop Tiger Hills upland (Prest et al. 1967). However, Welsted and Young (1980) questioned designation of Brandon Hills as an end moraine, because they found that much of the hills consists of stratified drift. They also considered, but rejected, the possibility of glacier thrusting, apparently because deformed bedrock is not

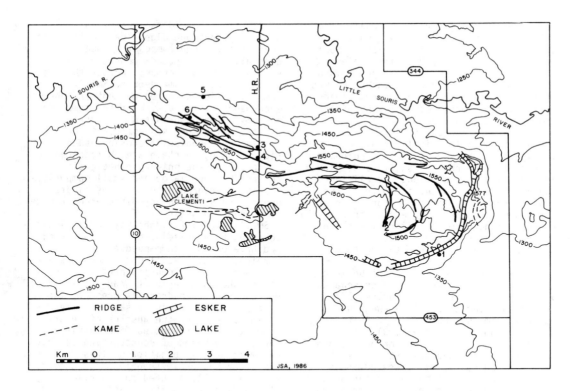

Fig. 4. Topographic map of Brandon Hills vicinity, Manitoba, showing glacial features. Based on interpretation of aerial photographs. Provincial highways shown by numbered circles; H.R. = Hydraulic Road. Elevations in feet; contour interval = 50 feet (app. 15 m). Locations of exposures shown by solid dots.

present. However, the stratified drift is strongly disturbed in certain exposures. Brandon Hills are, in fact, ice-shoved ridges consisting entirely of drift (Aber 1987).

Brandon Hills occupy a rectangular area roughly 10 km east-west and 4 km north-south (Fig. 4). The hills rise up to 100 m above the Little Souris River valley imme-diately to the north. Brandon Hills include three distinct morphologic types: composite-ridges, esker ridges, and kame-and-kettle moraine. Closely spaced, sub-parallel, composite ridges make up the northwestern, central, and eastern por-tions of Brandon Hills. The ridges are covered by a thin veneer of till, and they resemble a giant fishhook in overall plan view.

A high, single esker ridge meanders over the eastern end of Brandon Hills. This ridge extends with a couple of interrup-tions to the southwest and then northwest to form a large loop around the eastern half of Brandon Hills. The esker ridges

are composed predominantly of sand and gravel, and rest on a continuous substra-tum of till (Welsted and Young 1980). The southwestern flank of Brandon Hills con-sists of low kames and lake-filled kettles such as Lake Clementi.

The internal structure of Brandon Hills is fairly simple at most sites, consisting of 2 to 3 m of stratified sand and gravel overlain by 1 to 2 m of till. The till is banded and consists mainly of material eroded from the underlying stratified drift. Deeper structures are exposed at site 6, a large gravel pit cut into a northwest-trending ridge. Pit walls reveal two dislocated stratified drift units, each 15 m thick, which are faulted together, tilted southwestward, truncated and capped by 1 to 4 m of discordant till. Various associated thrust, reverse, and normal faults are in complete agreement; all strike west-northwest subparallel with the ridge.

The creation of Brandon Hills is most easily explained by a single ice advance, which pushed up ridges consisting of displaced masses of older stratified drift. Overriding ice then truncated the ridges and deposited a veneer of till composed largely of material reworked from the ridges themselves. Upon ice stagnation, a subglacial tunnel system developed in which meltwater laid down the esker ridge on top of till. The association of eskers with ice-pushed hills is a common one, as documented by Bluemle and Clayton (1984) in North Dakota. Kames and patchy ablation drift were finally deposited as the ice melted away. This sequence of events took place during the last glacial advance to reach Tiger Hills vicinity.

The Lostwood (late Wisconsin) glaciation completely covered Manitoba and reached far southward into the United States (Fenton 1984). Deglaciation was accomplished by stagnation of large areas of marginal ice, alternating with readvances or stillstands of active ice. The Assiniboine Sublobe of the Northeastern Ice was responsible for the final readvance--Marchand phase--into Tiger Hills (Klassen 1975).

Brandon Hills were created by this final glacier advance in material that was probably thawed and saturated with meltwater at the time of deformation. The existence of large proglacial lakes and wasting stagnant ice masses throughout this region during late Wisconsin deglaciation argues against permafrost as a factor in the glaciotectonic genesis of Brandon Hills.


5 CUPOLA-HILL

Many conspicuous hills, both small and large, have the general characteristics of ice-shoved masses, but lack the hill-hole relationship or the typical ridged morphology. Bluemle and Clayton (1984) placed such hills into the category of irregular hills that Clayton, Moran, and Bluemle (1980:46) defined as "... an irregular jumble of hills with no obvious transverse ridges and no obvious source depression." The glaciotectonic origin of such hills can only be proven with evidence of subsurface deformation of bedrock or drift, although such hills may be suspected from other evidence.

Perhaps the most common form of these irregular hills is the type that Smed (1962) first called cupola-hill (kuppelbakke in Danish). Cupola-hills have an internal structure similar to composite-ridges; however, unlike composite-ridges their morphology was substantially modified by the action of overriding ice. Cupola-hills possess three

basic attributes (Smed 1962):
1. interior structure -- thrust glacial and interglacial deposits, plus floes of pre-Quaternary strata;
2. external form -- long, even hill slopes with overall domelike morphology, unlike marginal moraines; varying from near circular to elongated ovals in form, 1 to 15 km maximum length, most 5 to 10 km long, from about 20 to >100 m high;
3. discordant till -- overridden by ice that truncated deformed structures and laid down a capping till layer.

The hills north of the village of Hjelm, including Hvideklint on the southern coast of Moen, Denmark are an ideal example of a cupola-hill (Fig. 5). At Hvideklint, upthrust floes of Upper Cretaceous, Campanian, and lower Maastrichtian white chalk, covered and separated by deformed drift, form a cliff nearly 1 km long and up to 20 m high. Undisturbed bedrock underlies southern Moen at elevations of -25 to -35 m and presumably consists of chalk. At the surface, deformed Quaternary strata and allochthonous chalk floes are displayed. Hvideklint has been examined by several geologists (Haarsted 1956; Berthelsen et al. 1977; Aber 1979; Berthelsen 1979).

The cupola-hill north of Hjelm has an oval outline and contains many irregular small hillocks and closed depressions (Fig. 5). It rises to a maximum elevation of 44 m at Bavnehoej. Unlike the higher cliffs of eastern Moen, the chalk masses here do not form distinct ridges inland from the cliff. Only one subdued, linear ridge, Glinsebanke, is present. Chalk is exposed at several places on Glinsebanke, so it is probably the upturned end of a thrust block. There are otherwise no obvious linear trends in the cupola-hill morphology; land slopes are moderate throughout the hill.

Three tills along with associated stratified drift are exposed at Hvideklint (Fig. 6). A discordant till overlies upper and lower dislocated tills. The drift covers detached floes of chalk in which chalk breccia and chalk-till melange are locally developed. The discordant and upper dislocated tills are generally the same, characterized by a large amount of local material and about equal contents of crystalline and Paleozoic types typical of northeastern derivation. Conversely, the lower dislocated till has noticeably less chalk and more Paleozoic limestone, indicative of a Baltic source.

Hvideklint displays a progressive increase in deformation from a minimum at the northeastern end to a maximum toward the southwest. Thus, the cliff is con-

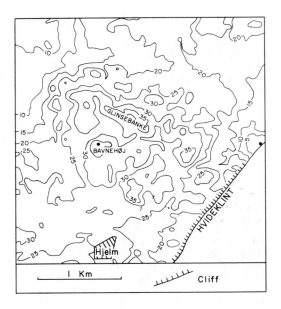

Fig. 5. Topographic map of cupola-hill at
Hvideklint. Contour interval = 5 m; small
closed depressions not shown. Small dot at
right edge indicates starting point of
measured section (Fig. 6) at fishing
hamlet.

veniently divided into four portions (Fig.
6):
   1. Eastern chalk floe (200-410 m) -- A
slightly deformed chalk mass, tilted
toward the southwest, is overlain continuously by discordant and upper dislocated tills. Small thrusts and brecciated
zones are present within the chalk, particularly near its southwestern end.
   2. Central drift sequence (410-640 m) --
Discordant and upper dislocated tills
carry across this portion of Hvideklint.
Large bodies of deformed stratified drift
are found near the base of the upper
dislocated till, and no solid chalk is
present.
   3. Central chalk floe (640-855 m) -- The
largest single chalk body at Hvideklint
forms a broad anticline in which a conjugate thrust system is developed. The
chalk is thrust over isoclinally folded
and sheared stratified drift at its southwestern end.
   4. Western chalk floes (855-1080 m) --
Highly disturbed chalk and drift display
extreme stretching and thinning toward the
southwest. Penetrative deformation has
caused intermingling of chalk and drift in
isoclinal folds and shear bands. A body of
lower dislocated till is enclosed by

deformed chalk near the eastern end of
this portion (860-890 m).
   The orientations of glaciotectonic
structures are consistent along the cliff
section. Fold axes are mostly near horizontal, and all trend either southeast or
northwest. Thrust and normal faults also
strike northwest-southeast. These structural data correspond to southwestward ice
movement at Hvideklint. The orientation
data are clustered in two groupings: 1)
100 to 140 degrees and 2) 160 to 165
degrees. This suggests that there were
actually two episodes of structural
disturbance associated with ice advances
from about N30E and N70E. The chalk
megablocks were presumably derived from
the seafloor somewhere north of Moen.
   Southern Moen was subjected to the same
Weichselian ice advances that affected
eastern Moen. The discordant and upper
dislocated tills relate to the main
Swedish phase. The upper dislocated till
was probably deposited by the initial
Swedish advance, which also transported
the chalk masses and constructed the cupo-
lahill. The discordant till was then deposited by the first Swedish readvance,
which caused some additional structural
disturbance and smoothing of the hill.
Hvideklint was apparently little altered
by subsequent Young Baltic advances. The
lower dislocated till is problematic; it
could be the Weichselian Old Baltic Till
or it could relate to a still older
Saalian advance.

## 6 FLAT-LYING MEGABLOCKS

The terms floe, raft, scale, and megablock
have all been used for large masses of
glacially dislocated and deformed bedrock
and drift. The common occurrence of flat-
lying megablocks throughout glaciated
plains was not appreciated until fairly
recently (Stalker 1976; Sauer 1978). These
megablocks are more-or-less horizontal,
slightly deformed, and are often buried
under or within thick drift giving little
or no morphologic clue to their presence
in the subsurface. They may, in fact,
easily be mistaken for bedrock, if deep
exposures or drilling logs are not
available. In other cases, the megablocks
form flat-topped buttes, small plateaus,
or irregular hills, which have also been
mistaken for bedrock outliers.
   Most megablocks are composed of poorly
to moderately consolidated Mesozoic or
Cenozoic sedimentary strata, but some consist of harder bedrock or unconsolidated
drift. All megablocks exhibit some signs
of deformation--shear zones, folds,

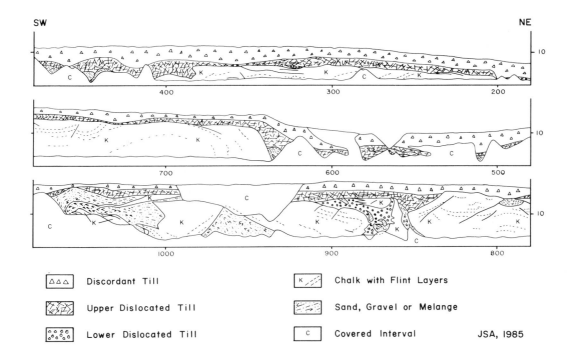

SW                                                                      NE

| △△△ | Discordant Till | | K ⟋ | Chalk with Flint Layers |
| ⧄⧄ | Upper Dislocated Till | | ⫶⫶ | Sand, Gravel or Melange |
| °₀°₀° | Lower Dislocated Till | | C | Covered Interval      JSA, 1985 |

Fig. 6. Measured profile of Hvideklint section as it appeared in 1979. Base of section at beach level. Note 2X vertical exaggeration slightly distorts geometry. Scale in m; adapted from Aber (1979, Fig. 3).

faults, brecciation--as a result of ice pushing. This criterion distinguishes megablocks from large, undeformed erratic blocks (Aber 1985). The sources of many megablocks are unknown or poorly iden- tified, but most were probably transported only a few km. Some megablocks were moved as much as 300 km, however (Ruszczyńska- Szenajch 1976).

All such megablocks have one trait in common; they are remarkably thin (typically <30 m) compared to their lateral dimensions (often >1 square km), although some are much larger. They are, in effect, "rock pancakes," which could only have been transported by freezing onto the underside of a glacier. The fact that such megablocks are scattered over broad regions behind ice-margin positions supports a subglacial origin for many.

During the course of geological and groundwater investigation in eastern Saskatchewan, a truly enormous megablock of Cretaceous shale was discovered at Esterhazy (Christiansen 1971). The megablock covers an area roughly 1000 square km in extent (Fig. 7). The area is part of the Saskatchewan Plain, a drift-

mantled region of relatively low relief underlain by soft Cretaceous sedimentary bedrock. The monotonous plain is broken by a large spillway channel, the Qu'Appelle Valley, which cuts through the middle of the megablock. The Qu'Appelle Valley is >100 m deep and about 2 km wide. Its bot- tom is drift filled, and bedrock outcrops are found along its walls.

Bedrock in eastern Saskatchewan is generally undeformed, dipping gently toward the south (Fig. 8). The Cretaceous formations are mostly clastic strata con- sisting mainly of siltstone, claystone, and shale. Hard, siliceous shale of the Odanah Member, Riding Mountain Formation is more resistant than other strata and forms the main mass of the megablock.

Highly folded and faulted bedrock of the Riding Mountain Formation is encountered west of Hazel Cliff, east of Tantallon, and south of the Qu'Appelle Valley in both surface exposures and drill holes. Breccia, slickensides, and mylonite are common microstructures. This deformation is confined to bedrock above about 1500 to 1600 feet elevation. The general plan of the megablock is an egg-shaped oval,

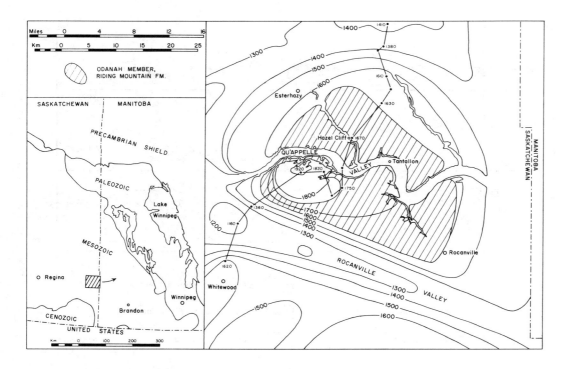

Fig. 7.  Bedrock contour map of Esterhazy vicinity, eastern Saskatchewan. The 1600-foot contour between Esterhazy and Rocanville defines position of megablock. Small dots show test wells used in constructing cross section (Fig. 8). Elevations of bedrock surface in test wells shown in feet; contour interval = 100 feet (approximately 30 m). Adapted from Christiansen (1971).

roughly 38 km long and 30 km wide. Bedrock in the west-central portion reaches 1920 feet elevation, indicating a maximum thickness of about 100 m. The megablock is thus at least 300 times wider and 380 times longer than it is thick.

A test well drilled near the western end of the megablock intersected 2 m of till after penetrating 80 m of brecciated and mylonitic bedrock. Similarly disturbed rock material rests directly on undeformed bedrock south of the Qu'Appelle Valley. The actual decollement zone is located in Riding Mountain Formation claystone below the Odanah Member, and some claystone above the Odanah Member is also part of the megablock. Assuming an average thickness of 60 m and an area of 1000 square km, the megablock's volume is estimated at 60 cubic km.

Ice flow in this part of Saskatchewan was generally from the northeast; however, neither the direction of emplacement nor the source of the megablock are known. The megablock may not have moved far (perhaps less lateral displacement than its own width) in order to produce the observed structures. The megablock is situated on the northern edge of a major buried glacial valley, the Rocanville Valley. Meltwater erosion and filling of the valley presumably took place during an earlier (pre-late Wisconsin) glaciation. It appears that the Rocanville Valley truncates the southern side of the megablock; thus, emplacement of the megablock must predate cutting of the Rocanville Valley.

The only conceivable means of displacing a megablock of such size was by freezing onto the bottom of an overriding ice sheet, in which case the megablock became the basal layer of the ice sheet. It is highly improbable, given its dimensions, that this megablock could have been pushed in front of an advancing glacier, whether it was permafrozen or not. Subglacial sliding of a permafrozen slab over a thawed substratum seems to be the most likely explanation for displacement of the Esterhazy megablock.

290

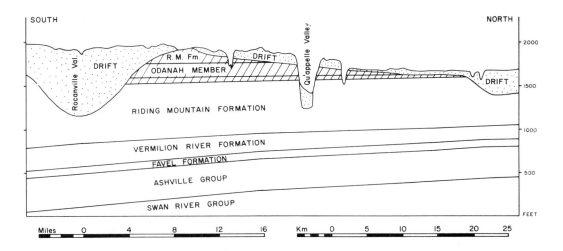

Fig. 8. Subsurface cross section showing megablock (diagonal lining) of Riding Mountain Formation at Esterhazy. Vertical exaggeration = 42X; adapted from Christiansen (1971).

ACKNOWLEDGEMENTS

Financial support for field work during 1986 in the Brandon Hills area, Manitoba, was provided by a research fellowship from Emporia State University. For discussions and information concerning case examples, I thank A. Berthelsen, E. Christiansen, M. Fenton, M. Houmark-Nielsen, and R. Klassen. A. Dreimanis and M. Fenton reviewed the manuscript and suggested several improvements.

REFERENCES

Aber, J.S. 1979. Kineto-stratigraphy at Hvideklint, Moen, Denmark and its regional significance. Geological Society of Denmark, Bulletin 28:81-93.

Aber, J.S. 1985. The character of glaciotectonism. Geol. en Mijnbouw 64:389-395.

Aber, J.S. 1987. Classification of constructional glaciotectonic landforms. XII INQUA Congress (Abstract), Ottawa, Canada.

Agricola, G. 1546. De Natura Fossilium. Basle.

Berthelsen, A. 1978. The methodology of kineto-stratigraphy as applied to glacial geology. Geological Soc. Denmark, Bulletin 27, Special Issue, p. 25-38.

Berthelsen, A. 1979. Recumbent folds and boudinage structures formed by subglacial shear: An example of gravity tectonics. In W.J.M. van der Linden (ed.), Van Bemmelen and his search for

harmony. Geol. en Mijnbouw 58:253-260.

Berthelsen, A., P. Konradi, and K.S. Petersen 1977. Kvartaere lagfolger og strukturer i Vestmons klinter. Dansk Geol. Foren., Arsskrift for 1976:93-99.

Bluemle, J.P. 1970. Anomalous hills and associated depressions in central North Dakota. Geological Soc. America, Abst. with Prog. 2:325-326.

Bluemle, J.P. and L. Clayton 1984. Large-scale glacial thrusting and related processes in North Dakota. Boreas 13:279-299.

Christiansen, E.A. 1971. Geology and groundwater resources of the Melville Area (62K, L) Saskatchewan. Saskatchewan Research Council, Geology Division, Map No. 12.

Clayton, L. and S.R. Moran 1974. A glacial process-form model. In D.R. Coates (ed.), Glacial geomorphology, p. 89-119. Binghamton, New York, SUNY-Binghamton Publ. in Geomorphology.

Clayton, L., S.R. Moran, and J.P. Bluemle 1980. Explanatory text to accompany the Geologic Map of North Dakota. North Dakota Geological Survey, Report of Investigation No. 69.

Fenton, M.M. 1984. Quaternary stratigraphy of the Canadian Plains. In R.J. Fulton (ed.), Quaternary stratigraphy of Canada--A Canadian contribution to IGCP Project 24. Geol. Survey Canada, Paper 84-10:57-68.

Fenton, M.M. and L.D. Andriashek 1983. Surficial geology Sand River area, Alberta. Alberta Geological Survey, map scale = 1:250,000.

Fenton, M.M., S.R. Moran, J.T. Teller, and L. Clayton 1983. Quaternary stratigraphy and history in the southern part of the Lake Agassiz Basin. In J.T. Teller and L. Clayton (eds.), Glacial Lake Agassiz. Geol. Assoc. Canada, Spec. Paper 26:49-74.

Haarsted, V. 1956. De kvartaergeologiske og geomorfologiske forhold pa Mon. Meddr. Dansk Geol. Foren. 13:124-126.

Hintze, V. 1937. Moens Klints geologi. Copenhagen, C.A. Reitzels. (Edited posthumously by E.L. Mertz and V. Nordmann).

Houmark-Nielsen, M. 1981. Glacialstratigrafi i Danmark ost for Hovedopholdslinien. Dansk Geol. Foren., Arsskrift for 1980:61-76.

Houmark-Nielsen, M. 1986. Glaciotectonic unconformities as marker-horizons in Pleistocene stratigraphy and their evidence for flow directions of former glaciers, with special emphasis on modelling the last Scandinavian icesheet. INQUA's Working Group on Glacial Tectonics, Field Meeting on Mon, Denmark, Abst.

Jessen, A. 1931. Lonstrup Klint. Danmarks Geologisk Undersogelse, IV raekke 2(8), 26 p.

Johnstrup, F. 1874. Ueber die Lagerungsverhaltnisse und die Hebungsphänomene in den Kreidefelsen auf Moen und Rügen. Z. Zeitschrift Deutsch. Geologische Ges. 1874:533-585.

Klassen, R.W. 1975. Quaternary geology and geomorphology of Assiniboine and Qu'Appelle Valleys in Manitoba and Saskatchewan. Geol. Survey Canada, Bull. 228.

Klassen, R.W. 1979. Pleistocene geology and geomorphology of the Riding Mountain and Duck Mountain areas, Manitoba-Saskatchewan. Geol. Survey Canada, Memoir 396.

Kupsch, W.O. 1962. Ice-thrust ridges in western Canada. Jour. Geology 70:582-594.

Moran, S.R. 1971. Glaciotectonic structures in drift. In R.P. Goldthwait (ed.), Till: A symposium, p. 127-148, Columbus, Ohio State Univ. Press.

Moran, S.R., L. Clayton, R.L. Hooke, M.M. Fenton, and L.D. Andriashek 1980. Glacier-bed landforms of the Prairie region of North America. Journal of Glaciology 25:457-476.

Prest, V.K. 1983. Canada's heritage of glacial features. Geological Survey Canada, Misc. Report 28.

Prest, V.K., D.R. Grant, and V.N. Rampton 1967. Glacial Map of Canada. Geol. Survey Canada, Map 1253A, scale 1:5,000,000.

Puggaard, C. 1851. Moens Klint section. Reproduced in International Geological Congress XXI, Session Norden (1960), Guidebook I.

Ruszczyńska-Szenajch, H. 1976. Glacitectonic depressions and glacial rafts in mid-eastern Poland. In S.Z. Rozycki (ed.), Pleistocene of Poland, vol. L, p. 87-106. Warsaw, Studia Geologica Polonica.

Ruszczyńska-Szenajch, H. 1978. Glacitectonic origin of some lake-basins in areas of Pleistocene glaciations. Polish Archiwum Hydrobiologii. 25(1/2):373-381.

Sauer, E.K. 1978. The engineering significance of glacier ice-thrusting. Canadian Geotechnical Journal 15:457-472.

Sjorring, S. 1981. The Weichselian till stratigraphy in the southern part of Denmark. Quaternary Studies in Poland 3:103-109.

Slater, G. 1927. The structure of the disturbed deposits of Moens Klint, Denmark. Transactions of the Royal Society of Edinburgh 55, part 2:289-302.

Smed, P. 1962. Studier over den fynske ogruppes glaciale landskabsformer. Meddr. Dansk Geol. Foren. 15:1-74.

Stalker, A.M. 1976. Megablocks, or the enormous erratics of the Albertan Prairies. Geological Survey Canada, Paper 76-1C:185-188.

Stephan, H.-J. 1985. Deformations striking parallel to glacier movement as a problem in reconstructing its direction. Geological Soc. Denmark, Bull. 34:47-53.

Surlyk, F. 1971. Skrivekridtklinterne pa Mon. In M. Hansen and V. Poulsen (eds.), Geologi pa oerne. Varv, Ekskursionsforer 2:5-23.

Wateren, D.F.M. van der 1985. A model of glacial tectonics, applied to the ice-pushed ridges in the Central Netherlands. Geological Soc. Denmark, Bull. 34:55-74.

Welsted, J. and H.R. Young 1980. Geology and origin of the Brandon Hills, southwest Manitoba. Canadian Jour. Earth Science 17:942-951.

*Genetic Classification of Glacigenic Deposits, Goldthwait & Matsch (eds)*
*© 1988 Balkema, Rotterdam. ISBN 90 6191 1988 Balkema, Rotterdam, ISBN 90 6191 694 1*

# List of contributors

Aber, J.S., Emporia State University, 1200
Commercial, Emporia, Kansas 66801, U.S.A.

Ashley, G., Department of Geological
Sciences, Rutgers University, New
Brunswick, New Jersey 08903, U.S.A.

Borns, H.W., Jr., Department of Geological
Sciences, University of Maine, Orono,
Maine 04473, U.S.A.

Dreimanis, A., Geology Department,
University of Western Ontario, London,
Ontario, N6A 5BY, Canada

Ellis, K., Ontario Geological Survey,
Ministry of Northern Development and
Mines, 77 Grenville Street, Toronto,
Ontario, M7A 1W4 Canada

Elson, John, Department of Geologic
Sciences, McGill University, Montreal,
Quebec, H3A 2A7, Canada

Goldthwait, R.P., Department of Geology
and Mineralogy, The Ohio State
University, Columbus, Ohio 43210, U.S.A.

Haldorsen, S., Department of Geology,
Agricultural University of Norway,
Aas-NLH, Norway

Hicock, S., Department of Geology,
University of Western Ontario, London,
Ontario, N6A 5B7, Canada

Hirvas, Heikki, Geological Survey of
Finland, SF-02150 Espoo 15, Finland

Juozapavičius, G., Lit NIGRI,
Schevtshenkos 13, Vilnius, 232600,
Lithuanian S.S.R., U.S.S.R.

Jurgaitis, A.A., State University, Dept.
of Hydrogeology and Engineering Geology,
Lithuanian S.S.R., Vilnius, Ciurlionio

21, U.S.S.R.

Kujansuu, R., Geological Survey of
Finland, SF-02150 Espoo 15, Finland

Lawson, Daniel E., U.S.A./C.R.R.E.L., Box
282, Hanover, New Hampshire 03755, U.S.A.

Levson, V.M., Department of Geology,
University of Alberta, Edmonton,
Alberta, T6G 2E3, Canada

Lundqvist, J., Department of Quaternary
Research, University of Stockholm,
Odengatan 63, S-11322 Stockholm, Sweden

Matsch, C.L., Department of Geology,
University of Minnesota-Duluth, Duluth,
Minnesota, 55812, U.S.A.

Mickelson, D.M., Department of Geology and
Geophysics, University of Wisconsin-
Madison, Madison, Wisconsin 53706, U.S.A.

Morawski, W., Institute of Geology, ul.
Rakowiecka 4, 00-975, Warsaw, Poland

Nenonen, Keijo, Geological Survey of
Finland, SF-02150 Espoo 15, Finland

Niewiarowski, W., Institute of Geography,
Nicholas Copernicus University, Torun,
Poland

Parkin, G.W., Department of Geology,
University of Western Ontario, London,
Ontario, N6A 5B7, Canada

Pedersen, S.A. Schack, Quaternary
Division, Geological Survey of Denmark,
Thoravej 31, DK-2400 Copenhagen N1,
Denmark

Raukas, A., Geological Institute, Academy
of Sciences of Estonian S.S.R., Estonia
pst. 7, 200101 Tallin, Estonia, U.S.S.R.

Rutter, N.W., University of Alberta,
  Department of Geology, Edmonton,
  Alberta, T6G 2E3, Canada

Saarnisto, Matti, Department of Geology,
  University of Oulu, 90570 Oulu 57,
  Finland

Salonen, Veli-Pekka, University of Turku,
  Department of Quaternary Geology,
  SF-20500 Turku 50, Finland

Shaw, J., Department of Geography, Queen's
  University, Kingston, Ontario, K7L 3N6,
  Canada

Stephan, H.J., Geologisches Landesamt,
  Schleswig-Holstein, PF 5049,
  Mercatorstr. 7, D-2300 Kiel, W. Germany

Warren, W.P., Geological Survey of
  Ireland, Beggars Bush, Haddington Road,
  Dublin 4, Ireland

White, O.L., Engineering and Terrain Geol.
  Section, Ontario Geological Survey, 77
  Grenville Street, Toronto, Ontario, M5S
  1B3, Canada